BY BROCK YATES

ENZO
FERRARI

ENZO FERRARI

THE MAN AND THE MACHINE

BROCK YATES

RANDOM HOUSE NEW YORK

2023 Random House Trade Paperback Edition

Copyright © 1991 by Brock Yates
Foreword and epilogue copyright © 2023 by Stacy Bradley

Published in the United States by Random House, an imprint and division of
Penguin Random House LLC, New York.

RANDOM HOUSE and the HOUSE colophon are registered trademarks of Penguin
Random House LLC.

Originally published in hardcover in the United States by Doubleday, a division
of Bantam Doubleday Dell Publishing Group, Inc., in 1991.

ISBN 978-0-399-58861-7
Ebook ISBN 978-0-399-58862-4

Printed in the United States of America on acid-free paper

randomhousebooks.com

9 8 7 6 5 4 3 2 1

Book design by Jo Anne Metsch

When *Enzo Ferrari: The Man, the Cars, the Races, the Machine* was published in 1991, it was an honest and sometimes raw biography that looked at the life of Ferrari in the context of his family, his temperament, his education and his socioeconomic standing in a country financially ravaged by two world wars and the Depression, and an intense vendetta against Fiat that would take almost fifty years to play out. Brock Yates examined the birth, life and death of Enzo Ferrari, sifting through the gossip and mythology of the man in order to offer a clear picture without the romanticism and awe of many biographers of the time.

When the book was released, Brock was celebrated and vilified in equal measure. Historians lauded his accuracy and strength of research, while acolytes of Ferrari condemned the biography for showing the human side of Enzo, warts and all, without the rose-colored glasses of blind adoration. This was not a rhapsodic reinvention of Ferrari's life but rather a realistic and surprisingly erudite look into what made Enzo tick, from birth to death, no holds barred.

Brock was forever amazed by the outrage of those who disagreed with his portrayal of Ferrari. Had Enzo been a weak man, his company would have failed almost immediately, earning not even a footnote in the annals of history. That being said, Enzo was not a dilettante. He was tough, iron-fisted, sometimes cruel and often despotic, and he created a company that survived chaotic times and continuously faced financial ruin as it tried to find its place within the incredibly fluid political climate of the times. Enzo's strength of will, his business savvy and even his boorishness all combined to give his company longevity, creating a foundation on which Scuderia Ferrari and Ferrari S.p.A. could thrive.

Brock's belief in journalistic integrity was sacrosanct, and while he never caved to criticism, he was open to self-reflection and honest enough with himself to step away from a topic in order to see what could or should be changed. This self-reflection began almost immediately following the initial publication of *Enzo Ferrari: The Man, the Cars, the Races, the Machine.*

As with all writers, Brock agonized over his syntax, phrasing and research. Had he done enough? Said enough? Said too much? These questions would linger and ultimately become the topic of conversation throughout our years together, as we often debated the changes he would make, the phrases he would polish and the information he would excise if given the chance.

As his daughter, I was raised in the industry, so I spent most of my life surrounded by cars, writers and racing. I debated the merits of his work and served as a sounding board and confidante, and often a buffer for his insecurities—cheerleading when necessary, but more often than not praising and admiring the brilliance of his craft and his incredible turns of phrase. As I grew older and more confident, I became an editor and assistant in his work, and after he was diagnosed with Alzheimer's, I became his memory, his archivist and his touchstone.

When the opportunity arose to republish Enzo's biography, I already knew, based on our prior conversations, what changes needed to be made to fulfill my father's legacy. Brock's illness had progressed to a stage at which he was unable to help make these changes, but given the closeness of our relationship and my firm grasp of his wishes, I feel confident that the book is now truly complete.

To those readers who are familiar with the first edition, the changes in the original text are minor, yet pivotal to the overall tone of this biography, all without changing the reality of Enzo himself. Simple errors have been fixed; antiquated language, which might be considered offensive by today's standards, has been refined; and great new photos have been added. Brock's biggest regret with the first edition was the inclusion of certain information that could not be verified and was therefore closer to speculation and rumor than fact. These few details have been excised completely, but in no way do these alterations impact or diminish the overall biography.

The final change to the book is my contribution. While Brock's first edition ended with the death of Ferrari, I have added an epilogue that offers a look at the company since Enzo's death. Brock's portrayal was concise at the time of publication, but no one was prepared for how the company would explode onto the world market in the years following Ferrari's death. In light of this growth, no biography of Enzo published today could be complete without a look at how the three decades since his death have unfolded.

In adding an epilogue to this book, I am not trying to reinvent the wheel. Ferrari's death in 1988 ends the biographical nature of this narrative as regards Enzo himself, but the company's progress and continued meteoric rise following his death offer further insights into the iron-fisted will and determination of the man behind the company, whether in life or in death.

In the years since Enzo's death, the eponymous company has faced the prospect of financial ruin and has been restructured (with a 90 percent stake going to Fiat), while its Scuderia division, dedicated to race cars, has seen marked changes, with incredible highs and lows in Formula One racing. In 2014 one of the biggest changes in Ferrari's modern history occurred when Luca di Montezemolo unexpectedly quit as chairman and Fiat Chrysler announced that Ferrari was being spun off as an autonomous company.

The wealth of information and the complexity of many of the players associated with Ferrari in the past three decades might threaten to overwhelm most biographies, but Brock's insight and direction set the ideal stage for a continuing look at the company in relation to the vision and structure Enzo engineered throughout his storied career.

In light of this, it makes eminent sense to look at the recent history of Ferrari, concentrating on Fiat's majority stake in the company, Luca di Montezemolo and Sergio Marchionne's complex and often contentious relationship, Ferrari's continued highs and lows in Formula One racing, and the explosive mystique of Ferrari not simply as an automobile but as a worldwide brand as well.

The new material will pinpoint the most important and defining moments of the past three decades without getting mired in the minutiae of information that has accumulated since Enzo's death. The intention is not to take away from the biography itself but rather to add the most important highlights that encapsulate and define the company that Ferrari has become.

Considering the worldwide reverence and marketability associated with Ferrari, this biography is more on point with today's readers than ever before, and these additions will breathe new life into what is already a relevant and respected biography, offering insight into the company for a new generation of acolytes who eagerly embrace any and all things Ferrari.

The process of republishing this book was not a solitary endeavor on my part, and there is a small group of people I would like to thank for enabling me to update Brock's work and make the small changes that he fretted about over the years. First and foremost, my mother, Pamela, Brock's muse, who worked diligently to reintroduce his work to new readers while recognizing that the timing was perfect for republishing this work.

To Michael Mann, for optioning *Enzo Ferrari: The Man, the Cars, the Races, the Machine* for a movie and for working incredibly hard to create a stunning screenplay that respects the merits of this biography while paying homage to Enzo's legacy. Brock was thrilled by Michael's dedication and determination to bring this story to the big screen.

To Carol Mann, the most amazing literary agent and a truly spectacular woman, who remained loyal to Brock for years and without whom this project could never have moved forward.

And finally, to Brock, who engendered a passion for the written word and trusted in my ability to be his voice when his was silenced by Alzheimer's disease.

Brock passed away on October 5, 2016, after a long and courageous fight with Alzheimer's. His impact on automotive journalism is without equal. He cherished his craft and loved every aspect of his career. The drivers, races, history and people associated with motorsports were his life's breath, and he felt blessed every day for his career and his friendships. He never took any of this for granted, and the hardest day of his life was the day he powered down his computer for the last time, knowing that he would never again speak truth to power, debate the merits of a race, or reach out across the miles from the pages of a magazine or book to a fellow devotee of the sport. As not only his daughter but also his fan, I am so proud that I got to play a small part in bringing his biography of Enzo Ferrari back to life.

STACY BRADLEY
2023

CONTENTS

For much of his life, Enzo Ferrari's public appearances were flaw-lessly choreographed events, designed to enhance the great man's image. He was a powerful presence, utterly confident of his ability to dominate the throngs of reverential customers, eager suppliers, oc-casionally pugnacious journalists and curious fans. As he grew older, the "Pope of the North," as he was often called, became increasingly reclusive, which only enhanced his regal reputation. It was no acci-dent that this simple man from Modena achieved a nearly superhu-man stature among his devoted followers around the world, which in turn brought enormous benefits to the Modena businesses he oper-ated for nearly sixty years.

I met Enzo Ferrari in private but once—in the late summer of 1975. I was in Maranello with Phil Hill, the great American racing driver, who had won the World Championship for Ferrari in 1961 following a tragic race at Monza that had seen his chief rival and teammate, Count Wolfgang von Trips, die in a bloody crash. Hill, a sensitive and thoughtful man, had left the Ferrari team a year later amidst loud recriminations. He had been gone for thirteen years when he and I arrived in the city to work on a documentary film. Without warning, Hill was summoned into Ferrari's dark, blue-walled office. For a reason that escapes me to this day, I was invited along as well, although, as an American journalist with no formal ties to either Hill or the Ferrari factory, my presence seemed utterly superfluous.

Enzo Ferrari was a more imposing man than I had expected. Nearly six feet tall, he stood half a head above Hill, who, like many expert racing drivers, was a slight, almost frail man. His host, then seventy-seven years old, moved with the brusque ease of a successful Italian, his prominent chin thrust forward, chest out, arms extended, his large hands opened flat to the sky, beckoning an approach. Ferrari's sweep of white hair, his imperious Roman nose and his drab suit were well-known trademarks and did not surprise me. The sunglasses, which had become increasingly common in his public appearances and which gave him the ominous persona of an aging *capo*, were absent, revealing a bright, riveting gaze. But it was his voice, soft and sonorous, that shocked me. I had expected stentorian power, soaring and dominant, or perhaps even Brando-like mumbles as popularized in the recently released *Godfather*. Anything but the velvet tones that issued forth in the quiet room. Ferrari's bellowing tantrums were the stuff of legend, but on this day—a day of peacemaking—he seemed muted and faintly subdued.

He moved around his vast, bare desk and swept Hill into his arms, nearly smothering the Californian as I stood by, a thoroughly uncomfortable witness to this strained, difficult and long-overdue reunion. I understood little Italian at the time but could decode enough of the conversation to realize that the exchange was stiff and formal. The old wounds were not yet healed. (As Ferrari and Hill grew older, the rift was repaired and Hill visited Maranello often.) The brief meeting ended with another embrace and the presentation of two signed copies of Ferrari's informal memoir known as "The Red Book." Quite obviously I had no idea that someday I would take on the prodigious chore of writing the biography of this powerful and contradictory man.

Ironically, Ferrari left very little in the way of a paper trail, other than a series of carefully edited, self-serving, informal autobiographies that were sporadically published—both publicly and privately—

from 1962 to the early 1980s. His private correspondence was for the most part limited to business matters.

I discovered rather quickly that there were two Enzo Ferraris: the private individual and the artfully crafted public version. My intent is to record as accurately and judiciously as possible the real Ferrari, who in reality was an amalgam of the two.

The resulting picture may offend many of his followers who have come to believe that he was a demigod, a genius who personally created marvelous automobiles out of bits of steel and aluminum. Sadly, this was not the case, and in dissecting this myth I will no doubt be accused of being the worst kind of muckraker and revisionist. Nothing could be further from the truth. I entered into this project with no preconceived notions. But the reader must be mindful that Enzo Ferrari has in death become more a mini-industry than a mere mortal. A massive financial structure has been built around the Ferrari mystique. Companies such as Fiat S.p.A., which now owns the business—as well as Italian industry as a whole—rely on the Ferrari myth to enhance their reputation for advanced technology. Moreover, a bull market for collecting Ferrari automobiles was generated after the death of the founder. Hundreds of millions of dollars were invested to elevate even the most ordinary Ferrari models into alleged works of art. Sustaining the aura of the man who created such machines descended to raw commodity speculation. (On several occasions I have been asked by concerned Ferrari owners if my book would adversely affect the value of their automobiles.) In addition to those who have a financial stake in the Ferrari mystique there are the multitudes of simple enthusiasts around the world who root for Ferrari racing cars, collect the team's memorabilia, glory in the exploits of the marque and in general propagate the Ferrari faith. There is a mindless fan worship that surrounds Enzo Ferrari which is as intense as any enjoyed by the long-departed Elvis or movie idols like James Dean and Marilyn Monroe. Therefore, any criticism of the man is likely to produce a firestorm of outrage. But so be it. The object here is to paint as accurate, dispassionate and fair a portrait of this complex man as possible. If certain investors or acolytes are offended, it cannot be helped.

Several years have passed in the preparation of this book. Elsewhere I have credited scores of men and women who aided in the research and who openly offered background material and personal recollections. To all of them I am grateful in the extreme. But several other individuals deserve even more credit. First, I owe thanks to my editor, David Gernert, whose fertile brain generated the idea for a Ferrari biography in the first place; second, to Gayle Young, my secretary, who typed and retyped this manuscript with continuing good humor; and last and most importantly, to my wife, Pamela. It was she who trekked across Europe and America with me, aiding in translation and source finding, but also acting at all times as my best friend and adviser.

To everyone who helped in this project, as well as to the hundreds of men and women who contributed so much over the years to establish the Ferrari empire and its mystique, I extend my sincere thanks.

BROCK YATES
Wyoming, New York
1990

FERRARI

1

The autoroute to Beaune lay ahead, an inkblot of asphalt. Behind him a few remaining lights of Lyon flickered in the rearview mirror as the big Renault sedan settled into a steady lope at 125 mph. He would be in Paris before dawn.

He was a driver. At eighty-two years, he still handled an automobile with the same authority and resolve that had carried him to three victories at Le Mans. Twenty-four hours. Around the clock. Those races were his specialties. He had never been a sprinter, never particularly fast in the twitchy, pitiless single-seat Grand Prix machines. But give him a full-fendered Alfa Romeo or Ferrari sports car and he could run like a marathoner, hour after hour, kilometer banked on kilometer, crushing the opposition with his sheer grit and the reserves of muscle coiled in his compact bricklayer's body.

He checked his watch. The hands were slipping toward three o'clock on the morning of August 14, 1988. It was a drive he had made a hundred times: Modena to Paris, now a headlong rush along the table-smooth Italian autostradas and the French autoroutes. For forty years, the mission had been the same. The central theme had been the cars, those low-slung beasts that had caused him so much joy and grief. Up and down the autoroutes, crossing and recrossing the spine of the Alps like some crazed tourist, back and forth over the bleak expanse of the Atlantic, always with one singular purpose—the glorification and enrichment of the most supremely outrageous, over-powered and oversexed automobiles of the modern era, the crimson bolides of Modena and Maranello—the magic machines of Enzo Ferrari.

Ferrari. There had been times, often lasting months, even years, when he wished he had never heard the name, but too much time had passed for that. Seventy years, he mused. He had known this in-credible man since 1918, when both had been young veterans poking around the postwar automobile business in Milan and Turin. Ferrari had made Luigi Chinetti a rich man while making himself richer. Ferrari had made him important and respected, while elevating him-self to world fame and becoming a demigod to the poseurs and nou-veaux who seemed prepared to sell their firstborn into slavery in order to obtain one of his automobiles.

It had all been an insane aerobatic display of emotion and ego warfare and now he was tired of it all. They had dueled too long and, if a winner had to be declared, it would have to be Ferrari. But did he not always win? Did he not always prevail, sometimes coming off the mat after repeated, bloody knockdowns to land a knockout punch? They had sparred many times, hugging between rounds, and he had often scored well, but now it was over, and the old warriors had fought their last fight. There would be no more confrontations, no more angry words, no more window-rattling arguments, no more lawsuits, threats, walkouts or vile insults, all generally forgiven or forgotten. Still, Ferrari towered over his life, having given it purpose and mean-ing, and for this he knew all the madness had been worthwhile. Yes, Ferrari had been at the center of it all.

Both men were hard, unyielding, classic Italians, easily tempted to let a relationship flare into wild boasts, vicious insults and bare-faced fabrications. For all the years he had known Ferrari, business negotiations with him remained an elaborate, protracted drama of artful parlays, suspended agreements, temper tantrums, operatic claims of impending bankruptcy, social ruin, family shame, incurable disease and violent death. It was in this Byzantine atmosphere that Ferrari had thrived and men like Chinetti had either learned to survive or been summarily defeated. At the core of any dealings with him had been the money—preferably Swiss francs or American dollars. Winning automobile races had been his obsession, but *only* if the money was right.

As the carefully preened image of Enzo Ferrari had risen to world prominence, Chinetti and those other old associates who had been with him since before the war had faded into the shadows. Ferrari shared the spotlight with no one, not even the man who had helped create a market for his automobiles larger and more lucrative than anyone could have imagined. Chinetti understood the crazed, comic-opera politics of the factory as well as anyone. On his numerous trips to Maranello he learned when to flatter, when to threaten, when to simper, when to sulk and when to primp and pamper the soaring ego of the man who rivaled the Pope as the most famous and popular personage in all of Italy.

This contentious friendship of the old warriors had often stretched to the point of shredding, yet the bond could not be broken. As for Chinetti, he came to know Ferrari too well—to understand his chameleonlike public persona, his ability to orchestrate the press and the public, to artfully and floridly articulate positions perfectly tailored for his audience, to play the lovable, beleaguered, poverty-stricken patriarch one minute and the ruthless, egomaniacal despot the next. He had seen the public Ferrari, the regal old don who oozed respectability, as well as the private Ferrari, the ribald, belching, farting, cursing, bragging, hectoring "Modenese *paesano*" who bore his lower-class background like a caste mark on his forehead. He knew him as the consummate manipulator of men, especially of the racing drivers who had driven the vermilion cars of the Scuderia, sitting in splendid isolation in Maranello monitoring his teams' successes and failures by telephone, telex and television but not attending a single automobile race for the last thirty years of his life.

There was no question that he was uncomfortable in large crowds, and as the years passed, his public appearances were restricted to flawlessly choreographed press conferences during which he traded verbal blows with the fickle and often contentious Italian sporting press. His true friends could be counted on the fingers of one hand and included a few longtime, utterly faithful business associates. Beyond that perimeter he tended to a flock of supplicants, sycophants, customers, adoring fans and a legion of racing drivers with baronial aloofness, handing out favors, slapping wrists, administering punishments and settling arguments from an invisible throne.

Once, when asked to comment on an observation made by another close associate that Ferrari cared little for his drivers, despite his public bleatings to the contrary, and that he in fact felt closer affinity to his mechanics in the shop, Chinetti paused for a moment, then replied, "I don't think he liked anyone."

Now he was driving away from that place in Emilia, that place occupied by that Goliath of a man whom he somehow knew he would never see again. He rushed on, calling up part of the reserves of endurance that had served him so well since he began to drive automobiles at unseemly speeds.

He was on a long, flat stretch when a sharp explosion shuddered through the Renault, jarring the steering wheel and causing him to jerk upright in the seat. He slowed to a crawl, checking the dimly lit instrument panel for a possible source of the detonation. It had been a thunderclap, spearing the car without warning. He groped for a reason. The car felt perfect. No strange noises or vibrations, no odors of burning wire, boiling oil or frying paintwork. Cautiously he rolled to a stop on the shoulder of the autoroute and climbed out. It was still. Moonless. Only the distant hiss of traffic intruded on the otherwise tranquil scene. He walked around the car, feeling the stiffness in his legs following the long hours behind the wheel. He poked under the automobile, then lifted the hood. He could find nothing that even suggested a possible murmur, much less the deafening salvo that had mysteriously rattled out of the night.

Nonplussed, but relieved that the Renault was undamaged, he crawled back aboard and resumed his journey. He reached the southern suburbs of Paris by dawn. Traffic was building and his pace was slowed by the notorious anarchy that was the daily Parisian commute. He turned on the radio. A newscast was beginning.

It was the lead story. In matter-of-fact tones a voice announced that earlier that morning death had come to Enzo Ferrari at his home in Modena, Italy.

Ferrari. Dead. It was expected, but somehow it stunned him. This old warrior, this curmudgeon, this masterful manipulator, this tireless competitor, this overwhelming presence, this imperfect but perpetually fascinating man was finally gone. A blizzard of incomplete thoughts rushed through his brain: What of the factory? The racing program? His son Piero? His mistress Lina? Would the aura of the automobile survive now that its life source was gone? Would the technocrats at Fiat destroy the personality of the company? What of the fabric that the Old Man had so carefully woven over the years? What part would be preserved and what would be shredded? He ruminated about these things, reaching no conclusions.

And then it struck him. He checked his watch in disbelief. No, not possible. He was a rational man. Yet it gnawed at him. The explosion on the autoroute. That great, thudding noise in the Renault. If the radio bulletin was correct . . . yes, if it had given the time of Enzo Ferrari's death accurately . . . that ominous burst of sound had come at almost the same moment as Enzo Ferrari's last breath. Surely there was no connection . . .

2

Winter bit hard at northern Italy in February 1898. On the eighteenth a blizzard swept down from the Alps and sluiced along the Po River valley in an icy assault, burying the small trading city of Modena in a foot of snow. Because of the bad weather it was two days before Alfredo Ferrari could dig out of the drifts and make his way to City Hall in order to register the birth of his second son, Enzo Anselmo. Therefore his official age would be recorded as two days younger than it actually was, not surprising in a nation whose commitment to record-keeping could at best be described as casual. Italians evidence no urges to imitate the prissy German and English need for detailed civil documentation outside their omnipresent Catholic Church. Therefore the official records of most Italians below the noble class during the nineteenth century are cursory at best, and in the case of Enzo Ferrari inaccurate as well.

His father, Alfredo Ferrari, was born in 1859 at Carpi, a farming town eighteen kilometers to the north of Modena, where the family had been small food merchants. Enzo's mother, the former Adalgisa Bisbini, came from Forli, in Romagna, sixty kilometers to the southeast along the ancient Via Emilia, that amazing, ruler-straight Roman road that had served as the lifeline of the region since its construction in the third century B.C. Both were members of the petit bourgeois merchant and landowner class that had developed in the region during the centuries since discharged Roman legionaries had settled there on land given them as a reward following twenty-five years of service. Those early veterans had drained the swamps and on the Padana Plain created the farms and vineyards of the area. From them had developed the traditional Emilian distrust of authority and of outsiders.

Modena, which the Romans called Mutina, enjoyed perhaps its greatest notoriety in 43 B.C., when Marc Antony pursued Decimus Brutus to the city following the assassination of Julius Caesar. Brutus was one of the major conspirators but not *the* Marcus Junius Brutus who is said to have administered the fatal wound. Following that brief action, the city fell into a thousand-year slumber before becoming a member of the Lombard League. The powerful Este family developed Modena into a first-rate manufacturing and trading center in the seventeenth century, but it began to molder again during repeated invasions by the French and the Austrians.

When Enzo Ferrari was born, it was a drab city of fifty thousand souls, noted more for its steaming summers, stifling humidity, voracious mosquitoes and grim, fogbound winters than for its local wine, Lambrusco, or for its culinary specialty, *zampone*, a spicy concoction of stuffed pig's feet. Its citizens were known to be rather dour, industrious people (by Italian standards) with legendary skills in all sorts of crafts, particularly those dealing with the casting, shaping, cutting, forming, molding and fabrication of metal. Such was the tradition of the city about which Enzo Ferrari was to develop an enormous, almost overweening pride.

His father's family was but one line of literally dozens of Ferraris that still dot the region. Even today, the name is as common in the Modena phone book as Smith or Jones in any midsize American city. Over nine hundred Ferraris are listed. If there was noble blood in Enzo Ferrari's veins, either he was not aware of it or he chose to manifest a mysterious and notably uncharacteristic modesty in concealing its presence.

Prior to setting up his own business, Alfredo Ferrari was trained in the metalworking trades and for a number of years was employed as a senior technical supervisor at the Rizzi Foundry in Modena. His academic background is unclear, but he was sufficiently sophisticated to enjoy the company of teachers, local university professors, clerks and fellow engineers while Enzo and his brother were growing up.

Ferrari later fostered the impression that his father was a poor worker, scratching out a living by fabricating crude buildings and other components for the Italian National Railway System. He wrote how he and his brother, Alfredo, who was two years older, "shared a room over the workshop and were awakened in the morning by the ringing of hammers. My father, who employed from fifteen to thirty workmen according to the amount of work on hand, made gangways and sheds for the state railways and acted as the manager, the designer, the salesman and the typist for his firm all at the same time."

The Ferrari home was located at 264 Via Camurri (now renamed the Via Paolo Ferrari) on the northern outskirts of the town and adjacent to the rail lines that were the source of income for Promiata Officina Meccanica Alfredo Ferrari, as the firm was called. It was a long, narrow brick building, with the machine shops dominating most of the space and the family's rather spacious and comfortable quarters tucked upstairs on the western end. It still stands, essentially unchanged today (at number 85), but is now a museum, Museo Casa Enzo Ferrari. Alfredo Ferrari was hardly a wealthy industrialist, but Enzo Ferrari reveals that his father's business created sufficient wealth to afford more than a modest inventory of amenities. By 1903 they had an automobile, an expensive single-cylinder French De Dion–Bouton, which was but one of twenty-seven private cars in Modena. Two more were to follow it into the family garage, and the Ferraris enjoyed the services of a combination handyman and chauffeur. As boys Enzo and his brother, Alfredo (who was familiarly known as Dino), were provided with state-of-the-art FN three-speed bicycles and a stable of pigeons, with which they competed. After young Enzo learned that he was not sufficiently athletic to win on the 100-meter running track that he and his friends had laid out on the wide front yard of the family home, he turned to shooting; he says he became quite an expert at skeet and at killing rats that lurked along a canal bank behind the house. He also received fencing lessons at the local gymnasium. None of this—the automobiles, the bikes, the leisurely hours at boyhood play—smacks of a youth drowned in poverty.

In fact, the general impression given by Ferrari in his oddly haphazard memoirs is one of a rather pleasant boyhood. Ferrari endured the heartbreak of losing a Great Dane who died after choking on a bone shard, but there is little to suggest anything in his early life that could be described as traumatic either in terms of ill health or family disorders. He does admit, however, that he hated school. While brother Dino was an accomplished student, Enzo brought home sufficiently poor grades to qualify for harsh strappings from his father, who was insisting that he enter technical training to ultimately earn an engineering degree. Ferrari writes that he simply wanted to be a "worker" and nothing else.

It is reasonable to assume that Enzo was brought up in a relatively normal Italian household, with Papa Ferrari operating as the undisputed despot, demanding respect from his children and silent, groveling subservience from Mama Adalgisa. The family unit was sacred to the Italian male of Ferrari's generation, within, of course, the rather wide latitudes defined exclusively by the man himself—who remained free to philander, spend freely and ignore the tenets of the Church, while in turn demanding goose-stepping obedience from his children and saintly fidelity from his wife. He adored his children, and treated his wife as a basically asexual helpmate modeled on the most perfect female of all time—his mother. Mothers were worshipped, wives tolerated and other women treated as objects of either scorn or lust, or both. Women, to men like Ferrari and his forebears, fell into two simplistic categories: the chaste females who produced them and lustful, low-life harlots from whom they selected their mistresses and concubines.

This idiotic double standard generated within men of Ferrari's generation (and their sons) perhaps a greater fixation with sex than any other civilized race. For example, during Enzo's lifetime, adultery by a man in Italy was not punishable, unless attended by scandalous behavior. However, if it was committed by a woman, she could be sent to jail. But for a man, having more than one woman was a measure of manliness, pure and simple. This, of course, presents contradictions, confusion and ultimately massive insecurities. In his seminal work on his countrymen, *The Italians*, the noted journalist and historian Luigi Barzini put it this way: "Most of them [men] harbor secret doubts and fears. A moment comes when every one of them is struck by the fact that most of the women he has had an affair with are somebody's wives and that it is not, therefore, materially possible for all the husbands in Italy to stray from marital fidelity while none of the wives do so. There is no escaping the fact that each day a substantial number of proud Italian males, jealous, suspicious, overbearing, proud men, are being made *cornuti* and become the object of scorn and ridicule."

Such was surely the mind-set that blossomed in the psyche of Enzo Ferrari. He was to become a classic Italian male: the haughty, tyrannical, sex-obsessed, posturing *padrone* who expected little of himself and his male children but everything from his women, either in terms of beatific, Madonna-like perfection or undemanding performance in the bedroom.

Until the end of his life Ferrari remained handcuffed to his late-nineteenth- and early-twentieth-century origins. He learned to communicate in the florid style of old Italy, always in broad terms of flattery and scorn, never directly dealing with an issue that could be treated obliquely and without revelation of his true feelings.

His first contact with automobile racing came on September 6, 1908, when his father took him and his brother to a race outside Bologna. He was ten years old as they cruised along the Via Emilia to Bologna to see two of the world's best drivers, Felice Nazzaro and Vincenzo Lancia, race their powerful Fiats over a long, flat 53-kilometer circuit composed of public roads outside the city, including the Via Emilia itself. Lancia had already begun the company that would bear his name, but remained on the Fiat team. He set a lap record of 84 mph, manhandling his monster machine across the Po flatlands at speeds exceeding 100 mph, but it was Nazarro who won the race, averaging a prodigious 74.1 mph.

The following year young Ferrari trekked two miles across open farmland to watch his second race. It was a minor event, on the Navicello straightaway of the old Modena–Ferrara provincial highway. Ferrari recalled watching local volunteers for the Modena Automobile Association sprinkle buckets of water on the dirt surface to keep the dust at bay between runs. A sportsman driver named Da Zara won, averaging 87 mph. Ferrari said, "I found these events immensely exciting."

Still, there is no evidence that his passion for automobiles and racing overwhelmed other interests in his youth. He retained a boyish enthusiasm for the local Modenese football team and at one point seemed to be headed for a career in sportswriting. He made several contributions to Italy's powerful national sporting newspaper, *Gazzetta dello Sport*, when he was seventeen years old, revealing a strong sense of drama and a purposeful hand with his prose.

Years later Ferrari would confess that as a boy he dreamed of three careers: sportswriting, opera singing and race driving. Ironically, he would make his mark at none of them, although it is likely that his skill with words and his stridently stated opinions might have led to success in the Italian press. As for singing, he openly admitted to possessing a tin ear and was never known to try an aria, even after a few too many Lambruscos. He warmly recalled dinners in the small family dining room with a tenor named Bussetti, an ex-mechanic who had sung in the chorus of the Metropolitan Opera Company. At one impromptu after-dinner concert, when Bussetti rattled the tiny room with a soaring high-C, Ferrari remembered that the lights went out!

Though he was raised, like virtually everyone else, in the arms of the Roman Catholic Church, Ferrari was never religious and would later lament, "Although a Catholic by baptism, I have not the gift of faith, and I envy those who find in it a refuge from their torments." He wrote of finding the catechism "repugnant" and recalled that the old parish priest at Santa Caterina was openly annoyed with him for his obvious lack of enthusiasm. He also admitted that the first and last time he confessed his sins was on the Saturday before his first communion—this despite years of being carted off to church with Dino, decked out in their matching blue sailor suits.

The recollections of Ferrari and other examinations of his youth supply nothing to indicate any extraordinary talents, no glittering, undiscovered genius, not even the singular talent for leadership and organization that was to blossom later in life. Enzo Ferrari, for his first twenty years, was simply another Italian boy who liked bicycles and car races and football but who was slowly being sucked into the gory maelstrom that was to engulf Europe after Gavrilo Princip fired his two pistol shots at Sarajevo on the morning of June 28, 1914.

From that moment the world was immersed in the most frustrating, gruesome, mindless mass slaughter in the history of mankind. Virtually every European male of Enzo Ferrari's generation would be affected, although nearly a full year would pass before Italy entered the war on the Allied side in May 1915. The mission was clear. *Italia irredenta* was a national obsession centered on lands held by the Hapsburg Empire in the Trentino salient and along the Adriatic coast. Italy, while nearly as populous as France, was the poorest of the so-called great powers, with the south locked in a medieval agrarian society and the north barely able to sustain a nascent industrial economy. Worse yet, Italy was weak politically, having united its warring regions and provinces less than fifty years earlier.

The youth of Italy, especially those northerners who had grown up with an endemic hatred for the Austrians, rushed to join the Army, duplicating the fatal naïveté that was surging through the young men of France, Great Britain and the Central Powers. Dino initially operated the family Diatto as an ambulance, transporting wounded from the Alpine fronts to hospitals around the Po Valley. When he reached nineteen, he joined the Italian Air Force, a decision that, by coincidence, was to have an impact on his brother long after the war. Dino became a member of the ground crew of Squadriglia 91a, whose insignia consisted of aviator's wings and a *cavallino rampante*, or prancing horse. This insignia was carried by the squadron's most famous ace and Italy's supreme flying hero, Francesco Baracca (thirty-four kills), before his Spad S13 biplane crashed at the front.

In February 1916, Enzo Ferrari had his first close encounter with death—an acquaintance that would stay with him for the rest of his life. His father died of pneumonia. The little business in Modena collapsed, and Enzo, untrained and apparently uninterested in maintaining the metal-fabricating operation, drifted through a series of menial jobs.

He apprenticed for a time with the Modena Fire Department, then took instruction in a lathe operator's school and became a trainee in a small factory manufacturing artillery shells. This was a confusing time for Ferrari. His father was dead, and with the convulsions of the war, plus his lack of serious career goals, he meandered through his late teens without direction, awaiting his call to the colors.

Then suddenly, shockingly, Dino was dead as well. Ferrari describes him as a victim of "a malady caught while doing military service." This could have been influenza, typhoid fever or any one of a dozen illnesses that ravaged the trenches and camps of the World War I combatants. Dino lay ill for a time before passing away at the sanitarium at Sortenna di Sondrio later the same year. Ferrari was devastated by the loss of his father and brother.

Dino had volunteered, but Enzo waited to be drafted two years later, at age nineteen. By then the Italian infantry was mired in a brutal mountain impasse, facing the Austrians along a front that stretched from the Isonzo River northeast of Venice into the craggy Alpine peaks of the Dolomites and the Trentino Alto in the west. Ferrari was assigned to the 3rd Mountain Artillery, at the time bogged down in the Val Seriana Mountains to the north of Bergamo. Claiming a background in mechanics, Ferrari expected to be assigned to one of the crews tending the big Breda howitzers that lined the highlands controlled by the Italians. But a Piedmontese second lieutenant chose instead to assign him to a detail shoeing mules—the beasts of burden employed to haul artillery over the mountain terrain. At that time the better billets, including the entire officer corps, were saved for the upper classes, and Ferrari, coming from a backwater city in Emilia and possessing no bloodlines with which to negotiate better duty, was quickly shunted into the menial labor thought to befit his origins. This lasted but three months before he was leveled by a near-fatal disease, thought in retrospect to be pleurisy. He was removed from the front and placed in a military hospital at Brescia, where two operations were performed. If the diagnosis is correct, it is possible the procedures were done to remove fluid from his lungs. Whatever the ailment and the proposed cure, he was then removed to a crude hospice of battered huts for the incurably ill near Bologna. He recalled being left in a chilly, darkened room on the second floor of a building from which he could hear the steady clatter of workmen's hammers as the lids of coffins were pounded into place. There, following more cursory treatments, much pain and a long rest, Ferrari was finally released, broken physically and mentally, but infinitely luckier than hundreds of thousands of young countrymen whose bodies littered the mountains to the north.

He was discharged in 1918 as Europe gasped for breath and men groped to justify the sacrifice of 30 million lives. While he makes no mention of it in his memoirs, and no student of his life has been able to ascertain the reason, Ferrari seems to have been seized with an epiphany of sorts during this period. A gap exists in the story that is hardly unusual in the Ferrari papers, but leads to intriguing speculation. Who was the mysterious person, or what was the pivotal event, that caused Enzo Ferrari to decide, between the time he entered the Army in 1917 and his release a year later, that he wanted to follow a career in automobiles?

In his autobiography he recalls sitting, prior to the war, outside the family home on a warm summer night with a friend named Pepino. Under the light of a gas lamp, they had examined a photograph of Ralph De Palma, the marvelously talented Italian who had moved to America in 1893, by 1916 had won both the Indianapolis 500 and the Vanderbilt Cup and was believed by many to be the greatest driver in the world. Ferrari claims that he told Pepino, the son of a Modenese importer of foodstuffs, "I will become a racer." Moreover, he often repeated, probably apocryphally, that the inspiration for his choosing V12-type engines when he started to build his own cars lay in his fascination with the big Packard V12 command cars he saw used by the American Army at the end of the war. These stories may or may not be correct—Ferrari openly employed history at his convenience—but there exists a thread of truth, implying an interest in cars that had become a central theme in his life by the end of the war and that took him far from home when the fighting stopped.

Following the Armistice, Ferrari surfaced in Turin, seeking work with Fiat. Unlike his hometown, Turin was a center of Italian manufacturing, a bustling, congested city along the Po River that had prospered from the war effort. When Ferrari arrived in the city in the winter of 1918–19, all of northern Italy was in turmoil. The Great War had seemingly solved nothing, and the powerful voices of socialism, Marxism and ethnic nationalism were ringing through the cities. Italy, its treasury ravaged and its feeble political contract of unity badly shredded by the rift between the industrial north and the backward, rural south, was aflame with social unrest. Gabriele D'Annunzio, the fabled war hero, poet and ardent nationalist, was assembling a noisy rabble of disenchanted veterans in hopes of capturing the Dalmatian coast, a prize denied Italy by the Treaty of Versailles. In Milan a pugnacious left-wing journalist named Benito Mussolini was making an artful switch from idealistic socialism to political opportunism. As editor of a fiery paper called *Popolo d'Italia*, Mussolini formed a virulently antiestablishment organization called I Fasci, which miraculously began to attract the disaffected, the frustrated and the malcontented from both the left and the right. For a while the Fates would smile on this bloated, preposterous poseur from Romagna as his black-shirted Fascistas took control of Italy and led it successively through a period of artificial pomp and prosperity and finally into shame and defeat.

Ferrari was seemingly unaffected by these swirling winds of social and political change that ripped across Italy. He had come to Turin with a single purpose in mind, to gain employment with the rapidly expanding Fiat works. Founded in 1899 by Giovanni Agnelli, Emanuele Cacherano di Bricherasio, and Count Carlo Biscaretti di Ruffia, FIAT (Fabbrica Italiana Automobili Torino) had been reorganized in 1907 and renamed simply Fiat S.p.A., with the Agnelli family firmly in control. The change signaled the expansion of the company from a small, regional automobile manufacturing firm to an industrial giant that produced not only cars but trucks, ball bearings, aircraft and marine engines, military aircraft, railroad locomotives, rolling stock and even complete cargo ships. By the time Enzo Ferrari appeared at the company's headquarters on Turin's Corso Dante, final plans were in place to mass-produce the 1.5-liter Tipo 501, a technologically advanced, low-priced passenger car that would become a staple of the company for years. But more to Enzo's liking was Fiat's continued interest in motor racing, where a small, elite crew of designers, mechanics and drivers were ready to contest the great continental races with the likes of Mercedes, Peugeot, Ballot and the upstart American company Duesenberg. The Fiat design team, led by chief engineer Guido Fornaca, enjoyed the services of a brilliant young Piedmontese designer, Vittorio Jano, in addition to a staff of experienced, university-trained engineers. In those wildly creative times, when little was known of the metallurgy, combustion-chamber science, suspension geometry, aerodynamics, and so on that would ultimately hold the keys to speed and durability in automobiles, Fiat was an acknowledged leader in the new technology. This prompted an Italian authority on the subject to proudly reflect, "Fiat did not copy; it taught, after having created."

It was into this exclusive milieu that the poorly educated, still unhealthy, totally unqualified former Army farrier poked his prominent nose on a winter day in 1918. He was bearing a letter of recommendation from the colonel of his regiment, the contents of which are unknown. Whether or not it was merely a form letter outlining Ferrari's vague skills—presumably revealed while hammering nails into mules' hooves—is lost to history. Whatever the message, it was useless. The postwar job market was glutted with veterans and Ferrari was doomed from the moment he entered the mahogany-paneled office of engineer Diego Soria. Soria, whom Ferrari described as "stalwart, with close-cropped reddish hair turning gray," politely but firmly described Fiat as too small a firm to absorb the services of the thousands of veterans who were seeking work in Turin. Worse yet for Ferrari, residents of the city were getting the available billets.

Ferrari wrote of this moment with poignancy, describing how he lurched into the wintry gloom outside the Fiat offices and wandered through the busy streets to a bench in Valentino Park on the banks of the Po. There in the shadow of the immense Castello del Valentino, he said that he brushed away a layer of snow and sat down. "I was alone. My father and brother were no more. Overcome by loneliness and despair, I wept."

Fiat owed him a debt of honor that would have to be repaid. If, as Nietzsche wrote, a capacity for protracted revenge is the sign of a noble mind, Enzo Ferrari was an aristocrat of the highest order. The slight by Fiat festered in his brain, creating an anger that blossomed with the passage of time. Revenge was a priority within him that would not be subdued, and repayment in kind to the Agnellis and their Fiat minions was a debt he swore to fulfill, no matter how many years it might take.

3

I t was only days after Enzo Ferrari's arrival in Turin that he began to display the gritty resourcefulness and powers of persuasion that were to become his trademarks. Operating on what he described as a small inheritance from his father, he began to poke into the thriving community of automobile manufacturers, drivers, mechanics and promoters that had sprung up as satellites to the Fiat operation. By frequenting a network of bars and restaurants along the Corso Vittorio Emanuele II where racing champion Felice Nazarro and the equally prominent driver Pietro Bordino ate and drank, this twenty-one-year-old upstart made friends with some of the very same daredevils he had seen race as a small boy in the Emilian outback.

The Bar del Nord, near the immense Porto Nuova railway station, was a center of racing gossip and deal making, and it was here that he hooked up with a Bolognese car dealer named Giovanni, who hired the eager Ferrari on the spot. Giovanni was purchasing war-surplus Italian Army trucks—primarily small Lancia Zeta trucks—and stripping them to the bare chassis. He was then transporting them to a coach maker in Milan, where passenger-car bodies were fitted and sold to car-hungry postwar motorists.

Ferrari was taken on as a general handyman, first helping to remove the truck beds and cabs, then driving the bare-boned vehicles the hundred-odd miles to Milan. When he learned to drive is unknown, and it is unlikely that he did much motoring while shoeing mules in the Army, but by the time he reached Turin he had developed at least minimal skills operating the crude, cranky, unforgiving cars and trucks of the day. For the rest of his life, Ferrari carried Italian driving license number 1363, implying that he was one of the nation's earliest motorists and was driving well before his release from military service.

Whatever his credentials as a driver, Ferrari found himself crashing over the bumpy roads of the Po Valley, ferrying skeletal surplus Army trucks to the tiny Carrozzeria Italo-Argentina in the center of the giant, grimy, industrial and political hotbed of Milan. But this was hardly work for a man who had aspired to employment with the elite firm of Fiat, and he quickly moved to improve himself. Once again he made for the trattorias where the racers and automotive sportsmen of Milan gathered—places like the Vittorio Emanuele Bar, where he made the first important decision of his young career.

Ugo Sivocci was a brash, mustachioed former bicycle racer who had moved into the world of cars, first with the small Milanese firm of De Vecchi, then as chief test driver for the new automaker CMN (Costruzioni Meccaniche Nazionalia). It had gone into business during the war manufacturing four-wheel-drive artillery tractors powered by Isotta Fraschini engines. Now the company, located on the Via Vallazze, was embarked on a full-scale campaign to manufacture high-performance sporting vehicles. And as was the universal custom of the day, the factory was entering a variety of races and hill climbs to demonstrate the prowess of its products. Parlaying his modest credentials as a driver, no doubt with considerable hyperbolic flourishes, Ferrari was hired by Sivocci as his assistant. He had begun his climb up the ladder.

Their initial joint assignment was to act as the support vehicle for ace motorcycle racer Marco Garelli as he ran a team in the Raid Nord-Sud endurance run from Milan to Naples. Sivocci and his new sidekick were to carry extra fuel, tires and spare parts for Garelli's twin-cylinder, two-stroke motorcycle during the event—hardly a glamorous assignment, but still Enzo Ferrari's first taste of competition.

His initial legitimate racing experience in an organized contest came on October 5, 1919. He and the tiny CMN team were entered in a regional hill climb called the Parma–Poggio di Berceto—an open-road contest leading into the Apennines over a distance of 53 kilometers. Ferrari was assigned one of the 2.3-liter Tipo 15/20 CMN tourers, its bodywork a mere shell for the engine, two seats and a pair of spare tires hung over the wooden-spoked rear wheels. Using a riding mechanic drafted by the team in Parma, Ferrari picked his way along the twisting route up the slope of Mount Piantonia to finish fifth in his class, but well down in the overall standings. It was hardly a dazzling debut, but considering the limited potency of his rather crude automobile and his rank-novice status, the fact that he figured at all in the final standings is worthy of mention. Sivocci, by contrast, finished second overall.

He raced again in November 1919, when he and Sivocci were entered in CMNs for the Targa Florio, an open road race around Sicily organized by Vincenzo Florio, scion of a powerful Sicilian clan that controlled immense tracts of land and a fleet of ships and wielded great political influence on the island.

In those days many racing cars were simply modified passenger cars bereft of bodywork and carrying more powerful engines. The CMN was nothing more than a hot-rodded tourer, unlike the elite of the sport, the Fiats, Peugeots, Ballots, Duesenbergs and Sunbeams that competed in the continental Grand Prix events and at such unlikely venues as the Indianapolis Motor Speedway in Indiana. Those cars were specially engineered, thoroughbred racing machines representing the best available technology—much of it fresh from the brutal but highly creative crucible of the war. They took advantage of the metallurgy, engine design and fuel formulas that had been learned during the feverish effort to build faster airplanes and tougher trucks and tanks for the front lines.

The tiny CMN firm could not afford to transport its race cars to

Sicily by rail or truck, as was the custom for the larger teams, and Ferrari and Sivocci were assigned the job of driving to the race. It was hardly a leisurely tour through the countryside. Ferrari recalled that they drove into a nasty blizzard in the Abruzzi hills. There they were set upon by a pack of wolves, which had to be dispersed by shots from the pistol he always carried under his seat. The pair struggled southward to Naples, barely arriving in time to board a Florio-owned ferry, the *Città di Siracusa*, for the night trip to Palermo—a journey he recalled featuring "a rough sea and the assaults of every kind of bug."

This would be a serious motor race. The Targa, as it was known, had been started in 1906 and was composed of a series of long laps around the craggy, rock-strewn, bandit-infested Sicilian countryside. (However, owing to the power and prominence of the Florio clan, all weapons were sheathed during the event and no one ever threatened the participants.) Ferrari's crude little CMN looked like a buckboard compared with some of the more exotic machines from Fiat and Peugeot, including a magnificent prewar L25 Peugeot Grand Prix car driven by the ace André Boillot. When Ferrari and his mechanic, Michele Conti, mounted up for the start, they faced four laps of 108 kilometers each (approximately 67 miles) and had little chance of success against the faster cars.

Ferrari drove as hard as his courage and his limited experience would permit, grinding through the gears as the little CMN bounced and juddered over the dirt and gravel roads. Early in the going the wagon-rutted surfaces jarred the fuel tank loose from its mountings and Ferrari and Conti lost forty minutes jury-rigging it back in place. Then as he thundered along the coast road east of Palermo he was flagged to a halt by a carabiniere on the outskirts of the village of Campofelice. Furious, he demanded to be let through. Impossible, said the policeman. The prime minister of Italy, Vittorio Emanuele Orlando, was giving a speech. The road was blocked until he finished. (Only the Italians, it seems, would schedule a motor race and a political address on the same day in the same place.) Helpless, Ferrari and Conti sat fuming in their car until the prime minister said his goodbyes. Only then were they permitted to move ahead, still being forced to straggle along behind the prime minister's giant black De Dion Bouton sedan until he turned off the main road.

The road cleared and Ferrari charged on toward the finish, sliding the wobbly machine through the gravel-strewn bends and thumping, full-bore, through the tiny, sunbaked villages that lined the circuit. With springs no more supple than those of an oxcart, and an open exhaust pipe blaring in the riders' faces, the CMN was a crude, hostile device that, like an unbroken horse, seemed to savor the opportunity of bucking its passengers into the hostile countryside. Still Ferrari pressed onward, brashly slewing down off the craggy hills toward the finish on the edge of Palermo.

By then Boillot had won, although he had clipped a spectator who had wandered into the road at the finish line and killed him. It was nearly dark when Ferrari finally straggled in. The area was deserted, save for a local policeman, who was doing what remained of the official timing with an alarm clock. Ferrari dismounted and chased down Florio to complain that he had been held up by the speech and therefore had been unfairly treated. A shouting match ensued. "What are you grumbling about?" asked Florio. "You were late, you risked nothing and we are even making you a present of including you in the classification!" So, in his second effort as a racing driver, Ferrari was "given" ninth place overall and third in his car class, which was, he said, "supposed to represent a success, albeit a small one."

Ferrari soon left CMN to seek his fortune in other realms of the Italian automobile business and to further his career as a racing driver. His exact movements are unknown, although it is documented that he appeared back at another running of the Parma–Poggio di Berceto hill climb in May 1920 with a car he at least owned in part. It was a prewar 7.2-liter Tipo 1M Isotta Fraschini of a type that had been run in the 1914 Indianapolis 500 (where one had blown a tire, flipped and caused a multi-car wreck). The big two-seater was noteworthy because of its four-wheel brakes. At the time most racing cars used brakes only on the front wheels, owing to the complexity of the cable-operated systems and the designers' inability to balance the stopping capacity between the front and back pair. In that sense the 130-hp Isotta was an advanced machine, and how and with whose funds Ferrari obtained it is unknown.

We do know the results of his three-race campaign in the Isotta Fraschini with riding mechanic Guglielmo Carraroli: a third overall and a second in the over-6-liter class at Parma–Poggio di Berceto, and two retirements with mechanical ailments at two minor events run in June of the same year. Shortly thereafter the Isotta disappeared into limbo and Enzo Ferrari began an association with Alfa Romeo that was to last for nearly two decades.

He was hardly aligning himself with a Fiat or a General Motors or one of the French or English auto-making giants. Alfa Romeo had been started in 1909 by Cavaliere Ugo Stella, who was formerly the manager of the Italian division of Darracq, the French carmaker. The plan was to take over the financially decrepit Darracq facility at Por-tello, on the outskirts of Milan, and to begin the manufacture of a line of all-Italian passenger cars under the name ALFA—an acronym for Anonima Lombardo Fabbrica Automobili (Lombardy Motor Manufacturing Company). By 1910 Giuseppe Merosi, a Piacenza native with an engineering background at Bianchi, and a surveyor by training, had designed a range of ALFA cars and production began. Four years later the company was sufficiently successful to warrant the fielding of a 4.5-liter Grand Prix machine.

When Italy entered the war ALFA began producing artillery shells and Caterpillar-type military trucks. At this point the company fell into the hands of a brilliant industrialist and politician named Nicola Romeo, a native of Naples and a professor of mathematics in Milan before entering the business world. Romeo was such a persuasive speaker that he later gained the name "the Siren" in the Italian Sen-ate. He also changed the name of his new car company to Alfa Romeo.

The details of the initial contact between Enzo Ferrari and Alfa Romeo are unclear except for two elements: Ferrari's purchase of a car and the presence of a brilliant young engineer named Giorgio Rimini, a crafty Catanian whom Ferrari remembered as having a cigarette seemingly glued to his lips. When Ferrari met him, Rimini was a jack-of-all-trades at Alfa, acting as the chief of the racing team and test department and also as the general sales manager. Whether Ferrari hooked on as a test driver with Alfa on the basis of his limited experience with CMN before or after he purchased a car from Rimini is unclear, but somehow a 6-liter G.I. Alfa touring car played a role in the connection.

In one of the few recollections in which Ferrari made himself the butt of a joke, he wrote that he bought the G.I. from Merosi with the loony caveat in the contract that the car would be delivered "as soon as possible and even sooner." When the delivery date fell hopelessly behind schedule, Ferrari protested to Rimini and this clause in the contract was produced. Ferrari learned, as he said, to thereafter read all contracts word for word. But the fact that Ferrari bought the G.I. Alfa Romeo in 1920, at the age of twenty-two, is intriguing. The Alfa was an expensive luxury model (based in part on the American Pierce Arrow—at least in terms of its production techniques) and seemingly out of reach of most young men from a self-described poor household. This being his second rather exotic machine, it is obvious he was a man of some means early on, either from his inheritance or from his activities with CMN and his limited motor-racing schedule—or more likely from a combination of all three.

There is no question that by this time Enzo Ferrari had become a compelling salesman and manipulator in the Milan-Turin car axis. He was already nearly six feet tall and towered over most of his contemporaries. His face was dominated by a large aquiline nose and baggy, hooded eyes that were almost reptilian when in repose. His hair was thick and black and combed pompadour style in shaggy waves. It was clear even then that his physiology doomed him to gain weight easily and, like his father, to gray prematurely.

Despite his obvious powers of persuasion and a sufficiently vivid personality to push his way into the inner sanctums of the Alfa Romeo test department, Ferrari remained gripped by his lower-class, bumpkin image. At one point he lamented, "I feel so provincial, so Emilian." But that sense of inadequacy was more than compensated for by his streetwise cunning and a brash ability to deal with social and professional superiors like a man twice his age. This skill was surely identified by Rimini, who made him a part of the Alfa team, although on a decidedly junior basis. It may be unclear whether the G.I. Alfa was purchased before or after his employment, but there is no question that he was at the Targa Florio on October 20, 1920, at the wheel of a prewar Alfa Romeo Tipo 40-60. He was a member of a three-car team led by twenty-eight-year-old Giuseppe Campari, a beefy, high-living sometime opera singer who was to become a close friend and associate of Ferrari's.

Over the years the term "Grand Prix" has been much misused and misunderstood. "Grand Prizes," as it were, had been awarded for decades for all manner of contests, products, works of art, and so on before the French term was applied to motorsports in 1906. Organized motorsports began in the mid-1890s, with the Automobile Club de France (ACF) having been formed in 1895. This organization was to assume a dominant role in international motorsports and to govern the rule-making for the next sixty years. In the beginning motor races were run as a sort of "Formula Libre"; that is, no specific engine or chassis specifications were required for entry. But as the industry developed, it was recognized that complex rules (which the French were masters at creating) would be necessary both to keep the competition even and to generate technical advances. Numerous major international events were staged in Europe and America during the first five years of the new century. Major race patrons like W. K. Vanderbilt, newspaper magnate Gordon Bennett, and the aforementioned Vincenzo Florio offered cups and trophies and international prestige for victory, but their rules were diverse and often contradictory. In 1906 the first Grand Prix of France was organized by the ACF and it formed the basis for international motorsports for years to come. From that point on, a pure "Grand Prix" was a race run according to detailed specifications for cars, race distances, and competition regulations. By the early 1920s most nations in Europe had a "Grand Prix" race run according to ACF rules. The term "Grand Prix" would often be bastardized and misapplied, but in its classic sense it refers to a series of international races run according to a set of stringent—and highly sophisticated—automotive specifications. In the beginning Grand Prix cars bore a faint resemblance to passenger automobiles, but by the end of World War I they had become light, powerful, open-wheel machines with streamlined bodies that were light-years ahead of their production counterparts in terms of raw speed and engineering.

That having been said, the Targa Florio was not run according to Grand Prix rules, but more as a Formula Libre competition. Hence the entry of a pure Grand Prix car for Campari and a more ordinary road machine for Ferrari. But this time he was more successful. With his heavy machine fitted with ungainly, Dumbo-eared front fenders to protect himself and riding mechanic Michele Conti from the mud that coated the 108-kilometer circuit, he managed a second overall in the difficult race. Although many of the faster cars had retired from the race, including that driven by his Alfa teammate Campari, it was a stirring finish for the young man and no doubt briefly convinced him that his future as a racing driver was assured.

But the modest firm of Alfa Romeo was still operating in the minor leagues of the automobile business. Chances of victory in the major events seemed a distant hope, and even with a brilliant driver like Campari on board, the might of Fiat, Mercedes, Sunbeam, Ballot et al. was simply too great to even consider a high placing in the major continental Grand Prix events. But Ferrari plunged into his job with typical vigor and in early 1921 he persuaded Ugo Sivocci to jump ship from CMN and join Alfa in the test department. Sivocci proved so adept at the job—both on the racetrack and on the highway—that Rimini quickly appointed him chief of the department.

During the early 1920s Ferrari was flitting back and forth between Milan and Turin, engaging in all manner of deals—selling and trading cars, buying parts, doing industrial spying on Fiat, gathering gossip, delivering cars to customers—as well as racing. As he grew older, he was to become anchored to Modena and Maranello, but during this period he bordered on the peripatetic, traveling at least once to

England and France and moving almost constantly around northern Italy and Switzerland. It was during one trip to Turin that he met a fair-haired twenty-one-year-old girl near the Porto Nuova railway station in central Turin. Her name was Laura Domenica Garello, and she was a native of the village of Racconigi, twenty miles south of Turin. She was the daughter of Andrea Garello and the former Delfina Parchetti. Both came from simple peasant stock and there is nothing in their backgrounds to provide a basis for later claims that Laura Garello came from landed wealth or that she was wellborn in any sense of the word. Laura was a rather thickly built woman with darting dark brown eyes and a quick, coquettish smile. Her heavy rural dialect was decidedly working class. Why and how the two met in Turin is unclear, although rumors persisted among those who knew them best that Laura was a café dancer who operated among the bars and hangouts frequented by the racing crowd. Whatever the circumstances, Ferrari seized on the relationship and swept her into his orbit. They began to travel together and she appeared regularly at the race meetings where Ferrari competed. By 1922 she was being referred to as Signora Ferrari although the two were certainly not married. A number of letters exist from the early twenties in which Ferrari corresponded with Laura in Genoa and Turin. They are addressed to "Mrs." Enzo Ferrari, but they were not married until 1923.

This charade may have been played for a number of reasons. It is possible that he wanted to spare the sensibilities of his mother, with whom he maintained a close and docile relationship. Moreover, while mistresses were accepted in appropriately shadowy circumstances among the upper classes, it was more difficult for a lower-class man like Ferrari to flaunt such a relationship. Even at that young age he was keenly attuned to the political atmosphere at Alfa Romeo and may have decided that his fortunes would be safer with a wife on his arm rather than a girlfriend with a questionable background.

Despite the addition of Sivocci to a driving team that already included Campari and the superb young Antonio Ascari, Alfa Romeo was hardly ready for prime time. The chief designer, Giuseppe Merosi, as noted, was a surveyor by training, and was referred to by the title "Geometra" for most of his life. His closest associate, Antonio Santoni, was a pharmacist. Neither had the necessary background to compete with the likes of Fiat and Mercedes, with their legions of highly trained mechanical engineers and technicians. The Merosi-designed automobiles were workmanlike tourers that could be converted to usable racing machines, and it was in equipment such as this that Enzo Ferrari saw limited competition, racing five times for the team in 1921, always in a backup role to Campari and Company, and recording a pair of fifth places at the Targa Florio and at Mugello. Admittedly the first-stringers were not doing much better (Campari was third in the Targa), but there is no question that Ferrari was never considered to be the equal of Ascari, Campari or Sivocci as a racing driver. But his skills at off-track organization were in the end to far transcend the value of the others to Alfa Romeo and the Italian automobile industry.

But while Ferrari immersed himself in the internal politics of Alfa Romeo, the nation was engaged in an infinitely more complex and desperate struggle of intrigue and turmoil. By 1919 Benito Mussolini had become a major political force, creating through his powerful oratory and inflammatory writing a Fascist Party which seemed to offer a morsel of satisfaction—wrapped in bravado and wild promises—for everyone. Once a socialist, Mussolini had switched horses to become a rabid anti-Bolshevist, in part to seduce the wealthy industrial classes. Ever the panderer, he pledged to the stolid *borghese* a return to law and order while the workers were wooed by a siren song of better wages and a liberal contract of cooperation between themselves and management for the glory of Italy. Backed by his black-shirted thugs, Mussolini organized a march on Rome in October 1922. (He let his minions make the rainy trek from Milan while he arrived a day later by train.) Mussolini took over the tottering government from the discredited prime minister, Luigi Facta, and diminutive, weak-willed King Victor Emmanuel II gave him total dictatorial powers.

Ferrari appeared untouched by these events, immersing himself instead in the world of motorsports and the internal politics of struggling Alfa Romeo. By 1922 he returned to Modena to form Carrozzeria Emilia, Enzo Ferrari & Company in a small rented garage on the Via Emilia a kilometer east of the walled gates of the old city. The use of the term *carrozzeria* in the title is mysterious, in that it relates to coach building. There is no evidence that Ferrari's tiny firm functioned as anything but a minor outlet for Alfas and used sporting machinery. It never engaged in the sort of custom automobile body fabrication the name implies. Carrozzeria Emilia did, however, create for him a base of operations on his beloved home turf, although his travels to Milan and Turin continued at a frantic pace.

His aggressive sales tactics permitted him to become the exclusive sales agent for Alfa Romeo automobiles in all of Emilia. By using his growing reputation as a racing driver he artfully enhanced the marque's image among the rich clientele in the region.

As his financial fortunes in the car business increased, his racing career entered into a severe decline. Rimini gave him but three rides in 1922, although two were at the wheel of an older Alfa Romeo 20-30ES. In July he defected from his corporate loyalty and drove an Austrian-built 6-cylinder Steyr-Puch sports car in a minor hill climb that ran up the Alpine St. Bernard Pass from the city of Aosta. The winner was a diminutive Torinese member of a prominent coach building family named Giovan "Pinin" Farina.

A development that surely caught Ferrari's attention in 1922 was the construction of the giant Monza autodrome in the Villa Reale Park on the outskirts of Milan. For years the leaders of Italian motorsports had been seeking a major venue where a world-class Grand Prix race could be staged to rival that being held annually in France. Several permanent, specially designed supertracks had already been built for racing and testing, such as Brooklands in England and the Indianapolis Motor Speedway in the United States. Moreover, it was openly discussed that immense tracks would be built at Linas-Montlhéry outside Paris and on the outskirts of Cologne. (The legendary Nürburgring would ultimately be built in the Eifel district, but not until 1927; the Montlhéry track would open in 1924.)

This massive, sometimes ominous track at Monza would enter into Enzo Ferrari's life repeatedly in the years to come. All manner of triumph and grinding tragedy would unfold for him there, but ironically he never rolled a wheel in anger on its smooth, tree-shaded surface. He tested a number of Alfa Romeos there in the 1920s and 1930s and surely made numerous laps in his own machines after the war, but he never actually competed in a race on that famed circuit. For all his attempts at reaching the big time—and Monza was to become the embodiment of major-league racing in Italy—Enzo Ferrari would be doomed to the role of spectator.

But whatever frustrations his racing career was generating, Ferrari's great leap forward in the world of motorsports came in 1923. Ferrari established a brief relationship with Dr. Albert Schmidt, the Genevan car builder and engineer, operating as his agent in Italy. He traveled with a party of Alfa Romeo senior executives, including Rimini, to the 1923 Paris automobile show and also made a trip to England, where he visited the immense new Brooklands supertrack near Weybridge and met Mario Lombardi, the British representative for Pirelli tires. According to Ferrari, Lombardi was "a man who came to count much in my life and my thoughts." Ferrari would race but four times during the year. Yet his stature as an organizer and leader of men would soar to unexpected heights, in the main because he was able to strengthen the Alfa Romeo team and aid in propelling it to the very pinnacle of the sport. It would be June before his luck would change. His new Alfa Romeo Tipo RL/TF had broken at both the Targa Florio and at Mugello, and he entered the Circuito del Savio outside Ravenna with limited enthusiasm. His riding mechanic was Giulio Ramponi, the nephew of the great Campari and generally the personal mechanic and riding companion of Antonio Ascari. Ramponi was a combative, capable member of the Alfa team who would also play an ongoing role in the Ferrari drama.

This time his fortunes changed. Running against amateur and semiprofessional opposition, Ferrari and Ramponi won the Circuito del Savio at Ravenna, a 25-lap, 359-kilometer race that led to a meeting that was to have an enormous impact on Ferrari. It is an incident that has been recounted dozens of times, being warped and twisted a bit more with each telling until the truth has been almost obscured. Let Ferrari tell it in his own words, with some observations to follow:

"In 1923, when competing in the first Circuito del Savio at Ravenna, I made the acquaintance of Count Enrico Baracca" (the father of Italy's number one flying ace, Francesco Baracca, who had been shot down after recording thirty-four German kills). "As a result of this meeting, I was subsequently introduced to the ace's mother, the Countess Paolina Baracca, who one day said to me, 'Ferrari, why don't you put my son's prancing horse on your car. It will bring you luck.' I still have Baracca's photograph, with his parents' dedication in which they entrusted the horse to me. The horse was, and has remained, black, but I myself added the gold field, this being the color of Modena."

So goes the Ferrari version of the legend of the *Cavallino*, the Prancing Horse of Ferrari, which has become one of the most famous logos of the twentieth century. But his story is rife with alterations and omissions. To begin with, the emblem was not Baracca's personally, but that of his own Squadriglia 91a. (It was still employed even after World War II by the jet squadron 4a Aerobrigata, and therefore was hardly his mother's to give away.) As has been mentioned, Ferrari's brother, Alfredo, was apparently a member of the ground crew for the squadron, and it is possible the countess suggested that the escutcheon be adopted in his memory. There is a reliable story that Baracca took the Prancing Horse after he shot down a German pilot over Tolmezzo in November 1916. The German, who was flying an Albatross BII, was from Stuttgart and was carrying the emblem of his city—also a prancing horse—on the side of his airplane. Therefore it is logical to assume that the *Cavallino Rampante* so revered by Ferraristas is actually of German origin and directly related thematically to that later carried by Ferrari's archrival from Stuttgart, Porsche! Examination of the two logos reveals a startling similarity. As for the countess giving Ferrari the emblem for "luck," one must puzzle over what kind of good fortune it brought to her son.

The final irony of the story of the Prancing Horse: While Ferrari sentimentalized the moment and the significance of the gift, he put it in a drawer for *nine years*, and it did not appear on the hood of one of his racing cars until the 24 Hour sports-car race at Spa, Belgium, on July 9–10, 1932!

The Italian automotive historian Luigi Orsini suggests in his excellent history of the Scuderia Ferrari that the Prancing Horse was actually presented to Ferrari in the 1930s just prior to its employment at Spa, but Ferrari himself repeatedly leaves the impression in his autobiographies that the gift was made shortly after his victory at Ravenna.

Whether the *Cavallino Rampante* was offered in 1923 or in 1932 remains a mystery, as do the exact origins of the emblem, although it is known that in later years Ferrari was unable to trademark the Prancing Horse due to complications arising from Stuttgart's proprietary claim on the animal.

Enzo Ferrari and Laura Garello finally made it official on April 28, 1923, after living together for at least two years. The marriage took place in a small Catholic church near the massive Fiat Lingotto works in a working-class neighborhood of Turin. The union had long been opposed by Laura's parents, whom Ferrari described as a "hard-working couple from modest circumstances." A small knot of guests, all from Laura's family, witnessed the ceremony. One can almost visualize the tiny cluster of swarthy men and women dressed unremittingly in coarse black, contrasting with the shimmering white dress of the bride, smiling woodenly, stepping into the hazy sunshine from the darkened precincts of the church. Surely Father Alberto Clerici, who performed the ceremony, had seen a hundred such couples blink uncertainly at the outside world, facing what he knew to be the drudgery and stylized artifice of the Italian lower-class lifestyle. Little did he know that the tall, barrel-chested groom with the distant gaze already harbored aspirations that would carry him far away, in terms of both geography and financial success. Ferrari's gift to Laura was a small gold purse. This entire episode was treated with shocking insensitivity by Ferrari. He later wrote, "I married young, somewhere around 1920 [*sic*]. I cannot remember the exact year, as I have mislaid the marriage certificate." He added in the third person, "This young man declared that nothing else mattered where there was love. I later came to realize that the rest did matter and matter a lot."

That Ferrari's marriage was to descend quickly into a legal arrangement is an understatement. While numerous photographs exist of him and his new wife at various races in the early 1920s, he soon became the traditional Italian husband, seeking sexual conquests not so much for pleasure but for simple gratification of the ego. Enzo Ferrari was to remain obsessed with sex for most of his life, and it was probably within months that his marriage vows to Laura were shattered. Years later he was to remark to Romolo Tavoni, a veteran racing manager and close personal associate, that "a man should always have two wives."

Ironically, Ferrari's business fortunes improved due to an event he did not attend. In June 1923 the French Grand Prix had been run at Tours and there had been a fight on the powerful Fiat team. Luigi Bazzi, a thirty-year-old expert technician and engine tuner, had argued with the factory's chief engineer, Guido Fornaca, and upon returning to Italy was approached by his friend Ferrari with a proposal. Because of the acrimony at Fiat, Ferrari suggested a move to Turin and a position with the Alfa Romeo racing department. After all, Merosi and company were in the final stages of developing a new 2-liter supercharged Grand Prix car—the GPR—that would be introduced on September 9 at the European Grand Prix to be held at Monza. Using the seductive powers that would become legendary, Ferrari lured Bazzi, a quiet, steady, tough-minded native of Novara, to Alfa Romeo. It would begin a relationship that would keep the two men together in the automotive trenches for sixty years.

Merosi's new racing car had potential. The GPR, or "P1" as it would become known in the years to come, was a well-conceived two-seater (a riding mechanic was still permitted by the rules), and with Ascari returned to full health to join Campari and Sivocci, Alfa Romeo appeared ready to join the front rank of Grand Prix manufacturers. But the team's hopes were cut short by tragedy.

On the day before the European Grand Prix at Monza, the field of cars, the Fiats, Sunbeams, Bugattis, Alfas, et cetera, joined by the great American Indianapolis champion, Jimmy Murphy, aboard a very fast, technically advanced Miller straight-8, engaged in some high-speed practice laps. All returned to the pits safely, save one. Ugo Sivocci, aboard his new Alfa P1, spun off one of Monza's ultra-fast bends and was killed. Sivocci was gone. The man who had brought Ferrari into the sport and given him his first chance to drive was dead. He would be the first of literally dozens of men who would fall around Ferrari, victims of this crudest of sports. Sivocci's death, plus the fact that the P1 did not seem quick enough to run with the fastest cars, prompted the Alfa team to withdraw. The race was to be dominated by a pair of Fiat 805s with Murphy third in the Miller.

The team drove the few miles back to the Portello works in shambles. Not only had they lost one of their ace drivers, and a close friend, but the P1 needed serious debugging and reengineering. Bazzi had a suggestion. There was a brilliant engineer on the Fiat staff who might be lured to defect. His name was Vittorio Jano. The Alfa Romeo staff of course knew of the thirty-three-year-old Torinese and were aware, through their Fiat contacts (some made by Ferrari), that the man was a legitimate genius with the new supercharged engines that were beginning to dominate the sport.

Vittorio Jano was a member of a family with a long and proud tradition in the technical fields. His father had headed one of the major Italian military arsenals located in Turin. Jano was a graduate of a technical institute in that city, the Istituto Professional Operaio di Torino, and within two years of obtaining his degree (not a full engineering diploma, more on the order of a technician), at age eighteen, he had moved into the prestigious Fiat engineering department.

While much has been made of Enzo Ferrari's role in persuading Jano to move from Fiat to Alfa Romeo, the eminent automotive historian Griffith Borgeson, following extensive interviews with Jano prior to his death in 1964, recorded the most accurate version. He says Jano claimed that Rimini, not Ferrari, instigated the contact that Ferrari was assigned to make in the late summer of 1923.

Because Ferrari was familiar with the automotive politics in Turin and because he owned an Alfa, it makes sense that Rimini chose him to undertake the Jano mission to Turin. It was a drive Ferrari had made hundreds of times and there is no doubt he had little trouble finding Jano's third-floor apartment on the Via San Massimo. He recalled that he was greeted by Jano's wife, Rosina, who said that it would be out of the question that her husband would leave his native city. Then Jano arrived home. Ferrari reported, "We had a talk. I told him the advantages of joining Alfa and the following day he signed up." Once again Ferrari embellished the story.

Jano told Borgeson that he received Ferrari's offer coldly, stating that if Alfa desired his services, more senior officials would have to enter the negotiations. Only after Ferrari transmitted this rejection to Milan did Rimini—and perhaps Nicola Romeo himself—bring about the transfer of Jano to their company. Jano told Borgeson his salary was raised from 1,800 lire per month to 3,500 lire per month and he was given an apartment as a further inducement to make the move to Milan.

Vittorio Jano was a quick worker. Within months he had transformed Merosi's P1 into one of the greatest racing cars of all time. His P2 was a masterpiece, if highly derivative of the work being done at Fiat on their latest racing machines. He knew the Fiat's weak points: feeble valve springs and engine blocks that became distorted when hot. He also knew how to correct them. His P2s were powered by 8-cylinder, 16-valve, 2-liter engines developing 135 hp thanks to their Roots-type superchargers. The brakes were hydraulic, similar to those introduced by Duesenberg two years earlier, and carried for the first time the "balloon" tires that had been introduced at Indianapolis.

The car was an immediate success. With Antonio Ascari back at peak form, and judged to be on the verge of rivaling the all-time greats of the sport, he and Bazzi teamed up to win at Cremona. Their P2 was clocked at 121.1 mph on the circuit's six-mile straightaway and set a record for a lap around the sprawling open-road course of 100.8 mph. Suddenly a new player was on the crowded Grand Prix field and the Alfa management responded by engaging a powerful team of drivers. Joining Ascari and Campari was the wily French veteran Louis Wagner, who, at forty-two, boasted a racing career that dated back to 1906, when he had won the Vanderbilt Cup on Long Island. Also on the squad, as a sort of junior varsity member, was Enzo Ferrari.

He bought from the factory a Tipo RLSS sports car and did rather well, winning the Chilometro Lanciato in Geneva and taking a second straight win at Ravenna. He won as well at the Circuito del Polesine at Rovigo and ran second at the Corse delle Torricelle. Perhaps his greatest moment behind the wheel came in July 1924 at Pescara on the Adriatic coast. He was entered with Campari as an Alfa team driver, except that Campari was aboard one of the P2 Grand Prix machines while Ferrari was assigned one of the 3.6-liter Tipo RL/TF sports cars—a heavier, less nimble car from the Merosi design school that was expected to serve as backup to the faster new model from Jano's drawing board.

But Campari broke his gearbox early, and Ferrari found himself in the lead, far in front of a pair of powerful Mercedes manned by amateurs Count Masetti and Count Bonmartini. Campari had cleverly pushed his broken Alfa into some underbrush on the side of the road, which left his competition thinking he was still running somewhere on the immense triangular circuit, which was 15.8 miles in length, with a pair of four-mile straightaways and a series of treacherous bends that snaked through a number of tiny villages. The race was called the Coppa Acerbo after Captain Tito Acerbo, brother of Professor Giacomo Acerbo, a Mussolini cabinet member and a powerful ally of Il Duce. In fact, that same year he had craftily maneuvered a bill through the Chamber of Deputies that had given the Fascists two-thirds of the seats and had handcuffed the nation to Mussolini and his party. Acerbo's Cup had been offered in memory of the late captain, who had been killed in World War I, and carried with it the award of Cavaliere della Corona d'Italia. Much has been made of this award by some Ferrari acolytes, but it was essentially meaningless. The Cavaliere award was so common that in the nineteenth century King Umberto was heard to scoff, "A cigar and a cross of Cavaliere—one does not deny to anyone." In *The Italians*, Luigi Barzini noted his countrymen's fascination with titles thusly: "It is in the use of academic or other titles that people affix to your name, as if to prove that you so visibly deserve such honors that it is impossible that you have not been awarded them." A middle-class man is called Dottore in his youth and becomes Commendatore, or Knight Commander, when over forty. Ordinary letters are addressed "illustrious," "celebrated," "renowned," "Signore," or simply "N.H."—the abbreviation for Nobil Uomo. The award of Cavaliere to young Ferrari was a simple formality automatically offered to the winner of the race and meant little in a nation obsessed by hyperbolic flattery and hollow titles.

But the victory at Pescara did produce more tangible dividends. He had driven well, and his performance sufficiently impressed the Alfa Romeo brass that he was promoted to the full Grand Prix team for the upcoming Grand Prix of Europe, to be held on a rugged network of public roads near Lyon, France. He was to join Ascari, Campari and Wagner in an event that was surely the most prestigious and most hotly contested of all the major European races. Again, adoring biographers have made much of this promotion, but in fact Enzo Ferrari was taken to Lyon as a substitute for Count Giulio Masetti—who himself was a second-rater and decidedly the junior member of the team. Nevertheless, this selection was a high-water mark for Ferrari as a racing driver and afforded him the chance to demonstrate his talents against the likes of his brilliant teammates, not to mention the great Pietro Bordino from Fiat, Sir Henry Segrave and Dario Resta with their Sunbeams, Robert Benoist and René Thomas aboard French Delages, and others.

This never happened, and the debate rages as to exactly why. Ferrari appeared at Lyon with the Alfa team and practiced briefly on the tricky, open-road fourteen-mile circuit, but without warning boarded a train and returned to Italy. In his memoirs he feebly explained, "I was not well all that year, being seriously run down. My indisposition, in fact, was grave enough to compel me to cut down and practically give up driving. This was the beginning of trouble with my health that was to afflict me throughout the years to come." What was the nature of this serious ailment? Some have suggested it was a nervous breakdown; others feel it may have been heart trouble arising from his wartime illness. A few darkly imply that perhaps it was the beginnings of his connection with a dreaded ailment, rumored to be syphilis—the reality, or the rumors of the reality, of which would dog him to his grave.

Giovanni Canestrini, the doyen of Italian motorsports journalists during the 1920s and 1930s, stated flatly that Ferrari was afraid. He believed that the young driver from Modena—who was then but twenty-six years old and had but twenty-seven races under his belt— was simply daunted by the level of competition and the nasty, blind-hilled Lyon circuit and went home. This remark by Canestrini was openly circulated and enraged Ferrari, who refused to forgive him until a truce was finally arranged thirty-five years later! Ferrari's long-time friend Gino Rancati made the following cryptic observation about Ferrari's race-driving abilities in his openly affectionate biography published in 1977: "He possessed certain limitations as a racing driver: an excessive respect for the machines entrusted to him [and] perhaps not the highest form of courage."

Whatever the actual reason, his defection at Lyon doomed his chances for further efforts with the Grand Prix team and he returned to Modena to concentrate on the automobile business. He simply stopped trying to further his sporadic racing career.

The Alfa Romeo team, meanwhile, hardly suffered from his absence at Lyon. Campari won after his teammate Ascari suffered a late-race engine failure. Sir Henry Segrave finished second in a Sunbeam, while Ascari had enough of a lead when he broke down to end up third. The once-dominant Fiats were so badly beaten that the Turin firm never raced seriously again. An effort was made a few years later to field a Grand Prix car, but the torch of Italian motorsports honor was effectively passed to the upstart Alfa Romeo team on that warm August day in 1924. While Ferrari could accept no credit for beating Fiat on the track, his contribution to their overall defeat by aiding in the defection of Bazzi and Jano was pivotal. Following the pair's flight to Milan, the once-unbeatable Fiat racing department was in shambles. Worse yet, another excellent engineer, Vincent Bertarione, had already moved to Sunbeam in 1923, leaving little technical reserve when Jano departed. Those losses, plus other internal policy changes, prompted Fiat to drop racing after the 1924 season. A short-lived return would be made a few years later, but for all intents and purposes they departed the sport forever. Considering his deeply rooted sense of honor and his need for revenge, one can be sure that this capitulation by the company that had rejected him six years earlier was lustily celebrated by Enzo Ferrari.

Ferrari maintained that he was uninterested in politics and, in view of his antipathy toward Rome and a general disregard for the Catholic Church—which was deeply embroiled in the Italian political scene at the time—there is little reason to doubt him. He did, however, make friends with numerous Fascist officials because most of those with power and influence in the domestic automobile industry maintained tight links with the party. Enzo Ferrari was learning to be a supremely political man, but only in the sense that this would advance his own fortunes, not those of any particular ideology. He met Benito Mussolini only once, in early 1924. Il Duce was at the height of his popularity at the time and stopped in Modena during a circuitous motor trip from Rome to Milan. He made stops along the way to meet with prominent Fascists and party sympathizers and to mend political fences in such backwater locales as Modena. There he met with a prominent Fascist senator for a long, typical Italian meal and talkfest. Because Mussolini remained loyal to Alfa Romeo during his entire reign—perhaps because the marque was manufactured in the city where he rose to power—he was driving a new three-seat sportster when he arrived in Modena. Being the local Alfa concessionaire and the city's most prominent motorsports personality, Ferrari was selected to escort Il Duce out of town. He tried to drive carefully on the rain-slickened cobbles, but still ran at a higher velocity than the less skilled Mussolini was capable of maintaining. After several wild skids by the proud but inept leader, Ferrari slackened his pace until the journey ended a few kilometers later.

Ironically, Alfa Romeo would follow Fiat to the motorsports sidelines. On July 26, 1925, their star driver, Antonio Ascari, was killed while leading the French Grand Prix at Linas-Montlhéry.

The death of the popular Ascari shook the team to its very core, somehow deadening the competitive zeal of all the participants from the surviving drivers down to the lowliest mechanic. Young Luigi Chinetti had traveled to Lyon as a team mechanic, but decided not to make the trip back to his home in Milan. He was uncomfortable with the increasingly bellicose posturing of the Fascists and was attracted to the easygoing French, who were enjoying the postwar exuberance that enlivened their nation during the 1920s. Chinetti, whose expertise with automobiles ranged from the arcane field of bearing fabrication to occasional race driving, moved to Paris, where he planned to open a small exotic-car repair shop.

Shortly after the race Alfa Romeo announced that they were permanently canceling their competition activities out of respect for their fallen champion. But there was more to the departure than pure grief. Griffith Borgeson speculates that the decision was based on a sagging economy and the intervention of the Italian government. Earlier that year the Fascists had begun to tug Alfa Romeo under the wing of the newly formed IRI, the Istituto di Ricostruzione Industrial, an immense government bureaucracy designed to finance ailing businesses and keep them commercially viable. It is likely that the IRI officials were influential in canceling the expensive and risky racing program.

With the Alfa Romeo motorsports operation inert, Enzo Ferrari concentrated on developing his retail car business. He now held the Alfa Romeo sales franchise for the regions of Emilia, Romagna and Marches and in April 1925 had begun the conversion of a large two-story house on the Via Emilia into a dealership and service center. He and Laura had been living in a tiny second-floor apartment above a nearby restaurant and now planned to transfer their quarters to the new building. Life in that small, steamy space was hardly tranquil. He and Laura were fighting constantly. Worse yet, Laura and his mother,

Adalgisa, had grown to despise each other, leaving Ferrari to act as arbiter in a bitter, futile intrafamily battle. He was consumed by the automobile business and was traveling either to Milan—where contacts had to be maintained with the Alfa Romeo executive staff—or to Bologna, where he had opened a satellite Alfa dealership on the Via Montegrappa. When in Modena his life centered on the noisy local bars and restaurants where the motoring enthusiasts spent endless hours discussing their beloved sport. So, too, in Bologna, where in places like the Café San Pietro and the Folia Bar along the Via Indipendenza nobleman and commoner alike swilled wine, swirled pasta and nattered incessantly about their universal passion. Ferrari was a celebrity of sorts in such crowds. After all, he was a winner of the Coppa Acerbo and had served on the Alfa Romeo Grand Prix team, albeit briefly and without distinction. Still, these were sufficient credentials to impress the locals and he used them to great advantage to promote Alfa Romeo sales. He began to spend more and more time in Bologna and openly discussed his desire to get rid of his wife. Divorce in Italy was utterly impossible, if not unthinkable, but during this period he made no effort to act the loyal husband. His womanizing, in both Modena and Bologna, reached a frantic level. By no means a handsome man, Ferrari still enjoyed great success with women, thanks for the most part to his towering sense of self-confidence and his growing powers of persuasion. In those days his conquests were mainly among the harlots and loose women who hung around the racing crowd, but as his prominence grew, so would his taste in females.

Always a frustrated journalist, he began to dabble in writing commentary for several motoring newspapers while maintaining close contact with the elite team of Alfa racing technicians and mechanics in Milan. Jano was completing the design of a new lithe roadster with a small, high-revving 6-cylinder engine that Ferrari assured his friends would be unbeatable both on the highway and in the endless round of minor events that were being staged across northern Italy and the south of France.

It was the prospect of this fresh line of sports cars that infused the company's old racers with enthusiasm and kept men like Ferrari from defecting to other brands. So, too, for Jano and Bazzi, who were slowly weaning themselves away from the pure Grand Prix machines and leaning toward the more lucrative, widely produced two-seater sports cars for the enthusiast market.

Alfa Romeo by no means left the competitive scene, but merely redirected their efforts toward great sports-car races, where more road-worthy, real-world automobiles competed, as opposed to the pure, open-wheel, single-purpose Grand Prix machines. The new 6C-1500SS sports car, with its racey Zagato bodywork, was shown at the Milan Motor Show in April 1925, but did not go into production until two years later. (As an indication of how primitive the car business remained in places like Modena, there was no formal automobile show in the city until 1925 and dealers like Ferrari were forced to exhibit their wares in a barn at the local horse show.)

His mysterious bout with illness notwithstanding, the mid-1920s were years of consolidation for Enzo Ferrari. His fitful racing career was on hold, and while he maintained close contact with the sport in 1926, he did no competitive driving whatsoever. He successfully expanded his car business in both Bologna and Modena, however, thanks to his compelling personality and a rakish line of new machinery from the drawing board of Jano.

By early 1927 his competitive fires were sufficiently rekindled to make a return to racing. He entered an older Alfa Romeo RLSS in the Circuito di Alessandria, a minor 160-kilometer sports-car race, and won the *gran turismo* class. A month later he won on his home turf of Modena at the wheel of a potent 6C-1500SS two-seater with old Alfa Romeo mechanic and compatriot Giulio Ramponi at his side as riding mechanic. Ferrari had the fastest car in the race and won the 360-kilometer event with ease, setting the fastest lap on the twelve-kilometer circuit at 113 kilometers per hour.

The race was one of hundreds of similar local events run across Europe by the automobile clubs of middling-sized cities. It turned nary a head among the sport's major-leaguers. But it was a promotional boost for Ferrari's dealerships and further enhanced his reputation within the region as a rising entrepreneur in the automobile business.

As his thirtieth birthday approached in February 1928, the Italian government, under the auspices of puppet King Victor Emmanuel III, gave him the title Commendatore (Knight Commander) in recognition—presumably—of his business successes and his sporadic efforts as a racing driver. The rather common title was wholly honorary and was vaguely akin to being designated as a Kentucky Colonel in the United States. Still, 1928 marked an acceleration of his motorsports activities. He returned to Alessandria and scored a class victory. A second consecutive overall win came at Modena, followed by a third place at Mugello—all aboard the same Alfa Romeo that he had campaigned in in 1927. His second victory at Modena was marked by the appearance of little Pepino Verdelli in the riding mechanic's seat—a copilot's position he would enjoy beside Ferrari on and off for another fifty years.

4

As the 1920s slipped away, Enzo Ferrari found himself in a surprisingly pleasant situation. He was established in Modena as well as Bologna, presiding over an Alfa Romeo dealership that enjoyed exclusivity in the Emilia, Romagna and Marches regions, all of which were prospering under Mussolini's Fascist regime. His career as a racing driver was hardly flourishing, but as a former member of the Alfa Romeo team—albeit as a second-ranker—he was able to embellish his association with greats like Campari and the late Antonio Ascari to a point where he became a celebrity in the racing-enthusiast bars and trattorias along the Via Indipendenza. Ferrari by now had gained confidence in his ability to impress and influence the well-born gentlemen sportsmen who composed his clientele. They were addicted to fast automobiles, and any association with a man like Ferrari, who had climbed considerably higher up the mountain than they had, and who had access to the latest Alfas, was to be cherished and cultivated. Ferrari understood this weakness and was able to exploit it to the fullest, not only in terms of increasing sales of Alfa Romeos but in enhancing his social connections.

If there were frustrations, they centered on his recurrent health problems dating from the war and his wobbly, inconsistent career as a racing driver, which had been reduced to a few fitful starts in minor races and hill climbs. An added irritant was his steadily declining relationship with Laura. It had turned into the typical lower-class Italian marriage with Ferrari operating as the household tyrant on his sporadic visits to their small apartment while confining his sexual conquests to the large collection of flashy, often trashy women who were attracted to the glamour and danger of motorsports. Increasingly he sought solace and companionship at places like the Café San Pietro, where talk of fast cars—and faster women—was carried on until the dim, still hours of the early morning.

Ferrari was also a regular at the late-night debauches held at the apartment of a likable veterinarian and part-time mineral-water producer named Ferruccio Testi. He was a fixture on the Modena sporting scene, possessed of a wild enthusiasm for motor racing and a keen eye for photography. Indeed, Testi's masterful photographs compose a riveting portfolio of Italian motorsports from this era.

Ferrari's domestic tensions were exacerbated by the acrimony that simmered between Laura and his mother. Daughter-in-law and mother despised each other almost from the first meeting, and Ferrari was caught in the middle. He was devoted to his mother, as is the case with most Italian men, and did his best to support her in Modena (as he would do for the rest of her life), but tried as much as possible to keep her separated from his wife. He would never speak publicly of these conflicts, but the tension between the two women bore heavily on him for years, and no doubt was a factor in his search for other female companionship.

But in spite of his contentious home life and his modest achievement on the racetrack, life seemed filled with opportunity for this thirty-one-year-old, rather portly Emilian as the decade came to an end.

Contrary to the comic-opera legacy left by Benito Mussolini, Italy was actually doing quite nicely, at least prior to the tsunami that crashed over the world economy following the Wall Street crash of October 1929. It has become legend that Il Duce's achievements were limited to draining the Pontine and Campagna marshes and regulating railroad schedules, but that is hardly fair. Between 1922 and 1930 over five thousand major public works projects were undertaken—including the construction of the first four-lane motorways in the world. In fact, by the end of the decade Italy had 320 miles of such superhighways and led the way in advanced road- and tunnel-building techniques. Moreover, major efforts were made to reduce the prewar illiteracy rate of over 40 percent, and that shameful number was in fact nearly halved by 1930. Other efforts, including a campaign to eliminate the Mafia in Sicily, were less successful, although in 1931 Prefect Cesare Mori of Palermo somewhat prematurely announced that his four-year crackdown had totally cleansed Sicily of the brigands.

Mussolini settled old territorial disputes in the Tirol region and signed the Lateran Treaty with the Vatican, which in turn resolved the "Roman Question," which he had described as "a thorn in the flesh of the nation." At roughly the same time the Fascists decided that the Italian population of about 41 million was too small to qualify the nation for great-power status. A goal of 60 million was set and a number of radical social-engineering schemes were instituted, including the prohibition of contraceptives, restricted emigration and an added tax on bachelors. The legal age for marriage was dropped to sixteen for boys and fourteen for girls, and in 1932 the railroads cut the fare to Rome for newlyweds by 80 percent. The first Italian "Mother's Day" was established in 1933, to be celebrated on Christmas Eve as further encouragement to childbearing. Whether any of these policies encouraged the birth of Ferrari's first and only child in 1932 is unknown, but the programs apparently had some impact nationally, because by 1934 the population had increased by nearly 1.5 million, although the Depression had by then caused the birth rate to shrink below the level of a decade earlier.

By the early 1930s Italy was making bellicose noises about grabbing territory from primitive, slave-trading Ethiopia and there was the matter of avenging the shameful defeat of its troops at the hands of a gang of barefoot Abyssinian tribesmen at Adowa in 1896. Much has been made of the fact that this expansionism into Africa, the pomp and circumstance of the Fascist poseurs and the rising military production were simplistic bread and circus acts to distract that population from the atrocious economic conditions, but that is hardly correct. The World Economic Survey of the League of Nations in 1934–35 ranked Italy at about the same level of prosperity (or lack of it) as the rest of the industrialized nations that had been shattered by the Depression. In the end Mussolini did delude himself and his countrymen with a brand of bloated, bombastic militarism, but for the first ten years following the celebrated March on Rome his odd alliance of socialism and freewheeling big business worked surprisingly well.

In 1930 the import duties on automobiles were doubled, which essentially closed out French, German and American manufacturers, although Henry Ford, whose sympathies with Mussolini and later with Adolf Hitler are well documented, quickly established an Italian subsidiary in Milan. But the overall effect of the tariff was to hand Fiat a monopoly on the mass-produced passenger-car market in Italy, which it essentially enjoys to this day. Alfa Romeo and a few other small manufacturers remained in business only as suppliers of sporting and luxury machines for the upper classes. As Italy steadily increased its military hardware, however, Alfa's efforts slowly shifted toward the manufacture of trucks, reconnaissance vehicles and aircraft engines.

Enzo Ferrari cruised through this national turmoil with his vision riveted on the tight little world of automobiles and the national craze of motor racing. No nation in the world embraced the sport more passionately, and even the smallest cities had booming automobile clubs which organized all manner of races, hill climbs and road rallies to bring honor and prestige to their region. By contrast the American entrepreneurial spirit had transferred auto racing to half-mile and one-mile fairground dirt horse tracks, where admission could be more efficiently charged. There the participants were for the most part working-class mechanics and tinkerers. In England, the stuffy

Edwardians had banned racing on public roads, and competition was restricted to a few special tracks where only the very wealthy gathered ("the right crowd with no crowding").

But in Italy motorsports crossed all social boundaries, rivaling football and cycling in popularity. While its center was in the industrialized north, races like the aforementioned Targa Florio in dirt-poor Sicily or the Coppa Sila, a race run deep in the agrarian toe of Italy, were the subject of national attention. The great and the near-great stayed close to the sport, with Il Duce himself offering financial inducements to keep Alfa Romeo powerful in international Grand Prix competition in the name of Italian prestige. His dashing son-in-law, Count Galeazzo Ciano, the Italian foreign minister and the former lover of the Duchess of Windsor, donated a cup in his own name for a race at Livorno. While it was a badge of honor for noblemen to race cars, many commoners sat on the starting grids beside them— Ferrari being but one—and the immense, all-encompassing love of fast automobiles injected a spirit of egalitarianism into motorsports that was unknown in other aspects of Italian life.

Ironically, it was the good fortune of another manufacturer that triggered the creation of the racing team which was to bear Ferrari's name and which would begin his rise to international prominence. Bologna was the home of the Maserati brothers, a family of five motoring enthusiasts led by elders Alfieri and Ernesto, whose racing experience dated back to 1907. They had begun building their own racing cars in 1926 and by 1929 had created a *sedici cilindri* (16-cylinder) Formula Libre machine by mounting a pair of 8-cylinder engines side by side (a layout first tried by the Duesenberg brothers nine years earlier).

The car was a brute to handle, but loaded with power and capable of blinding straight-line speeds. On September 28, 1929, baby-faced Baconin Borzacchini climbed aboard the Maserati and drove it along a ten-kilometer straight stretch outside Cremona at an average speed of 154 mph. Borzacchini was one of Italy's finest drivers and a popular hero, although he would later accede to Fascist objections to his first name—drawn from the Russian revolutionary Mikhail Bakunin—

and change it to Mario Umberto. In 1929 a terminal speed of 154 mph was prodigious, although two years earlier Californian Frank Lockhart had run a supercharged, intercooled Miller with an engine half the size of the Maserati's at 171 mph over the flying mile at Muroc Dry Lake and had turned a lap on the Culver City board oval at 144.2 mph. This is mentioned simply to show that in the early years before they became slaves to cushy rides, glitz and gadgetry, American automobile engineers were in the vanguard of technology.

Needless to say, Lockhart's accomplishment was ignored, if not unknown, in the afterglow of Borzacchini's run. The achievement was received with such impact that the Bologna Automobile Club threw a massive banquet to honor the driver and the hometown boys who had built the car. Enzo Ferrari, a fixture on the local motorsports scene, was in conspicuous attendance, and whether by coincidence or by plan, the seating arrangements were pivotal to his future.

Flanking him at the dinner table were two of the area's most fevered motoring enthusiasts. Alfredo Caniato was a member of a prominent family of textile and hemp merchants from Ferrara. He regularly traveled to Bologna on market days with his older brother Augusto. A few weeks earlier Alfredo had visited Ferrari's dealership on the Via Montegrappa and purchased, with cash, a new 1.5-liter 6-cylinder Alfa Romeo sports car. Though his only previous experience was in racing motorcycles, he chose to enter his new Alfa in the Circuito delle Tre Province, where he finished a surprising sixth overall. Taking third in the same race was a wellborn native of Bergamo and longtime resident of Bologna, Mario Tadini. Their high finishes in the long, difficult open-road event had fired their enthusiasm, and when they arrived at the table on either side of Ferrari they were open to suggestions about how to further their nascent racing careers.

Ferrari was a force to be reckoned with in situations like this. He had serious credentials, not only in the racing world but also in Alfa Romeo's engineering and sales departments. Surely it was he who choreographed their conversation, which had one central theme: with their money and Ferrari's connections and racing expertise, why not pool their talents and form a *scuderia*—a racing stable—to jointly advance their objectives in the sport? The deal would be simple: a limited partnership would be formed—a *società anonima*—in which the shareholders would set up an operation to buy, race and perhaps someday build high-performance cars. Fueled by wine, their dreams of an all-winning racing team soared as the celebration swirled around them. Ferrari assured Caniato and Tadini that with his connections at Alfa's Portello headquarters, favorable deals on the best new cars would be forthcoming. Moreover, his ties to the accessory business might afford additional funding from Pirelli tires, Shell Oil and Bosch ignition systems, among others. Better yet, the stable could, thanks to Ferrari's reputation in the upper echelons of the sport, attract top professional drivers who would compete in races across the continent, bringing added glory—and important revenue—to the fledgling operation.

The agreement to form the team was drawn up on November 15, 1929, and the corporate papers were filed with the government on the twenty-ninth. Total funding was 200,000 lire (about $100,000 in pre-Depression values). The Caniato brothers and Tadini contributed 130,000 lire; Ferrari himself the not unsubstantial sum of 50,000 lire; Ferruccio Testi anted up 5,000 lire; Alfa Romeo, 10,000 lire; and Pirelli, 5,000 lire. The final contract was signed on December 1 in the office of Ferrari's Modenese solicitor, Enzo Levi, officially forming the Società Anonima Scuderia Ferrari. Levi was to remain as Ferrari's lawyer for years, and in his memoirs Ferrari noted that Levi's maxim was: "Any compromise is better than a successful lawsuit."

Ironically, at roughly the same time the Scuderia was created, a number of old Italian automobile companies were plunging into a financial crisis that would erase them from business before the year was out. Victims of the Depression would be Ceirano, Itala, Chiribiri and Diatto—the marque with which Enzo Ferrari presumably first learned to drive.

Dozens of other rich men had formed similar arrangements to further their motorsports efforts, but the Scuderia Ferrari was unique. The team would carry Ferrari's name, but not those of the other partners. Enzo Ferrari, while a lesser contributor in terms of funds, was far and away the most important member. That the Caniato brothers, rich, proud gentlemen, were willing to subordinate themselves to a commoner like Enzo Ferrari in such a prestigious aspect of the operation is an indication of his critical influence on the success of the organization. A bouquet was tossed to Alfredo Caniato by naming him president of the Scuderia, but he was strictly a subordinate to Ferrari. Moreover, by having his name on the door, as it were, Ferrari was assured that he would be first among equals, no matter what future rifts or splits might sunder the partnership.

Additional funding would come from three distinct outside sources. First came sponsorship, quickly arranged through Alfa Romeo (which would come via low-cost racing automobiles and parts, plus technical assistance): Bosch, the German ignition manufacturers; Memini carburetors; Champion spark plugs; Pirelli, the giant Italian tire maker; and Shell. It was the personal contacts made by Ferrari himself that generated this support—a sum that some estimate would approach a million dollars per annum when the Scuderia reached the height of its power a few years hence. Second, the Scuderia would acquire capital by servicing racing cars for additional members of the team. Rich sportsmen would be invited to join, with the agreement that their racing cars, for the most part Alfa Romeos, would be prepared for competition and transported to various racing venues. The client would then merely arrive, don his goggles, and turn the key. Third, it was planned that a small coterie of top racing professionals would be hired to run in major road races and hill climbs around Italy and perhaps France and Germany.

In those days European motor races offered little in the way of prizes other than elaborate cups and trophies. For the most part cash awards to competitors came via "starting money," sometimes called "appearance money"—that is, a major team or driver would be hired to run, thereby attracting more spectators, who would then pay the organizers through admission fees, concessions, and so on. In America, the system was quite different. Starting money was unheard of and drivers were paid exclusively according to their finishing positions in the race. For a team like Scuderia Ferrari the potential for starting money was excellent, provided the new boss could attract several big names to drive and Alfa would cooperate by offering a selection of their best automobiles.

Alfa Romeo was interested. The company was having a hard time selling expensive sports cars as the Depression descended and more pressure was applied by the Fascists to redirect their efforts toward the design of military hardware. Therefore, if the new Scuderia could act as the company's quasi-official representative in selected races, while the official factory team concentrated on major international Grand Prix events, the corporate motorsports image could be maintained with less cost and effort. Enzo Ferrari was a trusted agent. With nearly a decade of loyal service, he could be relied upon to field well-prepared cars manned by competent drivers. Yet the new Scuderia lacked one final, critical ingredient.

A star was needed. Rank amateurs like the Caniatos and Tadini could hardly attract much starting money, and Ferrari himself—while a veteran of the racing wars—lacked the credentials to boost revenues much beyond the subsistence level. Enter Giuseppe Campari. The swarthy (his nickname was "Il Negher"), rotund, loquacious amateur opera tenor who had risen to the very top ranks of Italian racing stardom was openly frustrated with the Alfa Romeo competition program, which seemed to be operating in a series of directionless fits and starts. Ferrari knew of the great man's displeasure and approached him with a deal to drive for the Scuderia. (In an Alfa, of course.)

In today's motorsports milieu, where Grand Prix drivers tend to be spoiled, humorless technoids, Giuseppe Campari would have stood out like a circus clown in church. He was a large man, perhaps fifty pounds overweight during one of his eating binges, and enormously popular with his fellow drivers as well as the public. His love of opera was sufficient to get him invited during the 1920s to play Alfredo in *La Traviata* at a theater in Bergamo. For all his zeal, Campari's vocal cords were simply not up to the job; the higher registers were clear and lusty but the lower tones were feeble and unsettling. As Campari forged ahead, an irate member of the audience stood up and shouted a suggestion that he return immediately to the racetrack. Campari heard the complaint, stopped in mid-song and turned to the crowd. "When I race, they tell me to sing. When I sing, they tell me to race. What am I to do?"

Ferrari himself delighted in telling Campari stories, and recalled that while practicing for the second Mille Miglia in 1928 the two of them were hurtling over the Raticosa Pass in the Apennines when he began to feel warm liquid splashing against his face. Ferrari turned to Campari, announced that a cooling hose had ruptured, and suggested an immediate stop. Campari ignored him and rushed onward. Then Ferrari noticed that the mysterious liquid was issuing from the pant leg of Campari's coveralls. He pointed at the problem as they slewed, buckboard style, around a dusty curve.

Campari acknowledged the situation and shouted an explanation over the rumble of the exhaust and the whine of the wind. "Hey, you don't expect me to stop on a trial run, do you? When you're practicing, you've got to piss your pants, that's all!"

The deal to retain Campari's services for the Scuderia was concluded at a noisy dinner hosted by the great man at his house in San Siro, a patrician district in Milan. Ferrari recalled the evening with pleasure, noting that Campari, an excellent cook, brewed up a local pasta dish called *riccioline al sugo* while decked out in pinstriped pajamas resembling those worn by Italian convicts. The evening was concluded with Campari and his wife bellowing a duet from the first act of—again—*La Traviata*. But most important, the new Scuderia Ferrari had a name driver who could be used to pry impressive amounts of starting money from various race organizers as the 1930 season got underway.

It was understood from the start that Ferrari would base the operation in his hometown. From that point on, his peripatetic travel and his division of interest between Modena and Bologna would give way to a firm settlement in the steamy, mosquito-ridden little city of his birth. While he had never totally left Modena, save for the first few years following the war when he had sought his fortune in Turin and Milan, Ferrari symbolically viewed the formation of the new Scuderia as a triumphant homecoming to a place that—save for a few racing enthusiasts—had ignored his presence. "My return [sic] to Modena was a kind of mental revolt," he later wrote. "When I went away I had merely some slight reputation for being a strange young man keen on cars and racing, but did not seem to have any particular capabilities. My return to Modena after twenty years, in order to transform myself from a racing driver and team organizer into a small industrialist, marked not only the closing of what I might call an almost biological cycle. It also represented an attempt to prove to myself and others that during the twenty years [sic] I was with Alfa Romeo not all my reputation was secondhand and gained by the efforts and skill of other people. The time had come for me to see how far I could get by my own efforts."

Enzo Ferrari set up his temporary headquarters in the Gatti machine works on the Via Emilia, no doubt with an air of grudging satisfaction. While he had won several races in Modena, the vendetta was a theme that ran through his life, the notion that no slight would go unforgotten, that no insult would go unpaid, and that honor in such situations transcended all practical need. To his final days, he carried on a love-hate relationship with his native city, always harboring a childish suspicion that for all the accolades heaped upon him by his fellow citizens, and his final elevation to the status of a near-deity, he was not totally appreciated and that the lack of attention during his youth had not been atoned for. This same refusal to forget would cause him to patiently wait fifty years—*half a century*—before he believed Fiat's debt to him, incurred by their refusal to hire him in 1918, was fully repaid.

Certainly this desire to flaunt his new success in front of the Modenese was a prime reason for the full-time return to his birthplace. But there were other, more practical motivations, including the availability of experienced metalworkers, fabricators and machinists in the city and Ferrari's personal knowledge of the best local suppliers. Modena's location on the edge of the Po Valley and astride the Via Emilia gave it a strategic central location in terms of the racing venues across northern Italy, while being within a day's drive or train ride from both Turin and Milan.

The deep pockets of Tadini and the Caniatos permitted the purchase of three competition-type 6-cylinder 6C-1750s—the featherylight, wonderfully graceful supercharged sports cars developed by Vittorio Jano that remain among the most venerated high-performance automobiles of all time. Also in the new inventory was a collection of spare bits and pieces from Alfa Romeo, a few lathes and small machine tools, and a husky Citroën van to carry the necessary fuel, tires, extra parts, and the rest, to various motor races. (The cars themselves would be driven to the tracks until the Scuderia could afford large truck transporters.)

Ferrari hired a retinue of local mechanics, including Pepino Verdelli, the diminutive Modenese who had served as his riding mechanic since 1928 and would continue with the Scuderia for the rest of his life, serving out his final years as Ferrari's personal chauffeur and a man privy to most of the intimate details of his increasingly complex private affairs. The hiring completed, Ferrari and his small staff set to preparing the three shiny new Alfas for Italy's most important open-road race, the Mille Miglia. This epic contest, organized by the Automobile Club of Brescia, had first been run in the spring of 1927 and was a 1,000-mile race (the Romans measured distance in miles, not kilometers, hence the name), starting in Brescia, heading south along the Adriatic, then over the Apennines to Rome, north again over the Apennines a second time through Florence and Bologna and back to the finish at Brescia. Run rain or shine in the early spring, the Mille Miglia posed the threat of snow or sleet in the higher Apennine passes and offered the chance for 140-mph road speeds on the seacoast stretches north of Pescara and across the Po flatlands. The outrageous quality of the event, with high-powered cars screaming through tiny villages and major cities alike, captured the imagination of the Italian people and it became an instant success.

The two previous races in 1928 and 1929 had been won by Campari for Alfa Romeo, and by arrangement with the factory he was to remain on the official Alfa team. By contrast, the Scuderia's personnel consisted of three rank amateurs—Tadini, Alfredo Caniato and the prominent Fascist politician Luigi Scarfiotti. This trio was hardly in a position to contest for the overall victory with the likes of the top professionals from other teams.

The fourth Mille Miglia was run on April 12–13, 1930, with Ferrari located at a fueling stop south of Bologna. There he was to direct the team—as he would for years to come—receiving reports of his cars' progress via telephone calls from observers and race officials around the route. But he had little need to employ his growing skills as a race-team organizer and strategist in the Mille Miglia. The Scuderia's three cars retired early in the going after making no impact on the leaders. The race was won by Tazio Nuvolari after a duel with Achille Varzi that was to become part of racing lore. Legend has it that Nuvolari stalked the leader, Varzi, through the predawn with his headlights out, in order to lull his rival into slowing down. At the last moment, it is said, he pounced on Varzi, flashed on his lights, and roared by to victory. Historians have long since discounted the tale, but it remains part of the portfolio of Nuvolari's exploits—all sufficiently impressive in reality to qualify him in the eyes of many as the greatest racing driver of all time.

Two weeks later a chastened Scuderia went to Alessandria for a race on a city course to be called the Circuito Bordino. The appellation was in memory of Pietro Bordino, the old Fiat team leader who had crashed his Bugatti into the Tanaro River during the 1929 event and died. Prior to the start the entered drivers, including Ferrari himself and Alfredo Caniato, laid flowers at the base of a monument erected at the site of the disaster. While Caniato again flagged in the face of the high-powered competition, Ferrari drove with steady determination to finish third in his Alfa Romeo 1750SS/TF.

His obligations to the official Alfa team fulfilled, Campari was free to join the Scuderia, where he linked up with Tadini in 1750s to run a pair of minor events—without great success. The second, the

Reale Premio (the Royal Prize race run on Rome's Tre Fontane course), saw them facing some of the finest cars and drivers of the day, Nuvolari and Varzi among others, in the new, totally revised Alfa Romeo P2s—the old 1924 Grand Prix cars that had been modernized and updated by Vittorio Jano. The Scuderia was badly outclassed, competing with what were little more than fenderless sports cars against specially designed full-race automobiles. Campari managed a fifth-place finish, while Tadini, who was quite fast for a gentleman amateur, took seventh.

If there was a prototype operation for what Enzo Ferrari envisioned for the Scuderia, it had to be that of the legendary Ettore Bugatti in Molsheim, Alsace-Lorraine. This part artist, part engineer, part entrepreneur, part sculptor had since 1910 created an automotive fiefdom in the tiny village located a few kilometers west of Strasbourg and held forth there in baronial splendor. The Bugatti estate included a small, elegant inn for the entertainment of guests and customers, a stable of Thoroughbred horses and the factory itself—a series of low buildings set among landscaped gardens with a trout stream meandering through the screeching factory machinery.

Ettore Bugatti was a Milanese from a family of artists. *Le Patron*, as he was known, was generally to be found conducting business dressed in riding breeches, boots, a red waistcoat and a yellow coat. His automobiles were (and remain) a stunning combination of industrial aesthetics and the jeweler's art—as if Fabergé had somehow been able to motorize an egg. The Type 35s and 51s that the Scuderia faced were hardly the most technically advanced but were simple, flawlessly fabricated and as reliable as eight-day clocks. (Bugatti eschewed hydraulic brakes until very late in his career, preferring instead cable-operated mechanical brakes. "I build my cars to go, not to stop," he once explained airily.)

Ettore Bugatti was just one of a bevy of colorful eccentrics, dissolute nobles, playboys, dreamy commoners and hard-eyed egomaniacs who populated the world of European motorsports in the 1930s. He certainly stood above the rest in terms of lifestyle: a feudal barony had been created around the spidery machines he manufactured in limited quantities and sold only to those he personally deemed worthy. By contrast Enzo Ferrari was then still a drab, simple journeyman laboring in a small garage in a fetid Italian backwater.

But the example Bugatti was setting surely did not escape him. The transplanted Italian, living in that sliver of territory mired in endless disputes between France and Germany, was a prototype for success. He was manufacturing cars for the very wealthy and fielding his own team of professionals and wealthy amateurs. Beyond that, mobs of pretenders, dreamers, part-timers, has-beens and dilettantes were flocking to Molsheim to have their Bugattis anointed by the master before competing, with inevitably modest success, in the myriad of minor races and rallies being staged everywhere in Europe. Surely, if Bugatti could succeed at this, a similar concept could be developed on a more modest basis for the Scuderia.

Despite the joyous, pasta-laden, song-filled evening that joined Campari with the Scuderia, the marriage was to be short-lived. The arrangement between Alfa Romeo and the Scuderia involved an informal farm system — cars and drivers were transferred back and forth on an as-needed basis. For important races the best machinery and the top drivers would run for the official factory team, but on occasions when the factory was not competing some of the same equipment and personnel would be sent east to Modena.

A trade was made in the early summer of 1930. Campari left the Scuderia and went back to Milan. However, in return Ferrari received a powerful P2 Grand Prix car from the factory. This was essentially the same machine that had been so successful during the 1924 season, but it had been radically updated and improved by Vittorio Jano to race against the more modern Bugattis, Maseratis and Mercedes-Benzes that were dominating major races. The Scuderia's new P2 had been shipped back to the factory from South America, where major modifications were undertaken to make it the equal of any in the official stable.

But who was to drive it? Surely none of the Scuderia regulars — including Ferrari himself — possessed the skill to manhandle the powerful 8-cylinder, 175-hp machine (roughly twice the horsepower of the 1750s). Luckily, a star was waiting in the wings. And, if anything, he would be an improvement over the departed Campari. Tazio Nuvolari was on the verge of becoming an Italian national hero, a man who would be compared to Niccolò Paganini in terms of his wild, inspired, maniacal virtuosity behind the wheel of an automobile. Nuvolari was among the most vivid characters of the age—no more than five feet four inches tall, with a lantern jaw and eyes that blazed inside his swarthy, high-cheeked face. He drove like a madman, crashing often and flogging his cars as if they were recalcitrant beasts of burden. He was, in the argot of the day, the classic *garabaldino*—a driver with the slashing, all-out style of a winner, a charger who drove with such abandon that rumors spread through the crowds that he was haunted by a death wish or, like Paganini, had a pact with the devil.

He had been racing motorcycles since the early 1920s and had shifted to cars on a full-time basis after the middle of the decade. In 1925 he was given a test drive by Alfa Romeo in one of their prized P2s prior to the Italian Grand Prix. Before half a lap was completed, the gearbox is said to have seized, sending the car tumbling off the road. One of Nuvolari's legs was broken and he was told that he was to spend a month in bed, bound up in plaster. But ten days later he was helped aboard his Bianchi and rode off at high speed to win the Monza Grand Prix for motorcycles.

He and Ferrari, both possessed of towering egos, did not get along. They had known each other since meeting at a race in June 1924, and Ferrari had found the little man to have a caustic wit and to suffer from near-terminal cockiness. At one point Nuvolari taunted Ferrari by questioning his resolve as a team manager. After engaging him to run in the 1932 Targa Florio for the Scuderia, Ferrari gave him a round-trip ticket to the Sicilian race. Nuvolari looked at the ticket and scoffed. "A lot of people say you are a good manager, but I can see that isn't true. You ought to have bought only a single ticket, because when anyone sets off for a race, it's just as well to bear in mind that he might be coming back in a wooden box."

This was the race, incidentally, where Nuvolari took along a young riding mechanic and warned him that he would shout if he took a bend too fast, in which case the youth was to duck behind the cowling to prepare for the forthcoming crash. When the race was over (during which Nuvolari set a record that stood for twenty years) the boy, named Paride Mambelli, was asked how the ride with the master went. He shrugged and said, "Nuvolari started shouting at the first curve and never stopped until the end. I spent the entire race under cover and never saw a thing!"

Jano considered Nuvolari beyond the fringe, a madman intent only on wrecking himself and the cars he battered. In the mid-1920s the senior engineer had used his considerable influence at Alfa Romeo to prevent Nuvolari from being hired for the factory team, although it was acknowledged that he possessed enormous talent. He was in his mid-thirties, but Jano still insisted on referring to Nuvolari as "the boy"—a pointed barb at his alleged immaturity, and a nickname the engineer used even after Nuvolari had reached the very pinnacle of the sport. This rejection by Alfa Romeo forced Nuvolari into an uneasy alliance with Achille Varzi, a chilly, aloof, wellborn son of a textile magnate from near Milan who had also begun his racing on motorcycles. Varzi was the antithesis of the volatile Nuvolari on and off the track, where he maintained an impeccably ordered persona.

Nuvolari raced with a wild wardrobe, often choosing knickers and argyle knee socks, but always in a yellow short-sleeved shirt with his initials embroidered over the breast and a tiny tortoiseshell brooch at his throat, a small replica of a good-luck charm given him by Gabriele D'Annunzio. Varzi, on the other hand, drove only in perfectly pressed linen coveralls and was seldom seen without a cigarette laconically drooping from his unsmiling lips.

Their driving styles could not have been more different. Nuvolari lashed his cars around the course in a series of wide-open-throttle power slides, and he is credited with developing the so-called four-wheel drift, an exquisitely balanced slide that was eventually adopted by every fast driver in the world. Conversely, Varzi never seemed to put a wheel wrong. His cars appeared to run on rails, unruffled, always in a disciplined gait attuned to its driver.

This delicious contrast between the two men who were rapidly becoming the two finest racing drivers in Italy, if not the world, fascinated the sporting public and created a special celebrity status for each. Nuvolari, because he was so *Italian*, was by far the more popular, but owing to his proximity to Milan, Varzi was much loved and supported within the precincts of that great city. The joint effort formed by the pair in 1928 lasted but two years before the rivalry became intolerable for both. They were rejoined on the official Alfa Romeo team in 1930, which set the stage for more fireworks between them.

It is likely that this tension was the cause of Nuvolari being "traded" for Campari. There is no doubt that the Scuderia got the better of the deal. While both men were the same age, thirty-eight, Giuseppe Campari was by far the more battle-worn of the two. He had been racing since 1914, and with his weight problem and his distraction with opera, he lacked the intensity of Nuvolari, who had been racing automobiles full-time only since 1927.

At the same time Enzo Ferrari went with Enzo Levi to the Modenese Bank of San Geminiano and arranged a loan for one million Depression-devalued lire in order to expand the Scuderia. This permitted him to purchase a new and larger two-story garage and shop around the corner from Garibaldi Square at 11 Viale Trento e Trieste. This was to become the headquarters of the Scuderia and the home of Enzo and Laura Ferrari, who took up quarters in the small second-floor apartment and who would remain in residence there for nearly thirty years.

Suddenly fortunes were on the rise for the Scuderia. In the new shop was the P2 Alfa Romeo, and ready to pop behind the wheel was Italy's brightest racing star. Nuvolari responded brilliantly. His first race was the hill climb at Trieste-Opicina, which he won in record-setting fashion, thereby bringing the Scuderia its first outright victory. He followed up with a win at the Cuneo–Colle della Maddalena climb, and completed the hat trick a few weeks later at the Vittorio Veneto–Cansiglio hill climb. Three races, three victories. Suddenly Scuderia Ferrari was undefeated!

Power begot power and quickly "Gigione" Arcangeli and Baconin Borzacchini joined the team for the vaunted Coppa Ciano, to be run on the fast, hilly 13.98-mile Montenero circuit near Livorno. They would run 1750 Alfas while Nuvolari remained aboard the P2. The factory team, no doubt with Ferrari's full knowledge, countered with Campari in another 1750 and Varzi in a second P2.

As expected, Nuvolari and Varzi hooked up in a desperate duel for the lead. Both P2s finally broke down under the flogging administered them, but not before Nivola, as he was sometimes called, added yet another page to his growing list of driving feats. At one point he came howling off a hillside toward a sharp bend with a gasoline station situated on the apex of the corner. It was clear that he had arrived at much too great a speed to make the curve in a conventional way, but still Nuvolari refused to lift, and aimed the big Alfa at the tiny space between the pumps and the station proper. He squeezed through, at full blast, with no more than a few centimeters to spare. Undaunted, he duplicated the maneuver on the next lap!

Race drivers are never quite contented with their lot and carry on an endless, nomadic search for the perfect car. This was the case with Varzi and Arcangeli, who defected from Alfa after the Coppa Ciano to join the rival Maserati team. With the important Coppa Acerbo coming up at Pescara, Ferrari himself stepped in to replace the departed Arcangeli, who, it turned out, was to return to the Scuderia after the event. Varzi confirmed the wisdom of his switch to Maserati, winning in his 8-cylinder 26M after Nuvolari's P2's spark plugs fouled and he retired. Ferrari, at the wheel of a 6C-1750, was not a factor and also failed to finish. (A few weeks earlier he had also competed in the relatively minor Circuito Tre Province race, run on a wide course between Bologna, Pistoia and Modena, and had also dropped out. These events, plus the earlier Alessandria race, were the only competitions that Ferrari would personally enter in 1930.)

The Scuderia's last major race of the season came at Monza on

September 2, where Varzi soundly beat the P2s of Nuvolari, Borzacchini and Campari (now back under the Ferrari umbrella, which was for all intents and purposes operating as a stalking horse for the factory).

The Scuderia ended the season on November 8 with a banquet at Modena's San Carlo Hotel, where all the luminaries of the operation, as well as old factory hands Bazzi (who was soon to join the Scuderia) and Ramponi, sat down with Ferrari, Tadini and Caniato to celebrate what had been an auspicious beginning. The team had entered twenty-two races and hill climbs and had won eight times. Admittedly, some of the victories and high finishes had come against minor-league opposition, but they nonetheless qualified for the win column. Surely the Alfa management was happy and Ferrari himself could count on their continued support as well as that of his major sponsors. Sales of 1750s to private customers were thriving and would continue to grow as the Scuderia's successes on the racetracks expanded in number and magnitude. And this would surely happen, because even as the celebration raged into the night, the dour, reclusive genius Vittorio Jano was hunched over a drawing board in Milan inking into place the final details of a grand new racing car code-named the 8C-2300.

And most importantly, Ferrari had in his pocket a contract signed two weeks earlier that formally linked Nuvolari to the team. It had been formalized and signed on October 20, and was to remain in effect for the entire 1931 season. The arrangement was simple: Nuvolari was to receive 30 percent of all prize money, sponsorship funds from tire, oil, spark plug manufacturers, and so forth, as well as any starting money paid him by race organizers. He was also to be paid for all travel expenses associated with racing and testing and to receive a 50,000-lire accident insurance policy.

This was a generous deal for the time, although Ferrari would never establish a reputation for largesse in such matters. In fact, as the prominence of his organization increased, he was to become a veritable tightwad, relying on a driver's career needs and his desire to be a part of such an august organization. But at this moment, in the fading daylight of 1930, he was grateful to even have the opportunity of engaging such a powerful personality as Tazio Nuvolari and surely did not haggle long over the price. But note that Ferrari craftily did not guarantee Nuvolari a salary. Other than an assurance of travel funds, all remuneration was based on a percentage of presumed winnings. Theoretically, if Nuvolari failed to win any prize money and his sponsorship and starting money dried up, he would make nothing. Ferrari was to use this same system with his other professionals. The gentlemen amateurs, on the other hand, understood that membership in the Scuderia was a matter of expense, not income. They paid for the privilege of joining, receiving in return technical assistance and the preparation and transport of their cars to selected races. For this Ferrari received a handsome profit, and it was to teach him that rich dilettantes would always be willing to pay large sums of money to rub elbows with professionals.

With Nuvolari and Borzacchini in the fold and Campari and Arcangeli available for major races, his roster of pros was as strong as any in Italy. Backing them up were Tadini, the Caniato brothers and Scarfiotti, among others, who could be relied upon to willingly pay up simply to bask in the reflected light of the superstars.

And in the background stood Enzo Ferrari, learning more each day about the delicate craft of keeping all these willful, cantankerous puppets dangling on the proper strings.

5

As the bleak, fogbound winter of 1931 gave way to an early Emilian spring there was every reason for Enzo Ferrari to believe that his Scuderia had survived its puberty and was on the way to a successful adolescence, although a great deal depended on Jano and the rest of the Alfa Romeo racing department. If the new 8C worked as well as early tests indicated, they had a winner in-house. Yet with each passing day the Mussolini government demanded that more funds and manpower be diverted to the design and development of military vehicles, crimping even more the automotive side of the business, which was now beginning to suffer heavily in the worldwide economic chaos.

Still, operations at the Scuderia continued, with the usual isolation from the more serious aspects of life. Everything centered on the racing operation, and Ferrari would begin each day in the main machine shop on the first floor, a fuming, grimy chamber crammed with rows of lathes, grinders and milling machines belt-driven from a series of shafts that hung below the ceiling like steel rafters. And he was still traveling a great deal, rushing off to Milan and the inner sanctums of Alfa in order to advance the cause of the Scuderia by gaining spares and technical assistance, as well as lobbying to obtain one of the vaunted new 8Cs.

Good news came when the factory decided that a pair of 8C-2300 sports cars would be assigned to the Scuderia for the upcoming Mille Miglia. This was a major coup for Ferrari. He felt singularly honored to be able to debut such a formidable machine, but the Alfa management had reasons that transcended pure generosity. The cars were hastily prepared, and Jano was not convinced they were race-ready. Therefore, if the Scuderia Ferrari entered them, any embarrassing failures could be laid at the doorstep of 11 Viale Trento e Trieste rather than on the official racing department in Portello.

Ferrari was of course ready to accept the risk, although the 8Cs to be driven by Nuvolari and Arcangeli were to be buttressed by no fewer than eight other older cars, including well-proven 1750s for Campari, Borzacchini and Tadini. Facing them were Rudi Caracciola's immense, supercharged 7-liter Mercedes-Benz SSKL and Varzi's brutish 5-liter Bugatti. As expected, the untested Alfas soon experienced tire troubles and fell back. But Varzi retired early and only Campari could stay within striking distance of Caracciola, who won at a record speed, becoming one of three non-Italians to win the arduous race. Campari staggered home in second place, eleven minutes behind the German.

A series of poor finishes and minor disasters continued to afflict the team until May 24, when a lightened, narrow-bodied, single-seat, open-wheel version of the 8C made its debut at the Monza Grand Prix. The race was to be ten hours in length, requiring Nuvolari to team with Campari in one of the new machines, while Borzacchini and "Nando" Minoia would drive a second 8C. Also on hand was a monster, twin-engined Tipo A Alfa—a hot rod with two 1750 engines mounted side by side and reputed to generate 230 hp. These three formidable machines were entered by the factory, with the Scuderia Ferrari personnel relegated to the role of spectators. During practice Ferrari stood by as the jovial Arcangeli climbed aboard the Tipo A for a practice run. Arcangeli was famous for his hilarious and notably irreligious off-track behavior, but on that day there was no joking. "Gigione" Arcangeli spun on the tree-shrouded Monza bend called Lesmo and was tossed from the Tipo A to his death. Although he was not racing under the Scuderia colors, this much-loved competitor was considered part of the family and the mourning was protracted and highly emotional. Of course, Enzo Ferrari was no stranger to racing death. The loss of Sivocci, his first mentor in the business, was hardly forgotten, nor was the shocking death of Antonio Ascari in France. He had been present at Monza in 1928 when the wealthy Italian sportsman Emilio Materassi lost control of his 1.5-liter Talbot on the main straightaway and crashed into the gathered throng, killing himself and twenty-three spectators. But the death of Arcangeli, who had been a member of the Scuderia and figured prominently in plans for the coming season, was the first to affect him personally.

Despite the shattering loss, Alfa Romeo managed to salvage a major triumph from the weekend. The 8Cs were fast and reliable, finishing one-two in the long race and thereby gaining the nickname "Monza" and inspiring numerous variations of the car.

As the season progressed and the factory's military obligations increased, the Scuderia became more closely involved with the racing program. In fact, the boss himself returned to competition. In June 1931, during the Bobbio–Monte Penice hill climb, Enzo Ferrari set a fast time up the thirteen-kilometer hill in a special, Zagato-bodied 8C-2300 Mille Miglia, thereby gaining the last victory in his somewhat truncated and inconclusive racing career.

But there was even more significant news for him on the home front. Laura was pregnant. It is difficult to assume this was a planned addition to the family, as he and Laura had been married ten years without any prior indication that children were in their plans. Yet with pressure to raise a family coming from the Fascists and abortion difficult if not impossible to obtain, there was little choice but to accept the presence of a new Ferrari in the tiny two-bedroom apartment over the shop.

Still, there was racing to be done, cars to be prepared and customers to be satisfied. Slowly the operation gained sophistication and a certain persona that separated its policies and its automobiles from those of the factory. The team's mechanics were beginning to make special modifications on their Alfas, although they were still being constrained from actually manufacturing new parts to replace those made in Milan. However, the initials "SF," along with serial numbers, began to be proof marked into all key components of the engine, chassis and drive train in order to make assembly simpler and to give a certain imprimatur to the automobiles being prepared in Modena. The Scuderia was also able to obtain a pair of proper trucks with which to transport the primary team cars to the races. A Lancia Model 254 and a Ceirano Model 45 were fitted up with special enclosed bodies fabricated by Carrozzeria Emilia in which two race cars, plus spares, could be hauled long distances by each truck. The bright red flanks of the big rigs were painted with the names of sponsors, with Pirelli enjoying the top marquee position.

Enzo Ferrari was to have one last try at race driving, and the outcome sent him an eloquent message as to why his future lay behind a desk and not a vibrating steering wheel. The race was the Circuito delle Tre Province, a regional event in which he had participated twice before without success. The course was a single 79-mile lap around the Apennines to the southwest of Bologna, crossing the 4,000-foot Abetone Pass and starting and finishing in the small city of Porretta. The Scuderia had no serious opposition. Borzacchini joined Ferrari with big 8C-2300 sports cars while Nuvolari was given a small 1750.

The race was a classic example of how difficult it was to beat Nuvolari, no matter the odds stacked against him. Up until the last minute he had been undecided whether to run in such a minor event, telling his longtime mechanic and sometime Sancho Panza, Decimo Compagnoni, that he was totally ignorant of the course and therefore at a substantial disadvantage to the boss, who had raced there twice before. Still, he instructed Decimo to prepare his car for the start the following morning.

By the time he and Decimo pushed the Alfa to the line, both Ferrari and Borzacchini had left. (As in the Mille Miglia, the race would be decided on elapsed time.) Urged on by the screaming crowds, Nuvolari sped off in hot pursuit. The potholed road paralleled a railroad track for a short distance, then traversed it via a level crossing. The Alfa hit the lumpy tracks at an absurd velocity that sent it flying into low orbit. Decimo saw the crash coming and grabbed a pair of handles inside the cockpit for support (no one wore seat belts in those days). But the impact was too much. The handles gave way and the hapless mechanic was pitched onto the tail of the car, barely avoiding being tossed into the wake of the berserk machine. Nuvolari skidded to a stop to assess the damage. The crash had broken the throttle linkage and bent the suspension.

Decimo quickly jury-rigged a throttle connection with his leather belt and the two set out again, driving the Alfa in tandem; Nuvolari working the brakes and gearbox, while Decimo yanked on the improvised throttle cable with one hand and held on with the other. But the grab handles were gone, meaning that he had to reach outside the car for a grip, which in turn flayed his hand with flying gravel and stones. In pain, he withdrew it long enough to wrap the bleeding flesh with his handkerchief.

Three miles farther on they saw Borzacchini's car parked at the side of the road. Now only Ferrari remained to be beaten. Decimo protested that it would be impossible to catch him, which only made Nuvolari drive faster. They snaked down the Abetone Pass, barely on the edge of control as the Alfa slewed near the fenceless drop-offs. At the Sestola checkpoint they were informed that Ferrari had a forty-second advantage. It was a mere twenty-two miles to the finish in Porretta and to make up such a differential against a larger car over so short a distance bordered on the impossible. But with Decimo stretching his belt to the limit and Nuvolari flinging the Alfa through the bends wide open, they gobbled up the distance. They shot across the finish line as the stunned and no doubt chagrined Ferrari saw his victory slip away by the margin of a few seconds. This would be the final race of Ferrari's career, although he did keep his Automobile Club of Italy racing license (number 16) active for several more years. He explained that the birth of his son was the reason for his retirement, but that must be viewed with skepticism. Years later he wrote: "I made the decision not to compete anymore in January 1932 when my son Dino was born. My last race of the preceding season had been the Bobbio–Monte Penice hill climb on July 14. On the hills above Piacenza I debuted a new Alfa Romeo 2300 designed by Jano and I brought it to victory. But that day I promised myself that if I had a son born to me I would stop driving racing cars and would devote myself to organizing and competing with automobiles. I kept my word. . . .

"I cannot claim, furthermore, that I would have been a great racer. Already, at that time, I was driving doubt away because I knew that I was carrying within me a great obstacle. I was driving the car and respecting it. When one wants to get spectacular results it is necessary to know how to maltreat [the car] . . . Summing it up, I wasn't capable of making the car suffer. And this kind of love, which I can describe in an almost sensual or sexual way within my subconscious, is probably the main reason why, for so many years, I no longer went to see my cars race. To think about them, to see them born and to see them die—because in a race they are always dying, even if they win. It is unbearable." This is typically florid Ferrari prose without much relation to the truth. There is little evidence that he cared about *any* automobiles, much less his own. From the beginning they were implements by which to gain gratification of his own ego. In contrast to Ettore Bugatti, who was gripped by the Bauhaus aesthetics of the machine, Ferrari never expressed any such sensitivity other than the foregoing, which was more than likely intended to excuse his humiliation at the hands of faster drivers like Nuvolari. Automobiles were tools, nothing more and nothing less, by which to glorify the Ferrari name on the racetracks of Europe. He kept no cars of his own, other than mundane passenger sedans, nor did he hesitate to send his most successful racing machines off to the scrap heap once they had become obsolete.

Ferrari's decision to retire was probably more related to the changing demands of his career and his realization of his limited talents as a driver than with any abiding devotion to his young son. Surely, being run down by Nuvolari with a battered, lower-powered automobile had to have sent the message that his time and energies would be better served by running the Scuderia than actually driving.

Therefore, as the members of the Scuderia, Alfa staffers and a burgeoning crowd of outriders gathered at Modena's Ristorante Boninsegna on November 21 for the annual dinner, it was understood that Ferrari had driven his last race. But this mattered little. An extensive tire-testing session for Pirelli had just been completed at Monza, with Nuvolari doing most of the driving. The session had further solidified the Scuderia's position as a top-rank racing organization and was sure to give it access to the best rubber being manufactured by that prestigious Italian company.

The annual banquet was becoming a tradition tightly integrated into the Ferrari operation. The restaurateurs prepared a special menu for the guests, and the meal was lubricated with many liters of Lambrusco. Ferrari presided with his usual flair for the dramatic, intoning lengthy and emotional introductions for the drivers, each of whom received a gold neck chain with his name engraved on a small panel. A short but eloquent eulogy was given for their fallen comrade, "Gigione" Arcangeli, before the team's mechanics received silver cuff links and bonus checks. Ferrari also distributed to the gathering the first of what was to be a long chain of personal memoirs regarding the progress of the team—a seasonal summary composed in his terse, disciplined, sometimes enigmatic style, which he would refine over the years to a minor art. Titled *Duo Anni di Corse (Two Years of Races)*, the small book contained illustrations by Dr. Testi, who remained a minor shareholder in the Scuderia and was becoming an avid and talented amateur photographer. His record of the early years of the team's campaigns has provided historians with an extraordinary account of those heady days.

The second season had been what Ferrari described as a modest success. The team had further solidified its position with Alfa Romeo as a trusted subaltern in the racing wars and had thereby gained access to their best machinery and technology. Better yet, Italy's most popular driver was in the fold and sufficient funds were being generated to allow Ferrari to look ahead to 1932 with soaring hopes.

A new name was to be added to the team for the coming year, as well as an expanded dimension of the competition effort. Piero Taruffi was a twenty-six-year-old motorcycle champion with an impressive background. He held a doctorate in industrial engineering (which, even in title-crazed Italy, was a legitimate academic accomplishment) and was a skilled amateur tennis player, skier, oarsman and bobsledder. Taruffi had run in two minor events for the Scuderia in 1931, winning both. For the following year he planned to alter the direction of his career from motorcycle racing to automobiles. This led Ferrari to expand his racing operation to include a small stable of competition motorcycles. Ironically, Ferrari's increasingly fevered chauvinism was forced to give way in this instance and he bought a small fleet of English Norton and Rudge 500-cc Grand Prix motorcycles. At the time Italian motorcycle builders were simply not making machinery capable of competing with the best from England and the decision was made to go overseas. This effort in motorcycles lasted three seasons—1932, 1933, 1934—and is but a footnote to the story. Taruffi won the Grand Prix of Europe on a Norton 500 on April 14, 1932, and other riders, led by Giorando Aldrighetti and Guglielmo Sandri, won a number of regional events in the north of Italy, but overall the two-wheel campaigning is remembered as little more than a brief diversion from the increasingly serious and professional automotive side of the Scuderia Ferrari.

In keeping with the evolution away from amateurism, the original founders of the Scuderia, Mario Tadini and the Caniato brothers, found themselves displaced from the operation during the winter of 1932. Tadini had already sold his share of the business to Alfredo Caniato, who in turn sold out to the wealthy Piedmontese nobleman Count Carlo Felice Trossi. Trossi, he of the long nose, the omnipresent pipe, the brooding eyes and the pile of black swept-back hair, was a unique man. Enormously rich from banking and landholdings, Trossi was a serious lover of automobiles and the sport. His ancestral castle at Gaglianico, near Biella, had a racing shop, entered by a drawbridge, where the Count often joined his small retinue of mechanics to work on his cars, always wearing a pair of white linen gloves! "Didi" Trossi, as he was known to his social equals, replaced Caniato as the president of record of the Scuderia, but the title was essentially honorary. Enzo Ferrari ran the operation and Trossi acted only as a source of funds and as a driver who could be counted on to perform adequately in minor events and to serve as a backup in the more important races.

On January 19, 1932, Laura gave birth to Ferrari's first child. It was a boy, who had the heavy-lidded eyes of his father and what was to be the prominent, slightly downturned nose of his mother. He was named Alfredo, after Enzo's late brother, and immediately nick-named Dino. He was a frail, sickly child from birth. No doubt his father broke away from the routine of running the Scuderia long enough to celebrate the new arrival, but it is difficult to believe that during this hectic, intense period of growth Enzo Ferrari, a man who by his own admission was a poor husband and by implication an inattentive parent, spent a great deal of time doting on his new son.

The pressing business of the day involved the preparation of updated Alfa Romeo 8CMs for the upcoming Mille Miglia. The old connection with Memini carburetors had been dropped and a new link forged with Edoardo Weber's Bolognese firm. This association brought forth Weber's unique dual-throat design and would bring enormous rewards to both firms. Ferraris were to carry Weber carbu-retors until fuel injection displaced them in the 1980s, and much of the high horsepower produced by the marque was attributable to the excellence of the Weber design. In early 1932 Ferrari arranged a joint research project between Weber and Shell Oil to learn more about the relationship among fuels, induction, and combustion-chamber design that was to pay long-term dividends to all three firms.

At this point the Mille Miglia had been elevated from a simple motor race around Italy to a national craze. Millions of spectators, kept in order by legions of Blackshirts, militia and carabinieri, lined the course, even in the remotest mountain stretches. Fame and for-tune awaited the winner—and the builder of the winning car. Alfa Romeo had dominated the race until the shameful defeat at the hands of the Germans the year before, and much of the responsibility for reclaiming the checkered flag was laid on the broad shoulders of Enzo Ferrari.

Once again the Alfa Romeo management had called Nuvolari, Borzacchini and Campari back to the official team, while Ferrari was to provide massive backup with five more 8Cs, headed by Trossi, plus three 1750s and a 1500 to be driven by the Caniato brothers. The race began in the early-morning hours in Brescia and finished in the same city almost fifteen hours later. During that period all of Italy was either riveted to their radios or huddled along the shoulders of the route, their eyes squinting through the dust and flying stones of the passing cars, their ears battered by the howl of the unmuffled engines. Borzacchini won, breaking Caracciola's record of the year before, and thereby erasing the shame of the Mercedes-Benz victory. Trossi was second, with his friend and aristocratic neighbor the Marquis Antonio Brivio Sforza, riding as co-driver and navigator.

As the late-spring warmth began to spread across the Po Valley, the Scuderia opened its customary multi-pronged campaign across Italy, with members of the team competing in all manner of races, small and large, long and short. None were sufficiently important to bring the first string back from Alfa Romeo, where Nuvolari and Borzacchini awaited the final touches on Jano's latest creation — a single-seat *monoposto* Grand Prix car officially named the Tipo B but immediately dubbed the P3. This lean, lithe, 215-hp, twin-supercharged racing machine was to become one of the milestone designs of all time. It would, in the hands of Tazio Nuvolari, win its first race at Monza on June 5. That stirring victory in the Italian Grand Prix was a triumph for the Alfa factory, with the Scuderia playing no role whatsoever. The new P3s were the exclusive property of the factory, which was now deeply indebted to the Mussolini government, both financially and philosophically, and the closest Enzo Ferrari and Didi Trossi got to the fleet new machines was as wide-eyed spectators in the Monza pit area.

While Alfa Romeo headed off to compete with Maserati and Bugatti on the major-league Grand Prix circuits across the continent, Ferrari was left to maintain a kind of farm team for the company. His

retinue of cars consisted of now outdated 8Cs and 1750s and his drivers remained eager amateurs and aspiring former mechanics. This
was to be a frustrating time for Ferrari, who the year before had been
on the very edge of becoming the official factory racing team. But
now, with the P3s showing so much potential, and the erratic, on-
again, off-again racing policy of the factory at full-tilt, he had little
choice but to sit back and patiently marshal his brood of second-raters
to gain as many minor-event victories as possible.

In 1932 the most famous endurance race in the world remained
the 24 Hour race at Le Mans. The organizers of the luridly fast and
difficult Spa-Francorchamps circuit in the Belgian Ardennes Forest
had long been attempting to create an endurance event equal to that
of Le Mans, and in July invited the Scuderia to participate in their 24
Hour contest. The Alfa Romeo factory, having tasted major success
with the P3, tightened its grip on the services of the Italian Big Three
(now joined by the brilliant German Rudi Caracciola) and left their
Modenese "division" to field a pair of 8C-2300s, fitted with lightweight Zagato bodies. They were to race with Taruffi and newcomer
Guido D'Ippolito in one and Brivio and Eugenio Siena (who had
become the team's chief test driver) in the other. The race is noteworthy not because these teams finished one-two in what was a minor
event and doomed to remain so, but for the initial appearance of the
Prancing Horse escutcheon on the hoods of the Scuderia Alfas. As
mentioned before, the exact origins of the emblem remain unclear,
but it is agreed that Spa marked its first appearance in public. Why
Ferrari chose to introduce it at such an obscure race in a small nation
far from home is again a source of confusion, but it was to appear on
Scuderia machinery from that day onward, in direct contrast to the
four-leaf-clover logo that was a common sight on the cowling of the
official factory cars. It is possible that Enzo Ferrari, sensing that
the success of the new P3s would revitalize Alfa's official interest in
motorsports to a point where the Scuderia would be squeezed out of
its future racing plans, decided to establish a stronger, more independent image for his operation by instituting his own distinctive look.
The Scuderia cars were painted a different hue of Italian racing

red—a darker burgundy shade, as opposed to the vermilion of the Portello machines—and the Scuderia Alfa 8Cs also carried slightly different headlight mountings than the factory automobiles, a further visual differential.

But this zany game in which cars and drivers were passed back and forth between the two operations like chattel was to continue. The team had barely returned from Belgium to be met by Ferrari— who did not make the trip—before they were informed that a Tipo B P3 was on its way from Portello, with Nuvolari assigned as the driver. Again, it is possible that intense politicking by the boss was the deciding factor, but ostensibly as a reward for its victory at Spa the Scuderia was presented with one of the new machines, which it would enter in the upcoming Coppa Acerbo. Nuvolari won this event without apparent effort, although Taruffi crashed hard after trying to outrun Brivio in the early stages of the race. It is apparent that an intense rivalry had been festering between the two and had resulted in Taruffi overcooking his car on a fast bend outside a small village along the course. This led to some tense moments with Ferrari, who tolerated wildness only on the part of Nuvolari. But to his relief, Taruffi was kept on the team. Ironically, it had been Brivio who had been considered the top candidate to bend a car in such a situation, but he had managed to maintain his composure and finish the race, albeit far back.

Despite the mad round of races in which the team participated, there was a certain leisurely routine that governed things at 11 Viale Trento e Trieste. Accounts from the day indicate that the staff worked at what might be described as a leisurely Italian pace, with plenty of time for joking, long lunches at the nearby trattorias, ogling the women who passed by on the street, getting together for high-speed test drives on the open stretches of the Via Emilia, and regular games of dice whenever the Commendatore—as Ferrari was then called— was out of the office or had his back turned. Bazzi, who remained on the factory staff, and Ramponi were on regular runs between Modena and Milan, transporting parts and the latest technical innovations. The shops hummed with the racket of the machine tools, the hiss of compressed air, and the clang of hammers, while the acrid smells of oil, heavy grease, gasoline for cleaning parts, lacquer and paint thinner were omnipresent. The mechanics' hands were perpetually filthy, owing in part to the use of lampblack for testing the sealing quality of gaskets.

Ferrari's small office was generally filled with visitors or Alfa representatives, leaving him little time to compose letters on his typewriter, which were always signed in purple ink. He used that particular hue, he explained, in memory of his father, who always signed his name on legal documents with an indelible pencil which left an impression of violet on the carbon copy. He was to use that color until the end of his life. Upstairs was the other component of his life: Laura and the baby, Dino, both of whom were never to stray far from the maelstrom of noise and confusion that swirled below them.

The persona of the man who was running the operation was beginning to take a shape that would remain consistent throughout his life. While still only in his mid-thirties, Ferrari was firmly in control of the operation, and given to fits of temper that would descend on the place like summer thunderstorms. A misplaced cigarette, a badly fabricated part or the late arrival of a worker would send his temper soaring and strong men scurrying for cover. At the same time he could be a model of decorum, transforming himself into a charming maître d' when the moment demanded it, as when a highborn noble or Fascist official arrived at the Scuderia or a wealthy customer expressed interest in spending extra for a special-bodied Alfa. Ferrari was learning the art of manipulation with his drivers as well: how the subtle suggestion, the offhand remark, the critically timed slight, might make a man drive all the harder. Enzo Ferrari was on his way to becoming the consummate manager of men—not docile, soft-willed men, but proud, fiercely competitive, egocentric men whose livelihood, if not their very reason for living, depended on this most demanding and unforgiving of sports. If any man understood the dimensions of that unique human weakness the Greeks called hubris, it was Ferrari.

The season dribbled to an end with the second-stringers winning or placing well in a series of minor races and hill climbs. It was again capped by a banquet at Ristorante Boninsegna on November 19, with the Commendatore holding forth during a long evening of speeches and award giving. Ferrari handed out a pamphlet titled *The Third Year of Racing*, in which he summarized the season. Fifty races had been entered by the car and motorcycle teams and no fewer than twenty-six victories had been gained. But the inference was clear in the listing of events; there were no Grand Prix races, no international sports-car contests, no prestigious European hill climbs, no races other than minor club outings centered for the most part within a day's drive of Modena. This was hardly the level of competition envisioned by Ferrari when the Scuderia was formed. Hopefully 1933 would raise the operation from the depths of regional obscurity to prominence on the continental racing scene. Or would it?

The new year had barely been celebrated when devastating news arrived from Milan. For months Ferrari had heard rumors that Alfa Romeo was about to drop out of Grand Prix racing. Mussolini was moving closer to an aggressive colonial expansion into Africa, with Ethiopia a prime target, and Alfa Romeo was steadily shifting its engineering and production staff to war materials. Worse yet, the Depression was wiping out demand for high-performance sports cars. This made the image-building efforts of an international Grand Prix team seem irrelevant. It was announced in the bleak days of the Emilian winter that not only would the official Alfa Romeo team be disbanded but the powerful *monoposto* P3s would be locked in a shed at Portello and retired forever. Ferrari made repeated trips to the factory, practically lowering himself to his knees to obtain the six magnificent machines that now sat under tarpaulins, unwanted and unused. For all his skills at persuasion and his unquestioned loyalty to the firm, the Alfa Romeo management was unrelenting. The racing cars would not be released and the Scuderia Ferrari would have to be content with its inventory of eleven 8Cs and two supercharged 1750 sports cars.

Furious at the slight, Ferrari and Trossi considered buying a pair of potent Maserati 8CM Grand Prix cars and even a trio of British-built MG K3 Magnettes—tiny roadsters that were dominating small-displacement sports-car racing. At the same time, Didi Trossi arranged through connections with Champion Spark Plug representatives in Europe to take delivery of a two-man Indianapolis car that was erroneously labeled a Duesenberg. The monster had in fact been constructed by a former Miller engineer named Skinny Clemons in association with August Duesenberg, who was no longer connected with the family firm. He was the younger brother of the brilliant Fred Duesenberg, who had been killed at the wheel of one of his own magnificent roadsters while descending Ligonier Mountain near Johnstown, Pennsylvania, the previous August. Trossi's car, fitted with a *monoposto* body, carried a single-overhead-camshaft engine based on the earlier (1920–27) Duesenberg Model A passenger-car power plant. This one-off would be raced with little success, although it would play a role in a tragedy that was to rock Italy later in the season.

While Mario Tadini, who was becoming a minor master at such competitions, won an early-season hill climb at La Turbie on the French Riviera, the Scuderia's first serious outing came, as usual, at the Mille Miglia. Nuvolari and his faithful Decimo Compagnoni won easily after Borzacchini's sister 8C-2300 broke its engine. The team also won in the 2-liter and 1.5-liter classes.

But the limits of the aging 8C Alfas were obvious. Two weeks later Ferrari entered four of the cars in the prestigious Monaco Grand Prix, a marvelous event, run to this day, through the sinuous streets of the picture-book principality. Varzi was there with a flawlessly prepared Type 51 Bugatti, while his tormentor, Nuvolari, led the Scuderia with a wildly overstressed 2.6-liter 8C Monza. The duel between the two masters covered 99 laps and over three hours under a broiling Mediterranean sun. The lead was exchanged countless times, with Nuvolari hanging on by a car length into the final lap. But finally the old Alfa broke less than a mile from the checkered flag and Varzi went on to victory. Borzacchini salvaged some of the Scuderia's honor by finishing second, but the message was unmistakable. With the P3s now sitting inert at Portello, not even the great Nuvolari would be able to hold off the newer, more powerful Bugattis and Maseratis in the upcoming Grand Prix races.

After an uncontested one-two-three finish for the Scuderia at Alessandria, with Nuvolari besting Trossi and Brivio, the team sailed across what was then rather pretentiously being called "Mare Nostrum" by the Italians. A major race was scheduled at the spectacular rebuilt Mellaha circuit outside Tripoli in the protectorate of Libya. This fantastic 8.14-mile road course, which surrounded the steaming desert oasis of Tagiura a few miles from the Mediterranean coast, was to gain the reputation as the fastest natural circuit in the world, with straightaway speeds commonly approaching 200 mph.

But in 1933 speed was hardly the big news from Tripoli. What was to transpire was one of the maddest interludes in international motorsports and a moment of high comedy that is still often swept under the rug in official histories. Giovanni Canestrini was at the time the editor of *Gazzetta dello Sport* and the nation's most prominent motoring journalist. It was he who devised the scheme of a national lottery—"a Lottery of the Millions," as it was called—built around the upcoming Tripoli Grand Prix at the new track. The idea was warmly received by Italo Balbo, the new governor of the colony and a man responsible for the creation of the Mellaha course with its giant, cantilever-roofed grandstand, its formal gardens and acres of cutstone landscaping. Italo Balbo was perhaps the most charismatic of all Mussolini's cohorts. A World War I aviator, he had been one of the organizers of the March on Rome and had been rewarded with an appointment as air marshal of the booming Italian Air Force. By the early 1930s Balbo was being supplied annually with over 1,250 advanced fighters and bombers. In July 1933 he led a flight of twenty-four seaplanes across the Atlantic to the Chicago World's Fair and was acclaimed as a hero at home. But Mussolini, irrationally threatened by men who might rival his leadership, quickly banished Balbo to the governorship of the Libyan colony, which led the former air marshal to grouse, "As soon as he sees too much light shining on us, he turns off the switch." Balbo was a motor-racing enthusiast of the first order, which prompted him to construct the Mellaha racecourse at Tripoli. Always a realist about Italy's place in the sun, he understood full well that his nation—bereft of natural resources and a strong industrial base—could not fight a major war. Enzo Ferrari recalled asking Balbo if Italy was going to war with the great powers. "What would we fight it with, peanuts?" he replied.

Canestrini and his lottery also had the enthusiastic support of Augusto Turati, the secretary of the Fascist Party and another motoring aficionado. His plan was simple: twelve-lire tickets would be sold across the nation, with thirty finalists to be taken to Tripoli for the race. There each would hold a ticket on one of the participating drivers, with the holder of that on the winning car assured a prize of about $500,000. But Canestrini was a man who hated long odds. It is said that he contacted the owner of the ticket with Achille Varzi's name on it—a lumber merchant named Enrico Rivio—and proposed a plan. (Others claim that Rivio called Canestrini, but no matter, the result was the same.) If Varzi was assured of a victory, he and Rivio would keep half the money and split the rest with certain interested participants—namely, Varzi, Canestrini and all of the contenders for the victory, among them Nuvolari, Borzacchini, Giuseppe Campari (who was driving for Maserati) and Louis Chiron, the ace from Monte Carlo.

The deal was made. The fix would work perfectly. A phony early race would be staged, then Varzi would move his Bugatti into the lead for the big payday. Presumably Varzi was chosen for two reasons. First, his Type 51 was a potent car with a legitimate chance for victory, fix or no fix. Second, a victory by a French car might eliminate suspicion on the part of Balbo, who knew nothing of the arrangement. It must be presumed that Enzo Ferrari was part of the conspiracy, or at least privy to it, considering that all of his drivers were participants, but there is no concrete proof. What is known is that one driver, the Englishman Tim Birkin, was not involved, due either to outright ignorance of the cabal or to a spasm of high conscience. Being a knight of the realm and a rich sportsman of note, Sir Henry "Tim" Birkin was probably not considered either a threat to win or a willing candidate by the conspirators and presumably was kept in the dark.

This turned out to be a disaster for all concerned. At the start Campari led for a bit, before being overtaken by an enthusiastic Birkin, who chose to run at a much friskier pace than those who did not want to strain their selected champion. Nuvolari and then Borzacchini were forced to speed up to dampen the ardor of the Englishman while Varzi languished in third and fourth. Because Birkin had been quick during practice, a late-hour rule had been instituted to impede him. An absurd regulation was imposed, permitting only one mechanic to service each car. Because Maserati had sent along but one man to care for both Birkin's and Campari's 8CMs, Birkin found himself without a pit crewman prior to the start. He enlisted the aid of a local garage man, who turned out to be a hopeless drunk and spent the entire race dozing under a palm tree.

Then the farce began. Campari cruised into the pits and began a fueling stop that consumed about the time required to fill the tanks of a Savoia-Marchetti SM-81 trimotor bomber. Borzacchini then sauntered into the pits on foot. He had driven off the course, whacked an oil drum and declared the Alfa unfit to drive. He left baffled on-lookers peering at the pristine car and walked away, a wide smile on his cherubic face.

Nuvolari had the lead, but Birkin was pressing on with classic Edwardian resolve. However, the hapless Englishman was forced to stop for fuel and tires—which were being ground smooth by the blowing sand. He set to changing the wheels and fueling up alone, but unfortunately seared his arm on the hot exhaust pipe before climbing back aboard and continuing, now far back and out of contention. Nuvolari cruised on in the lead, with Varzi second. But as the Bugatti passed the immense stands, thousands of ears perked up. That noise! That evil rattle issuing from the Bugatti's exhaust! The engine was misfiring. If Varzi failed to finish, the entire plan was doomed.

Nuvolari immediately ducked into the pits for fuel. Decimo seemed mired in sand as he filled the tank. Campari staggered in. His mechanic began to examine the front suspension, which appeared to be in perfect shape. A catcall echoed from the stands. More followed. The crowd was on to the scheme. Nuvolari rushed out of the pits, attempting to look serious. Next time around his crewman signaled him that he had already made up twenty seconds on the reeling Varzi. Resourceful to the end, Tazio began sailing the 8C around the corners in mad, sand-slewing, time-consuming slides. But try as he might to go slower, he gobbled up the distance between himself and Varzi, whose sputtering Bugatti wobbled around the circuit at a near walk. Finally Nuvolari gave up the charade and merely tucked in behind Varzi and followed him across the line amidst torrents of jeers.

Balbo was furious. A commission was created to investigate the scandal. Varzi, Nuvolari, Campari, Borzacchini and the Monégasque Chiron were banished from racing in Tripoli for "life." Canestrini presumably escaped Balbo's wrath, as did Turati. Ironically, the man who paid the greatest price was the single innocent. Tim Birkin contracted blood poisoning from his burn and died three weeks later.

The same lottery scheme would be tried a year later, with a more subtle (and successful) plot. Varzi won again, despite the safeguards instituted by Balbo, but the race essentially ended the lotteries—and the only known attempts to fix a major Grand Prix. It might also be said that only the Italians, with their innate sense of the absurd, could have executed the attempt with quite the same combination of high comedy and good-natured larceny. It goes without saying that the "lifetime" sentences imposed on the guilty were quickly forgiven and all returned to race at Tripoli the following year.

Whether or not the plot in Tripoli had any effect on the ongoing operations of the Scuderia is unknown, but it did mark the beginning of a period in which the entire operation seemed to unravel. Nuvolari and Borzacchini took their manifestly slower 8Cs to Berlin for a race on the AVUS circuit, a gaudily fast twelve-mile superspeedway formed by linking two six-mile stretches of four-lane motorway with 180-degree corners at either end. Average lap speeds of over 170 mph would become common by the end of the decade, and in this environment the old Alfas were hopelessly outclassed by the more modern and powerful Bugattis and Maseratis. The two drivers returned to Modena angry and frustrated. Ferrari, on the other hand, engaged in a series of threats, friendly persuasions, tantrums, visitations, homages, attempted defections and open begging in order to wrestle at least two of the moldering P3s from the factory. The reality that the Scuderia was doomed without the P3s affected morale to a point where the once-tranquil shop was swirling with acrimony and dissent.

Nuvolari succeeded in winning a few races by sheer willpower and his uncanny talent, but by mid-season he had determined to break with Ferrari, contract or no contract. A week prior to the Belgian Grand Prix at Spa, Nuvolari signed a secret pact with Ernesto Maserati. He would drive one of the potent 8CMs in the race, despite the fact that Ferrari had sent along an Alfa for him. There was a colossal contretemps. Ferrari was apoplectic that his star was defecting, while Nuvolari, the obsessed competitor, loudly justified his departure on the basis that his aging Alfa Romeo was incapable of winning against the opposition. In the end a face-saving compromise was worked out whereby Nuvolari theoretically drove the Maserati (to an overwhelming victory) under the aegis of the Scuderia, but the arrangement was pure window dressing. He carried no Prancing Horse on the Maserati. Nuvolari was gone, along with his sidekick, Borzacchini. Then Taruffi walked out. Suddenly Ferrari was bereft of top-flight drivers. While rumors swirled through the sporting press, he quickly responded by hiring Campari—now universally acknowledged to be past his prime—and a feisty, muscular native of Abruzzi named Luigi Fagioli. Said to be an accountant by some, a pasta merchant by others, or simply a rich sportsman, Fagioli was, at thirty-five, a veteran driver who had had considerable success. But he was known to be a willful, difficult personality and it was predicted he would quickly butt heads with the boss of the stable.

The mutiny by Nuvolari and Borzacchini finally brought action from Milan. Alfa Romeo responded to Ferrari's cries for help and one day in late June a convoy of trucks creaked to a stop on the Viale Trento e Trieste. On board were six of the magical Tipo B P3 *mono-*

posti and mountains of spare parts. Better yet, Luigi Bazzi arrived to take over engine development, having left his post as head of the experimental department under Jano. Also transferring to Modena was Attilio Marinoni, a likable mechanic who had been Alfa's chief test driver and who would assume similar duties with the Scuderia. Vittorio Jano would remain in Milan, where he was assigned to develop a new series of aircraft engines for an Italian Air Force that Mussolini could brag was the most numerous, if not the most powerful, in the world.

In a moment's time, the Scuderia had been transformed from a loser into a potential powerhouse. This providential act may have saved the operation from oblivion. It would not be the last time a gift of this magnitude would reverse the fortunes of Enzo Ferrari and his Scuderia.

With the new cars in hand, Fagioli won the Coppa Acerbo at Pescara and a minor Grand Prix at Comminges, France, while the newly recruited Louis Chiron took the victory at Miramas, near Marseilles. The team was winning again, and none too soon. This set the stage for a pair of pivotal races at Monza on the second weekend in September. On the morning of the tenth, the Italian Grand Prix was run over the 6.2-mile combined-road-and-speedway course. It was a grim struggle between the Alfas and the Maseratis, with Fagioli wrestling a victory from Nuvolari when the latter was forced into a late-race pit stop to change a tire. The result, of course, was a tremendous disappointment for Nuvolari, who viewed second place as oblivion.

Unlike the Italian Grand Prix, the Monza Grand Prix was to be run on the so-called *Pista di Velocità*, a high-speed oval that lay within the perimeter of the road course. It was an immense, Indianapolis-style oval featuring a pair of very fast, semi-banked corners. The day was chilly, with the threat of rain lingering over the brooding speedway. A number of the drivers in the earlier race, including Nuvolari and Fagioli, chose not to compete in the Monza Grand Prix, which was to be run in three short sprints. A wealthy amateur, Polish Count Czaykowski, won the first heat with his immense 4.9-liter Type 53

Bugatti. The only memorable incident involved the Trossi Duesenberg Special, which retired early with engine failure. Some said it left a puddle of oil in the middle of the dangerous south curve.

The second heat opened with Campari, on board a Scuderia P3, and Borzacchini, driving a Maserati, charging off the line (the race began with a standing start) and roaring into the gloom side by side. But they never returned. Only three of the seven starters dribbled past the pits a few minutes later. Campari and Borzacchini, as well as two other drivers, Nando Barbieri and Count Castelbarco, were missing. Word quickly passed through the pits that there had been a giant crash on the south curve. It was confirmed. Campari and Borzacchini had touched wheels at full speed and pinwheeled off the track. Both men were flung onto the pavement while their cars tumbled crazily through the wire infield fence. Borzacchini's Maserati bounced to a stop on its wheels, looking practically undamaged. But its driver was mortally wounded and would pass away a few hours later. Campari, the beloved "Pepino," was killed instantly. The other two drivers, although their automobiles flipped over, escaped with minor injuries.

Standing in the damp gloom of the pits, Enzo Ferrari reeled at the news. Suddenly the naked brutality of the sport he loved pummeled him and tore at his insides. Here were two men, his old *compagnos*, now lying torn and lifeless like rag dolls in a muddy ditch. He had steeled himself for such moments. He knew of the flimsy cars and their propensity to buck men off their backs like Wild West broncos. He understood the risks, having accepted them himself. But for Campari and Borzacchini to go down together was almost more than he could bear. This would force him to separate the cruel, heartless business of automobile racing from the amiable side of motorsports—the casual talk in the bars, the bargaining over the sale of an Alfa to a wealthy client, the workaday routine in the shop—for until that day he had not dealt with the death of one of his own. Yes, Arcangeli had fallen, and so too had Sivocci and Ascari, among others, but they had not been in Scuderia cars. No blood had been spilled on the Prancing Horse until that evil afternoon, and it would mark a turning point for him. From that day onward Enzo Ferrari would draw a thin, invisible psychic shield between himself and his drivers. On rare occasions he would let that barrier be penetrated, but for the most part the men who would thereafter drive for Enzo Ferrari operated outside his most intimate emotional boundaries.

This black day ended with further tragedy. In the third and final heat Count Czaykowski lost control of his Bugatti and crashed in the same place. He burned to death in the wreckage. In the wake of the carnage the entire world of Italian motorsports was plunged into mourning and a spate of rumormongering. Trossi's Duesenberg was (and still is) labeled as the culprit, after allegedly spewing oil from its sump on the track. But Giovanni Canestrini, who, despite his foray into choreographing races, was a generally reliable journalist, reported that he had made a postrace inspection of the Trossi car and found its oil supply to be intact. Its retirement had been caused only by a burned piston. No matter, the oddball car remained the scapegoat for this terrible racing tragedy, although it is more likely the three fatalities were caused by overzealous driving and the damp surface combined with the hard-compound tires employed for the high-speed track.

Nuvolari was particularly affected by the loss and remained in the Monza hospital for the entire night with the dead drivers' grieving wives. When asked about how the death impacted upon his own driving, he replied philosophically, "It happens when you least expect it. If we were to drive eternally preoccupied with the dangers that beset us, we would never complete a single lap."

Enzo Ferrari dutifully attended the funerals of Campari and Borzacchini, but the racing continued without abatement. Louis Chiron, known as the "wily fox" because of his silky, almost inconspicuous driving style, was retained to replace Campari, and a week later, at Brno, Czechoslovakia, he beat Fagioli as the Scuderia recorded a one-two finish. But there was to be one more tragic surprise in this year of gyrating fortunes. The final race for the Ferrari team was held outside Naples, a Mille Miglia–like 480-mile open race around the boot of Italy called the Princess of Piedmont Cup. The event was hardly of the magnitude of the international Grand Prix races in which the Scuderia had become a common presence, but its amateur team was well presented. Barbieri and Comotti won easily, but that victory hardly counterbalanced the death of the widely liked and valued team member Guido D'Ippolito. His 8C-2600MM, which he was sharing with Francesco Severi, tangled with a local farm cart (public traffic was never completely stopped for races such as this) and he was killed instantly in the ensuing crash.

When the Scuderia gathered for its annual season-ending dinner, it was a more subdued reunion than in years past. Three of their cohorts were dead, and Italy's greatest driver, who had begun the year as one of them, had moved to Maserati. Worse yet, word was filtering out of Germany that the Hitler regime was funding a massive assault on international Grand Prix racing that could make obsolete all existing machinery, including the P3s. On the positive side the Scuderia could now count on being the fully authorized representative of Alfa Romeo in big-time motorsports, with the talented Bazzi and Marinoni on staff. For the first time since its formation, the Scuderia could be considered at the top level of the business, able to compete head to head with the likes of Bugatti and Maserati—and with the expected German onslaught. For all the death and acrimony, Enzo Ferrari could look back on the past season with a certain satisfaction, befitting only a man with a diamond-hard will to win at all costs.

6

Enzo Ferrari knew enormous change and uncertainty lay ahead for the Scuderia in 1934. Alfa Romeo's sagging economic fortunes had in early 1933 caused the Italian government to completely absorb the company into the IRI (Istituto di Ricostruzione Industriale) and placed strong-willed Ugo Gobbato at the helm. It seemed that the new chairman's policies dovetailed with Ferrari's desire to commandeer the entire factory racing program. Gobbato was a young, intense engineer who had risen to the top on the basis of his extraordinary administrative skills. He was in many ways the antithesis of Ferrari, one being thorough and methodical, the other inclining toward constant improvisation. Gobbato was a classic organization man, which included his enthusiastic involvement with the Fascist Party. Ferrari tended to be the lone wolf, relying on his own wits rather than the collective wisdom of committees. Gobbato had spent 1931 in Russia, aiding the Communists in setting up the world's largest ball-bearing factory. (At the time, the Stalin regime had actively solicited the help of numerous Western engineers to aid in the industrialization of the state. Some, like Gobbato, accepted their lavish enticements. Others, like Ferdinand Porsche, did not.)

Now back in Italy and heading a fully nationalized factory that was almost totally devoted to the military armaments business, Gobbato made the decision in November 1933 that Alfa Romeo would cease competition on its own part and turn the racing department over to the Scuderia Ferrari. Gobbato no doubt made it clear that fealty to the Fascist Party was critical if one had hopes of doing business with Alfa Romeo. While Enzo Ferrari never revealed any strong political orientation, he was a pragmatist of the first order and realized that Gobbato's sentiments would have to be acknowledged, if not openly supported. Like so many Italians, Ferrari's diet of Fascism was based on a thin broth, not the thick steamy soup of Hitler's National Socialists or Lenin's chilly Marxism. Mussolini seemed to be a powerful, resourceful leader and the masses adored him, at least from a distance, where his bombastic exhibitions could be enjoyed as pure jingoistic theater. Despite the gripping poverty in the south and the grumblings of the Communists in the north, the nation was sailing along on his coattails, unconcerned about where the trip might take them. Much of the Fascists' apparent success—the public works projects, the heavy-handed social engineering and the military buildup—appeared to evidence the reemergence of a strong, united Italy reminiscent of the Roman Empire. But it was a self-delusionary charade. Beneath the parades, the oratory, the strutting about in flashy new uniforms, the same old regionally fragmented, cynical, isolated, opportunistic, faintly bemused Italy chugged onward, unchanged and unaffected. Yes, the people cheered Il Duce, they poked the skies with the Fascist salute, they dared to dream of power and glory, but in the end, as the night settled around them outside their shuttered windows, the people understood that it was yet another game and that if they were to survive, it would be thanks to their wits and their ability to play the ancient game of life, Italian style.

Few men understood this game better than Ferrari and, thrust forward by his realism, he joined the Fascist Party in 1934. His business relationship with Gobbato and with Alfa Romeo demanded it. There is no evidence that he was a particularly devoted party member, although the Scuderia's newsletter did assume for the next four years a bellicose tone attuned to party doctrine. Fascist slogans were laced through the text and Ferrari's traditional sniping at his rivals—the Maserati brothers, the German teams and certain drivers and sponsors who had fallen from favor—became more pointed and strident. Ferrari—always the cold-eyed businessman—was not about to disturb a cozy arrangement with Alfa Romeo because of some silly, essentially irrelevant political doctrine. If Fascism was in fashion and it meant extra racing successes for the Scuderia, so be it; Enzo Ferrari would be a good Fascist.

With the official factory racing team of Alfa Romeo once again in one of its fitful states of dormancy, Jano and his engineering staff would be assigned other duties, including the creation of new high-powered fighter aircraft engines, while Ferrari would have to make do with the six P3 *monoposti* currently in the garage. There was one caveat: Two wealthy Algerians, Guy Moll and Marcel Lehoux, had already ordered new P3s and the Scuderia would be required to admit them as team members in order to fulfill the factory obligation.

This would create a complication, but one that was more than offset by the signing of no less an eminence than Achille Varzi. Now that Nuvolari was gone, the severe, unsmiling Varzi was open to advances from Ferrari. It was widely known that this regal, calculating professional had refused to drive on the same team as the temperamental, ebullient Nuvolari. But with both Maserati and Bugatti seemingly falling deeper into the ruck of outdated technology and with the thriving Scuderia in control of the vaunted P3 *monoposti*, Varzi joined up. Ferrari knew him as a difficult man, but one who could be motivated by pride. And surely, with a young tiger like Moll to prod him along, Varzi might attain unimagined heights during the season.

Although he was a relative newcomer, Moll's skills were surely known to Ferrari. A fresh-faced man-child of twenty-three, Moll had been given his first major drive two years earlier by the older, more experienced Lehoux in the Marseilles Grand Prix. He finished third behind Raymond Sommer and Nuvolari while turning numerous heads along pit row with his charging, heads-up driving style. The son of a French father and a Spanish mother, Guy Moll was the kind of cocky, headstrong, fanatically brave individual that Ferrari believed to be the prototypical race driver. He would be hard to handle, especially when teamed with the proud Varzi, but Ferrari was gaining more confidence in his ability to manipulate and control the soaring egos of his drivers. Moreover, he could identify in Moll the qualities of a future champion. As for Lehoux, he would be relegated to secondary status with Marinoni. Therefore, the front-line team would be Varzi, Chiron and Trossi in that order, with Moll operating as a kind of background agitator—an eager substitute who would be tossed into the fray at the moment one of the first-stringers faltered.

This situation with cars and drivers placed the Scuderia in the very top rank of Grand Prix teams for the 1934 season. In four hectic years the Ferrari operation had risen from a tiny regional racing entity to the exclusive representative of a renowned factory whose automobiles were the fastest and most powerful on the international racing scene. Virtually everyone in the motoring press predicted that the Scuderia would dominate the coming season. To be sure, the Bugatti team had Tazio Nuvolari and its new 240-hp Type 59, but the car was classic "old school" with beam axle suspension and mechanical brakes (the Alfas had similar antiquated components, but advanced independent suspension and braking systems were already in the works, while Ettore Bugatti would refuse to update his cars). The Maserati brothers had fine 2.9-liter 8CMs, complete with four-wheel hydraulic brakes (the first time a European racing design had employed such a system after the American-built Duesenberg introduced them at the French Grand Prix eleven years earlier) but the firm was small and relied on sales to private teams to finance its cash-poor racing campaign. Surely, went the current wisdom, the powerful, reliable, sweet-handling P3 Alfas in the hands of virtuosos like Varzi and Chiron would have little trouble disposing of the opposition.

But there was a new, unknown element looming to the north. The 1934 racing season was to introduce a new formula that the doyens of the sport had instituted to slow the cars to a more rational pace. As early as October 1932 the French-based AIACR had announced that from 1934 to 1937 Grand Prix races would run under a so-called 750-kilogram formula.

This alleged improvement in the regulations was motorsports' counterpart to the Smoot-Hawley tariff or the disarmament fantasies of the 1928 Kellogg-Briand Pact: it accelerated exactly that which it was intended to discourage. But whereas the tariff and the pact were concerned with peace and prosperity, the intent of the 750-kilogram formula was the reduction of rising speeds. As it turned out, the direct opposite was the result. The logic of the new formula, which mandated that a Grand Prix racing car weigh a maximum of 750 kilograms (about 1,650 pounds) minus tires and all liquids (coolants, lubricants, brake fluid, etc.), was original in concept but utterly bogus in practice. It was believed that the engines of the day, the 3-liter supercharged units of Alfa, Bugatti and Maserati, were about the maximum that could be carried in automobiles of the new weight. Therefore to build larger, more powerful engines would require beefier, heavier chassis and running gear. By holding down the weight, went the reasoning, power would automatically be limited.

This seemed to be a perfectly reasonable proposition, considering the twenty-year-old metallurgy and chassis technology being employed by the current competitors. These builders were obsessed with horsepower, believing that races were won by sheer straightaway speed. They paid little attention to such niceties as cornering power, weight saving, braking, steering, suspension design and other engineering esoterica. Their cars, even the somewhat advanced P3, were merely crude four-wheeled platforms upon which to plant engines. Within this mire of stagnant technology the 750-kilogram formula appeared to make perfect sense.

But if someone decided to install a large, very powerful engine on an advanced featherweight chassis with superior road-holding qualities, the new notion of speed control could be contravened. This is precisely what occurred to two vibrant engineering minds. Coincidentally, they were both located in the Swabian industrial city of Stuttgart. There Dr. Hans Nibel was the director of the central design office of Daimler-Benz AG. A few miles away, in the office of his private consulting firm, sat the renowned Austrian engineer Dr. Ferdinand Porsche. Both had essentially the same idea, although their approaches were reversed—at least in terms of where they located the engines in their startling new automobiles.

By 1932 the Depression had hammered hard at German industry. Employment at Daimler-Benz was halved to the level of 1928 and thousands of skilled German laborers were thrown on the street. This economic collapse was only part of the reason the nation lapsed into a mass psychosis and elevated that mad artist, architect and sometime paperhanger Adolf Hitler to power. He had taken over Germany's chancellorship in January 1933 and was about to start the nation on a twelve-year march to shame and near-oblivion. But as with all despots, there were flashes of rationality in his megalomania. For example, Hitler was a devotee of personal automobile transportation, which would ultimately lead to the mass-marketing of the Volkswagen long after his demise. He was quick to recognize the need for an advanced road system and by 1934 had appointed Dr. Fritz Todt of Munich to oversee the construction of 2,500 miles of the most advanced highways in the world (he managed to complete 1,310 miles before construction was halted in 1942). The idea was to utilize the massive road system for civilian transportation as well as the movement of military supplies during wartime. This, of course, was the same rationale for the creation of America's 40,000-mile Interstate network in the 1950s and 1960s.

Hitler was a motor-racing enthusiast of minor magnitude and in March 1933 acceded to a plan proposed by the management of Daimler-Benz to produce a world-beating Grand Prix car. The justification was simple: The machine would showcase German technology to the European masses and thereby function as a powerful propaganda tool for the New Order. The program would operate under the auspices of the newly created NSKK (Nationalsozialistische Kraftfahrkorps, or National Socialist Motoring Corps), a semi-comical home guard for motoring activities headed by the popinjay Korpsführer Adolf Hühnlein, an early Hitler supporter who had been jailed while trying to capture a telephone station during the ill-fated 1923 Munich Beer Hall Putsch. The NSKK, which ultimately attracted 500,000 members, was an essentially useless organization widely ridiculed as *nur Säufer, keine Kämpfer*—"only drinkers, no fighters."

During the Berlin Motor Show of 1933 it was announced that Hitler would support a major German Grand Prix racing effort and that a subsidy of 500,000 reichsmarks would be awarded to the company fielding a racing team. It was known that Daimler-Benz already had a design prepared and a de facto arrangement existed between the honored old firm and the government. Jacob Werlin, the company's Munich manager and a fanatic Nazi, was a close associate of Hitler's and had made the arrangements for supplying funds directly to the Stuttgart headquarters. Daimler-Benz, for all its later protestations of innocence, was a major contributor to the Nazi war effort, and was to be a prime beneficiary of the racing program thanks to lucrative contracts to build trucks and the awesome DB 601 aircraft engines.

But to everyone's surprise, a second firm stepped forward to claim a share of the subsidies. Auto Union had been formed in 1932 from the ruins of four depression-racked companies—Audi, Horch, DKW and Wanderer—and was now prepared to field a racing car as well. The machine had been independently created by Ferdinand Porsche and adopted by the new company. After a meeting between Hitler, Porsche and two Auto Union representatives following the Berlin Auto Show it was decided to split the subsidies between Daimler-Benz and Auto Union, thereby creating a rivalry and an added incentive for success.

Much has been written about the Nazi support of the blitzkrieg mounted by these two firms. It has been often repeated that they were lavishly subsidized by the government, but responsible historians claim that the funding—given through the German Transport Ministry—was far less than that needed to run a winning racing operation. Both companies spent perhaps ten times as much as the government subsidies. But surely they were additionally compensated through war contracts, crackdowns on labor unions, and generous loan terms.

Even as astute an observer of the racing scene as Enzo Ferrari could not have had the vaguest notion of the crushing blow that was about to be dealt to him and his now outdated P3s. The engineering departments under both Porsche and Nibel had already completed prototypes that would render the Alfas obsolete, once the development bugs were worked out. Their designs were staggering in terms of engineering creativity and potential performance. The Daimler-Benz Model W25 (to be raced under the company's brand name, Mercedes-Benz) featured all-independent suspension, enormous hydraulic brakes, a four-speed transmission mounted with the rear differential and a 3.3-liter straight-8 supercharged engine developing 314 hp, or 100 more than the P3s about to be fielded by the Scuderia. The Auto Union's car was even more impressive. Porsche's ultraradical Type A design carried a larger engine (4.4 liters) with 16 cylinders mounted *behind* the driver. It was twin-supercharged and developed about 300 hp. The so-called P-wagen utilized a five-speed gearbox and all-independent suspension like the Mercedes-Benz.

While the cars were radically different in terms of engine placement, the engineering philosophies that created them were nearly identical. Both the Mercedes-Benz W25 and the Auto Union Type A employed masses of lightweight alloys to package powerful engines within the 750-kilogram weight limit. The two designs also utilized supple, independently sprung suspensions that were the antithesis of the old Italian and French versions that were more appropriate to nineteenth-century oxcarts than to modern automobiles. Such advanced components, coupled with hydraulic braking systems, streamlined bodies and practically unlimited budgets for research and development, doomed the old establishment teams—including the Scuderia Ferrari—from the start.

The international motoring community had been chattering about the rumored Auto Union and Mercedes-Benz efforts for over a year, and there is no doubt that Ferrari was familiar with the projects.

But surely he was not overly alarmed. The wildly chauvinistic Italian press—which he read avidly—was openly confident about the vaunted Alfas and not a little disdainful of the rumors concerning the revolutionary machinery that were drifting out of Germany. With the race-hardened P3s perfectly prepared by Bazzi and Marinoni there was little reason to expect anything but a winning season. However, the team did suffer two serious losses on the technical staff. Young Eugenio Siena left, with Ferrari's blessing, to join in partnership with Swiss sportsman Walter Grosch as a mechanic and driver in a new racing enterprise. Ferrari viewed the likable, aspiring professional with a kind of fatherly benevolence and, contrary to many who abandoned the Scuderia, Siena was the beneficiary of a lavish going-away party. Not so with Ramponi, who departed for England to serve as chief mechanic for the independent team of Maseratis being fielded by American expatriate Whitney Straight (who, once he had vented his youthful exuberance on motorsports, went on to become the chairman of Rolls-Royce). Ramponi was not to be missed. He had made a great deal of money during the Tripoli betting scam and, according to his compatriots, had become insufferably cocky. Whether or not he offended the Fascists, which in turn caused his defection, is not known, but he was to remain in England until moving to South Africa later in life.

Also gone was Luigi Fagioli. In February, Mercedes-Benz team manager Alfred Neubauer brought prototype W25s to Monza for early testing. The driver was to be the impetuous, arrogant Manfred von Brauchitsch, a member of a prestigious military family from Hamburg and the nephew of Walther von Brauchitsch, who would rise to the rank of field marshal in 1940, only to be relieved of his duties a year later when the Wehrmacht failed to capture Moscow. Manfred, whose zeal often transcended his skill, crashed one of the cars, and Fagioli was called in as a substitute. The old "Abruzzi Robber," as he was called, was thirty-six years old at the time and had won three of the seven races he had run for Ferrari in 1933 and finished second three times. He was a tough-minded middleweight of a man, pugnacious, headstrong and perpetually convinced that his skills were underrated by oafish team managers. And Fagioli could drive. He demonstrated his skills to Neubauer at Monza and was hired by Mercedes-Benz as a team driver for the coming season. Ferrari was publicly furious over what he considered a traitorous act, but his histrionics surely veiled an understanding that Fagioli was a professional like himself and would race for the team—or the country—prepared to pay him the most money. Enzo Ferrari fully understood that he was engaged in a hard and demanding business that with each passing day had less to do with the romantic images of sport that had drawn him to automobiles as a young man.

Antonio Brivio, Trossi's friend and neighbor, had transferred to Bugatti, but the loss was not comparable to that of Fagioli. Tonino Brivio was a steady hand and could be counted on in sports-car races, but with the Scuderia now concentrating on major international Grand Prix competition, his gentlemanly approach to the sport was simply inadequate. The season would be decided by gritty pros and inflamed semiprofessionals like the young Moll, while the dilettantes would have to find other outlets for their competitive urges. At the same time Ferrari was becoming increasingly disenchanted with his *direttore sportivo*, Mario Lolli, a wealthy amateur gentleman who was expected to see to the routine details regarding entries, travel arrangements, contracts, and so on. Lolli could speak only Italian, which further handicapped him for work outside the country. Ferrari, on the other hand, had gained a workmanlike command of French, then the universal tongue of the continent, and would before the season was out replace Lolli with the better-traveled Nello Ugolini—yet another example of the increasing intensity and seriousness of the Scuderia's racing effort.

The opening Grand Prix of the season was set for Monaco, and Ferrari himself led the entourage west to the tiny principality. The course was much as it remains to this day, although the shimmering Romanesque architecture that lined the twisty 1.98-mile course has long since been replaced by high-rise apartments, condominiums and luxury hotels. The track, first laid out in 1929, was a simple network of harborside avenues, featuring a turn in front of the world-famous Casino and a fast, open quayside stretch along the Boulevard Albert Premier.

For the first time in history the starting grid would be set by qualifying times rather than the old practice of drawing positions by lot. Louis Chiron, who was from Monaco, was the sentimental favorite. The son of a former hotel maître d', Chiron had first gained his reputation as a lady killer while working as a *danseur moderne*, a sort of elegant pay-by-the-dance escort, in the Hotel de Paris. He became Marshal Foch's personal chauffeur during the war. At the time of the race he was involved in a sensational love triangle that was about to rock the continental sporting scene. For years he had been carrying on an affair with the lovely Alice "Baby" Hoffman-Trobeck, a multilingual beauty born in Hartford, Connecticut, to a German father and a Swedish mother. In the late 1920s Chiron had raced for a team owned by Baby's husband at the time, Hoffman–La Roche drug empire heir Freddy Hoffman. By 1932 the lovely lady had left her playboy husband for Chiron, but marriage for the obsessively vain Chiron to a woman several years older was out of the question. "Chiron's mattress," as she was called in private, continued as his loyal companion until the winter of 1933. At that time Chiron's friend and teammate, the German ace Rudi Caracciola, lost his devoted wife, Charly, in an avalanche while skiing in Switzerland. Baby stayed at Caracciola's side trying to comfort him and was so successful that the two fell in love. The three-way affair was an open secret to everyone except Chiron, who did not learn of the situation until two years later. Caracciola and Baby Hoffman were married in 1937.

Despite his eagerness to perform well in front of his home audience, Chiron was not as fast as his teammates Varzi and Trossi, who set the quickest times in qualifying. A number of other drivers, including Nuvolari and the rising French ace René Dreyfus, were quicker, while the brash young Moll equaled Chiron's time of two minutes flat. However, in the race itself the rest of his teammates faltered and Louis Chiron found himself with a wide lead over Guy Moll. With two laps remaining, Chiron, perhaps composing his victory speech or waving at a pretty lady in the crowd, overcooked his

Alfa in a corner and spun. By the time he got restarted the imperti-
nent Moll had slipped by and taken the victory. Chiron was furious,
but Enzo Ferrari was more than pleased. His Scuderia had won the
race, which was the first concern, and Moll had proven himself to be
a *garabaldino* of the first order. Though Louis Chiron would sob in
frustration, the Scuderia would return to Modena a winner—and it
mattered little which of the drivers carried the laurel wreath around
his neck.

A week later the team was ready to mount a massive attack on the
Mille Miglia, now the most important among a limited schedule of
sports-car events that would be entered. The primary concentration
was to be on single-seat, open-wheel Grand Prix competition, but
owing to the prestige of Italy's most famous open-road race, Ferrari
entered five Alfa sports cars, including four 2.6-liter Monzas for Varzi,
Chiron, Tadini and Caracciola. The opposition was Nuvolari and
Siena aboard a powerful 8C-2300 prepared by the Siena-Grosch
team. Carrying special bodywork constructed of aluminum by the
Brianza Coachwork Company, Ferrari's front-line Alfas also enjoyed
a significant advantage in the arduous contest. Pirelli had provided
new Stella Bianca rain tires for the team, specially designed units
with extra treads, or "sipes," which offered greater traction in the rain
that teemed down on the entire 1,000-mile course. Varzi, always stubborn, objected to using the new rubber, and during a refueling stop
at Imola he and Ferrari engaged in a shouting match over the advantages of the Pirellis. Finally Varzi relented and fitted up a set of the
Stella Biancas. Their added traction was considered a decisive factor
in his defeating Nuvolari and Siena, who finished far back in second
place on more conventional rubber.

No doubt seeking to avenge his defeat two weeks later, Nuvolari
lost control of his cranky 8CM Maserati on the soaked cobbles of
Alessandria's Pietro Bordino circuit and smacked into a tree at the
side of the road. He lay in a hospital bed, his broken right leg wrapped
in plaster, while Scuderia Ferrari, led by Varzi and Chiron, dominated the race by taking the first four places.

Guy Moll returned to the team for the Tripoli Grand Prix, and it was hardly a happy reunion. Once again the Italian penchant for larcenous mischief could not be avoided in the face of the mounting pile of lottery money. This time a more subtle plan was conceived at the bar of the elegant Hotel Uaddan among a small group of drivers (including all of the Scuderia team) who had the best chance of winning. The pot would be split equally; Varzi was selected to win again, with Chiron allotted second. Piero Taruffi, who was driving for Maserati, refused to cooperate, and Englishman Hugh Hamilton, who was driving another Maserati for Whitney Straight, was not apprised of the plan for fear that pangs of Etonian schoolboy sportsmanship would expose the conspiracy.

All seemed to be in order as the cars gridded on the wide main straightaway beneath the huge concrete-and-marble timing tower. As usual, a torrid gibleh wind spread furrows of sand across the giant, sweeping bends, making the race all the more dangerous. But Italo Balbo, decked out in his ceremonial finest, was in a marvelous mood as he moved among the drivers to wish them *"In bocca al lupo!"* and to stand at rigid attention while the Italian national anthem, "Giovinezza," was played by an Army band. This time, thought Balbo, his sand-swept colony would be untainted by any betting scandals.

Taruffi almost threw a wrench into the careful plan. He surged to the front when the flag dropped and kept the big Maserati in the lead for four laps, despite the determined efforts of the Ferrari stalwarts. Then the brakes on the heavy Maserati began to fade and Taruffi lost control as he tried to haul the car down from 180 mph for the Taguira hairpin. The pinwheeling car knocked down a large sign and came to rest with its driver suffering a broken arm and leg. His only consolation was having the corner named after him. Hamilton motored along valiantly as well for a while, but his engine exploded and Moll took up his assigned order behind Chiron and Varzi, who was now leading easily. All that was left was an easy drive to the finish and another big payday.

But Guy Moll would have none of it. He was a rich man to begin with, and the notion of winning the biggest race on his native continent was much more appealing than a docile third-place finish and the covert collection of some tainted gambling funds. Moll upped his pace as the Scuderia pit crew, up to then enjoying the African sunshine, noted with concern his lowered lap times on their watches. Chiron spotted the Algerian approaching in his mirrors and sped up as well, so the two Alfas began to close in on the cruising Varzi. Chiron managed to hold off Moll's bull-like charges until the final lap, when sinking oil pressure forced him to slow down and let his rebellious teammate past. Moll then reeled in Varzi and almost got the lead as the pair of Alfas skidded around the last corner. But Varzi was too experienced and too tough for the youngster. He cut in front of Moll under braking, forcing the Algerian to dive across the rutted apex to avoid crashing. Before he could recover, Varzi sped away to win the race by less than a car length.

The Scuderia's discipline was in tatters. The "Moll incident" was a cause célèbre in motoring circles, although it was considered simply a case of unruly behavior by members of a team, with little mention being made of the betting scandal. Ferrari's involvement in this incident is unknown. But as in 1933, one must presume that a man so intimately involved in the goings-on of his team would have been aware of the cabal, if not before, then certainly after the race. Balbo got wind of the game and altered the betting system to ensure that no repeat of the first two races would be possible. (His measures would be superfluous because the all-winning German teams were the most dominant in the remaining Tripoli races and their disciplined, well-paid drivers had no interest in engaging in any such games.)

At this point Guy Moll had run two races for the Scuderia Ferrari and had won one and finished second by a nose.

After Varzi triumphed in a lightly contested Targa Florio and Chiron took first against minor opposition in Morocco, the team headed north for its first confrontation with the vaunted Germans. The track would be the ultra-high-speed AVUS circuit outside Berlin. This four-lane superhighway, where on non-race days rich Berliners could pay a few marks for a high-speed drive, had been the site of some blistering record runs by Hans Stuck in his Auto Union car and he was clearly the favorite. In desperation Ferrari had arranged for some engineers at the Breda aviation works to design streamlined bodywork for one of the P3s. In those days automotive aerodynamics was a black art, and an improper shape could make a car behave like a berserk airplane.

The streamlined P3 was thought up by Ferrari because Alfa Romeo was rapidly losing all interest in motorsports. This situation would finally reach a point where the Portello staff refused to supply any spares to the Scuderia, other than unfinished engine castings and drive-train forgings which Ferrari had to pay for! It would then be up to Ferrari to complete the final machining and finishing, which he arranged to do through an independent firm in Porretta Terme, a small town thirty-five miles south of Modena.

The strange, high-tailed Alfa was tested on the Milan–Bergamo autostrada with Varzi at the wheel. The various bits of aluminum bodywork flapped alarmingly in the high-speed airstream and the ever-present frown on Varzi's face deepened as he climbed out of the cockpit. He would not drive such a monstrosity. But Moll jumped at the opportunity, perhaps aware that the extra streamlining would give him a major advantage on the six-mile-long straights at AVUS. Moreover, the impetuous young man no doubt enjoyed the psychological advantage gained by volunteering to drive a car that the team's Number One had rejected as unstable and dangerous.

Although it is certain that Adolf Hitler and his warped but inspired propaganda programmer, Joseph Goebbels, recognized the value of German racing cars tearing up the tracks of Europe as noisy, intimidating vanguards of the Third Reich, neither attended motor races after reaching power. Hitler was an occasional spectator at the Solitude circuit outside Stuttgart and at AVUS in the late 1920s, and each year during the 1930s he opened the Berlin Motor Show with a ceremonial look at the new Mercedes-Benz and Auto Union racing machines, but that was the extent of his personal involvement. Yet the massive funding that his government provided these two firms increased until 1939, when the last race was run on the eve of war.

It must be understood that in the spring of 1934 relations between Germany and Italy were hardly as cozy as they were to become following the creation of the Rome-Berlin Axis two years later. While Hitler openly admired Benito Mussolini, and in fact kept a bust of the Fascist leader in his Munich party headquarters, the compliment was hardly returned. Mussolini considered the Austrian corporal to be a vulgar mountebank. When the Germans repudiated the Treaty of Versailles in March 1935, an outraged Mussolini formed the so-called Stresa Front with France and Great Britain. But the flabby appeasement policies of the Western powers in the face of Hitler's boldness ultimately turned him into an ally. As the Scuderia Ferrari was trekking northward across the Brenner Pass toward Berlin and AVUS, Mussolini was saying the Nazis were "drunk with stubborn bellicosity."

Enzo Ferrari and his teammates would learn too late that the Nazis were elevating international motorsports from a simple contest of speed and daring to an instrument of government-backed propaganda. The Fascists would later respond to this threat by encouraging Alfa Romeo to build machinery sufficiently potent to counter the German threat, but funding and manpower would be minuscule compared with that being supplied to Mercedes-Benz and Auto Union. This was already evident when the Ferrari team sighted the arrowlike Auto Unions at AVUS. Their ace driver, the tall, ebullient Hans Stuck, had just set a series of international records in the new car and there was no possible way the aged P3s of Varzi, Chiron and Moll could stay with him. The shrieking, supercharged power of the Auto Union was stunning, even to race-hardened veterans. When the farcical mismatch got underway, Stuck completed the first twelve-mile lap with a lead of over a minute on Chiron! Stuck was miles ahead, cruising to an easy victory, when the clutch failed. Moll, in the twitchy Breda-Alfa streamliner, which could run 180 mph, then inherited the lead and won over Varzi, but the victory was a hollow one and everyone in Modena—especially Enzo Ferrari—knew it.

The intrepid Nuvolari, not surprisingly, finished fourth, with his broken leg still encased in a plaster cast. His ever-loyal mechanic, Decimo Compagnoni, had rigged the pedals in his Bugatti so they could be operated with Nuvolari's one good foot. The little man, in great pain, soldiered through the race, with his still-mending bones being endlessly jarred by the rough surface of the track.

Mercedes-Benz had entered at AVUS, but had withdrawn due to mechanical troubles prior to the start. However, their powerful W25s then appeared in full force for the Eifelrennen race at the supremely beautiful and challenging fourteen-mile Nürburgring road circuit. This time both German marques were faster than the Alfas, and Chiron drove well to finish a distant third behind Brauchitsch's Mercedes-Benz and Stack's Auto Union.

Still, these two German races were not considered full-fledged Grand Prix events, and the wildly jingoistic Italian press continued to delude itself that the full force of Alfa Romeo and the Scuderia Ferrari would not be brought to bear until the prestigious French Grand Prix at Montlhéry on July 1. This would be the first real showdown, with all of the front-line teams represented in their full fighting regalia. The titans were gathering at the splendid hilltop circuit in the lush countryside outside St.-Eutrope on the Paris–Orléans highway. Ferrari had sent his entire team to France with three immaculately prepared P3 *monoposti* with their race-proven Jano-designed 2.9-liter engines.

But the Commendatore remained in Modena. Years later the Ferrari legend would be enlarged by the fact that he refused to attend races. This would be attributed to many factors, from the death of his son in 1956 to his fear of crowds, but even in 1934 he was becoming increasingly rooted to his Modena home. Ferrari never traveled much after the Scuderia was formed and after 1934 attended almost no races outside Italy, even one as important as the French Grand Prix. Monte Carlo, at the beginning of the season, was to mark one of his final crossings of the Italian border.

Part of this bondage to Modena surely involved simple logistics. Ferrari refused to fly, was suspicious of trains and presumably employed only the automobile for travel, which, on European roads of the 1930s, meant endless hours behind the wheel. He would also not use elevators. Overall, this man who would call himself "engineer" maintained a less than wholly analytical view of life's technical complexities.

Thirteen of the finest Grand Prix cars and drivers were entered for the French Grand Prix: three Mercedes-Benzes for Luigi Fagioli, Manfred von Brauchitsch and Rudi Caracciola; two Auto Unions for Hans Stuck and the skilled amateur August Momberger; a trio of Bugattis for René Dreyfus, Robert Benoist and the still-wounded Nuvolari; two Maseratis for Freddy Zehender and Philippe Etancelin; and the three Scuderia Ferrari Alfas, in shimmering red, for Varzi, Chiron and Trossi, with Guy Moll listed as a substitute. In the event anyone doubted that the sport was being politicized, the Auto Unions (but not the Mercedes-Benzes, for some reason) carried swastikas on their long, pointed tails.

The three Mercedes-Benzes were shockingly fast in practice, and it was a depressed Ugolini who, despite the crude international long-distance phone system of the day, reached Ferrari in his Modena office with the lap times. This would be the beginning of a tradition that would last until the final days of Ferrari's life—the telephoned reports from trackside to Modena in which the team manager would convey the details of practice to the boss. In one of his masterful strokes of psychology, Ferrari told Ugolini to put Moll into Trossi's car for the final practice session. The young tiger had been pacing the pits and Ferrari reasoned that his fevered driving might inspire the regulars to go even harder. This appeared to work, and Moll was rewarded for his motivational efforts by being assigned as co-driver with Trossi.

The race started in front of packed grandstands, with Chiron making a gutsy charge from the third row to lead the first two laps. The suave stylist from Monte Carlo, decked out in blue silk coveralls, was driving one of his greatest races, but he could not hold off the more powerful Auto Union of Stuck, who took over until his engine failed. The attrition was shocking, due in part to the eagerness of every driver in the race to prove himself against a premier field of his peers. By the twentieth of the forty laps around the 7.76-mile combination speedway and road course, all of the German forces had either retired or were circulating at a crippled pace. When the checkered flag fell, Chiron was easily cruising in the lead, three minutes ahead of Varzi. Moll had replaced a somewhat overwhelmed Trossi early in the going, and was for once obediently motoring along in third. Only one other car, Benoist's Bugatti, was still running, four laps down.

It was a magnificent victory—on paper. The Italian press was delirious: a one-two-three finish against the might of Germany. Better yet, on-site reports assured the *tifosi* back home that the P3s could match the Auto Unions and Mercedes-Benzes in terms of sheer top speed and acceleration. This was, of course, nonsense, and Ferrari, Bazzi and Ugolini knew it. The German cars were still in their teething period, with engines, tires, suspension settings, gear ratios and so forth being developed, and the men of Scuderia, from the lowliest mechanic's apprentice to the Commendatore himself, knew that future victories with their old cars would be countable on the fingers of one hand.

Still there were dozens of minor races where starting money could be earned. The German teams were being selective about the events they planned to enter, restricting their activities to full Grand Prix or to meets where the organizers were willing to pay their stiff starting-money fees. Therefore numerous organizers of smaller races, mostly around the perimeter of the Mediterranean, were eager to welcome the Scuderia. In these races, where the competition was slim and the prizes were lavish, the Alfas were still able to compete against the equally aged Maseratis and Bugattis for minor-league status in the absence of the noisy leviathans from north of the Alps.

The next confrontation with the Germans came at the Coppa Acerbo at Pescara on August 15. Mechanical troubles bothered the Scuderia cars for most of the race, and in the latter stages only young Moll had a chance of catching Fagioli, who led in his W25 Mercedes-Benz. Moll, sitting bolt upright in the shining red number 46 Alfa Romeo, an odd plastic mask shielding his face, began to up the pace. This brash novice was taking on the gargantuan task of running down the doughty veteran Fagioli in a car with perhaps 100 more horsepower and an extra 20 mph top speed along the pair of four-mile straights. Still Moll would not back down, and he began to nibble two to four seconds off Fagioli's lead with each passing of the pits.

He rushed up on Ernst Henne, the former BMW motorcycle star who was driving a backup car for Mercedes-Benz. According to Henne in a 1985 interview with historian Chris Nixon, Moll caught him on a narrow stretch of Pescara's fast straight, where the road was no more than twenty feet wide. The two cars were running about 170 mph when Moll pulled alongside. Henne could see the Alfa's Prancing Horse emblem from the corner of his eye. Then, he recalled, Moll fell back and the Alfa began a gruesome yawing motion that sent it tumbling off the track. On the next lap Henne passed the crumpled Alfa Romeo leaning askew against a farmhouse. The broken body of Guy Moll was lying by the roadside. He had died on the seventeenth lap, which may or may not have any relationship to Enzo Ferrari's superstition about that number. Like many others, Ferrari attributed Moll's death to a collision with the Mercedes, although Henne firmly denied that the two cars ever touched. No matter, Guy Moll was to be honored by Ferrari in later years as one of the greatest. In 1962 he wrote, "I rank him with Stirling Moss as the only driver worthy of comparison with Nuvolari. In fact, he resembled Nuvolari in certain mental traits, in his aggressive spirit, in the calm assurance with which he drove and in the equanimity with which he was prepared to face death." When it is recalled that Enzo Ferrari was a witness to the efforts of many of the best racing drivers in the world over a span of nearly seventy years, this indeed is a true testimony of admiration.

The death of Guy Moll seemed to drain the last of the competitive juices from the Scuderia. Varzi was more sullen than usual, which tended to deaden the enthusiasm of the entire thirty-man staff, and Chiron, now called *le Vieux Renard* (the Old Fox) although he was only thirty-four, was in a slump from which he would never recover. Perhaps it was his increasingly chaotic personal life, or the fact that the cars were becoming so fast and unforgiving, but after his win at Montlhéry he never regained his early form. He was, however, to continue racing on and off until a final retirement at the age of fifty-eight.

The organizers of the Italian Grand Prix at Monza added a series of chicanes to the circuit in hopes that the superior speed of the German cars would be nullified on a tighter, twistier circuit. This did nothing except permit the Mercedes-Benzes and Auto Unions to showcase their superior braking and handling in the September race. A final insult was given on the hometown streets and avenues of Modena as Tazio Nuvolari debuted the new 3.7-liter 6C-34 Maserati and trounced the Scuderia before its own loyalists.

By the time the season had dribbled to an end it was clear that massive changes were in store for 1935. Varzi announced that he was leaving to take a seat with Auto Union. Joining him on the Chemnitz-based team would be Hans Stuck and—later in the season—the brilliant Bernd Rosemeyer, a pyrotechnic talent whose rise would be as meteoric as Moll's and whose career would be almost as brief. Chiron and Trossi announced they would return to duty, and Tonino Brivio had rejoined the Scuderia late in the year, but none of these men could be expected to challenge on the increasingly competitive Grand Prix scene.

To be sure, Jano was describing to Ferrari a new, independently sprung car he had on the drawing boards, as well as straight-8 and V12 engines, but what good would a faster car be without a superstar to drive it? The mannerly French star from Nice, René Dreyfus, was signed from the Bugatti team and he arrived in Modena to trade his Bugatti for a 2.3-liter gray Farina-bodied Alfa Romeo cabriolet, in which he and his new wife would travel to the races. Of course, the one man Ferrari needed had sworn never to return to the Scuderia. Yet Tazio Nuvolari had switched back and forth between Maserati and Bugatti without success during 1934 and could see that neither marque had the potential to field a German beater in the coming year. The door was closed at Auto Union with the hiring of Varzi (who again had stipulated that Nuvolari could not drive for any team while he was a member), and Mercedes-Benz, now that Caracciola was totally recovered from his 1933 crash and the loss of his wife, did not need or want the great Italian—who was now almost totally gray and, at nearly forty-three, considered by many to be reaching the end of his tether.

With few other options, Tazio engaged in lengthy secondhand negotiations with Ferrari. Jano acted as emissary, delivering by letter, by phone and in person a series of demands and counter-demands to the two headstrong parties. Nuvolari remained at his villa near Mantua and Ferrari in Modena. Finally, after a meeting on neutral ground in Piacenza, roughly midway between Modena and Milan, a deal was put in place. René Dreyfus, who was his teammate that year, recalls that in addition to a generous salary, Nuvolari received 50 percent of the winnings. Because he was the junior member, Dreyfus was given 45 percent of the purse but was required to absorb all his personal expenses. Nuvolari would lead the Scuderia Ferrari—and Alfa Romeo—into battle in 1935.

And this would be a battle. Gone were the elegant amateurs of the early days. Now the Scuderia was a bare-boned, hard-muscled professional racing operation. This immediately became apparent to outsiders like Dreyfus, who later wrote, "The difference between being a member of the Bugatti team and the Scuderia Ferrari was virtual night and day. . . . With Ferrari I learned the business of racing, for there was no doubt he was a businessman. Enzo Ferrari was a pleasant person and friendly, but not openly affectionate. . . . Enzo Ferrari loved racing, of that there was no question. Still, it was more than an enthusiast's love, but one tempered with the practical realization that this was a good way to build a nice, profitable empire. I knew he was going to be a big man one day. . . . Ettore Bugatti was *le Patron*. Enzo Ferrari was the Boss. Bugatti was imperious. Ferrari was impenetrable."

The beginning was auspicious. Nuvolari and Dreyfus opened the 1935 season with slightly modified and updated P3s by running one-two at Pau in the extreme southwest of France. But the opposition was mediocre and hardly representative of future trials. Then came a nearly two-month lull in the schedule, which gave Ferrari and Bazzi a chance to complete one of the most audacious racing cars in history. Several Formula Libre races were being run in Europe and Africa at the time, with major events scheduled for Tripoli and AVUS in 1935. At the annual Scuderia banquet in December 1934, Luigi Bazzi had suggested to Ferrari the notion of a special twin-engine machine—a *bimotore*—to be built by the Scuderia out of Alfa components. The prize money was lucrative, and besides, how else could the Germans be beaten? Ferrari liked the idea, and during the winter months of the new year Bazzi and shop foreman Stefano Meazza set to work on the radical machine.

A regular Alfa *monoposto* chassis was lengthened six inches and reinforced to carry a second engine, mounted behind the driver. Then Bazzi and engineer Arnaldo Roselli designed and built an insanely complex drive-train setup that linked both the front and rear 8-cylinder Alfa Grand Prix engines to the back wheels (in retrospect, a four-wheel-drive layout might have protected the car's Achilles' heel, tire wear). Bazzi, always the stern, spartan worker, was known to work around the clock and live only on occasional doses of pasta and wine. It was he who instituted the no-smoking rule in the workshops, a rule that remains in effect in the Ferrari factory to this day.

A pair of *bimotores* were to be built. One would carry two of the latest 3.1-liter supercharged Alfa Grand Prix engines, while the second was mounted with the earlier 2.9-liter units. Both cars weighed about 2,800 pounds (or about half a ton more than their German 750-kilogram rivals) and produced between 520 and 540 horsepower. These giants were expected to run over 200 mph on the straights at Tripoli, where their debut was scheduled. The larger-engined car was assigned to Nuvolari, and Chiron was given the 2.9-liter twin. The cars were fearsomely fast, but their prodigious weight ate tires as if they were made of angel-hair spaghetti. Nuvolari finished fourth after changing thirteen tires; Chiron was fifth. Caracciola, in his regular-duty Mercedes-Benz Grand Prix car, won easily. The team tried again at AVUS two weeks later. Again tires were the fatal flaw and Nuvolari could run no faster than 175 mph before the Belgian Engleberts literally shredded off the rims. The tires were so ill suited to the monster machine that the firm's president, Georges Englebert, is said to have personally thanked Nuvolari for not having made an issue of their failure when talking to the press after the race. Chiron, driving as if he were on Alpine black ice, feather-footed the other *bimotore* to a distant second-place finish. But the message was clear; the cars, while interesting technical exercises, were simply too heavy and cumbersome for any contemporary racing tires. But surely there must be another application in which the machines could save face . . .

Earlier that year Hans Stuck had set an international Class C record for the flying mile at 199 mph. To recapture this mark on Italian soil would be a wonderful publicity stunt and might help to erase memories of the fiascoes in Tripoli and Berlin. The attempt on the Stuck record was set for June 15 on the same highway, the Florence–Viareggio autostrada, along a level stretch between the two exit ramps for the village of Altapascio near Lucca. Now shod with tires made by Dunlop, the British company, which had been making high-speed versions for Sir Malcolm Campbell's Bluebird land-speed-record cars, the bigger *bimotore* was made ready for the ironclad foot of Tazio Nuvolari. Before the morning was over he had ripped up and down the autostrada in record time, breaking a number of international marks set by Stuck and attaining a top speed through the timing traps of 208.937 mph. Amidst the cheering someone noted that this made the Ferrari-Alfa and Nuvolari the fastest car-and-driver combination on earth, next to Campbell and his Bluebird (which held the outright record at 301.13 mph).

This machine was credited with being the first to carry the Ferrari nameplate, although historians dispute its lineage and some insist that it was merely a hot-rodded Alfa Romeo. Still, it caused a national furor and enhanced the reputation of the Scuderia at a time when its front-line Grand Prix cars were being trounced with depressing regularity by the Germans.

There was one final, monumental triumph for the Scuderia Ferrari before the capitulation to the German onslaught was completed. It came at the German Grand Prix at the Nürburgring, a circuit generally acknowledged to be the most demanding racetrack in the world. It was only proper that the greatest living race driver (and perhaps the greatest of all time), Tazio Nuvolari, should be the key to that astounding moment in motorsports.

By July 28, 1935, the Alfa Romeos had been relegated to also-ran status by all but the most witless and ill informed, blindly loyal Italian motor-racing aficionados. Even they understood that these lithe, but thoroughly antiquated automobiles were merely ambulatory cannon fodder for the all-conquering Germans. But Ferrari fought on, fitting the old cars with independent front suspensions and other variations in search of slightly improved performance. The engines of Nuvolari and Chiron had been stretched to 3.8 liters, a desperation effort in search of more power, but such a modification tended to further strain the already feeble transmissions. They had become so unreliable that the Scuderia mechanics named them "the Ponte di Bassano," after a bridge near Venice that had been endlessly bombed and repaired during the Great War.

The rise of the Germans had produced a national craze for motorsports, and by July it seemed like the entire population of Germany was descending on the Nürburgring, or simply the Ring as it was called, for the Grosser Preis von Deutschland, the richest, most prestigious European race of them all—and an event that was sure to fall to either Auto Union or Mercedes-Benz.

As was the tradition, the major teams booked rooms in the Eifelerhof Hotel in nearby Adenau, and there the tiny fraternity of brave men gathered prior to the race. Nuvolari was present with his long-suffering wife, Carolina, and his oldest son, Giorgio. Joining him were Chiron and Brivio, who replaced the ailing Dreyfus. Aligned against the three Alfa Romeos was a veritable panzer division of German machinery—five Mercedes-Benzes and four Auto Unions, all decked out in their aluminum bodywork (hence the nickname "Silver Arrows"), red numbers and ominous red-and-black swastikas. Five Maseratis, a single Bugatti and an English ERA filled out the field, but to the assembled throng (numbering over 400,000 scattered among the forests bordering the immense track) the question was reduced to which of the German teams would win. Although Nuvolari had been inspired in practice, equaling brash Bernd Rosemeyer's fastest qualifying time, he was not seriously considered to be a contender. The book on him was simple: an aging, forty-three-year-old, somewhat overwrought Italian who, when he was not crashing into some unyielding object, was ripping up the transmission or engine of his abused automobile.

Race day dawned with a drizzle blanketing the surrounding Eifel Mountains. But by the time legions of Nazi storm troopers escorted the twenty starting automobiles onto the vast starting grid, the sun was beginning to probe between the scattered clouds, although the track would remain damp for most of the day. The front row consisted of the two Mercedes-Benzes of Caracciola and Fagioli bracketing the Nuvolari Alfa—a mechanical sandwich with a red Italian morsel pinched between a pair of sure winners. But this was Tazio's day, and he launched the old Alfa off the line as if amphetamines had been mainlined into its fuel tank. He drove valiantly for the opening laps, but the lack of speed down the three-mile-long finishing straight (which featured a pair of nasty humpbacked bridges that sent the cars airborne) caused him to be steadily gobbled up by the Mercedes-Benzes and Auto Unions. Worse yet, Brivio and then Chiron retired with transmission trouble, leaving only Nuvolari to carry on the battle.

On the eleventh lap of the huge circuit Nuvolari made a routine stop for fuel and was delayed when a pump failed and the tank had to be hand-filled from cans sloshing with gasoline. By the time he jumped back aboard the Alfa, apoplectic over the delay, he had fallen far behind into sixth place.

At this point began what many believe to be one of the greatest feats of driving in the history of the sport—a titanic driver on a magnificent racetrack facing overwhelming odds. Nuvolari seemed at this moment to ascend into another sphere of skill entirely. Even the multitudes, lounging against the fences or huddled around the campfires in the bordering forests, could sense that a master was at work. He was hardly braking for the corners, those stomach-turning twists and hollows at the *Flugplatz*, at the *Bergwerk*, at the famed *Karussell* and through the dangerous kinks at the *Pflanzgarten*. He was charging into the bends flat out, then yanking the Alfa into a series of lurid slides, elbows akimbo, flailing madly to maintain control.

Manfred von Brauchitsch had assumed the lead in the later stages, as the other German aces had either faltered or stopped. Neubauer and the rest of the Mercedes team understood the situation. They could feel the mystical presence of this little terrier of a man as he snaked his way around the massive circuit. Their own driver was good, if a trifle unpolished and temperamental. He tended to have bad luck, and would finally gain the nickname *Pechvogel,* "unlucky bird," but on this day, in front of his homeland throng and possessed of the faster automobile, there was little doubt that he could hold off the mad thrusts of Nuvolari—who had now, amazingly, preposterously, surged into second place. But surely his old, upright relic of an Alfa could not stand the abuse.

Three laps remained and Nuvolari had cut Brauchitsch's lead to sixty-three seconds. Frantic pit signals were flashed at the German as he completed the twentieth of twenty-two laps. Hühnlein, the Hitler toadie, watched with increasing alarm from the press tower. In his hand was a speech, typical jingoistic boilerplate celebrating yet another German victory, which was to be presented before ten radio microphones and a mob of adoring Nazi journalists. On the penultimate lap Neubauer knew his man was in trouble. Observers on the circuit were reporting that Nuvolari was gobbling up the distance like a berserk hare. The crafty racing boss walked out onto the edge of the track as Brauchitsch thundered past. The right front tire was showing a white seam—a "breaker tread," which warned that the rubber was worn thin. Nuvolari sluiced by thirty seconds later. Decimo Compagnoni, his loyal mechanic, who had borne the brunt of Nivola's anger during the pit stop, reached for a small bottle of cognac hidden in a toolbox. The pressure was becoming unbearable. Could Nuvolari pull it off? One mad fourteen-mile lap remained.

The squad of NSKK troopers in black motorcycle helmets were standing by to hoist an immense swastika on a flagpole that towered over the grandstand. In the public-address tower the "Horst Wessel

Song" record was placed on the turntable in anticipation of the German victory. Meanwhile, out on the circuit, flitting among the bordering stands of pines now shielding the moist tarmac from a lowering afternoon sun, Nuvolari continued his banzai drive. The masses in the pit-row tribunes and the assembled teams heard the shocking news from the loudspeaker. It was the voice of the reporter stationed at the 180-degree banked *Karussell* about six miles from the finish. "Brauchitsch has burst a tire!" the voice screamed. "Nuvolari has passed him! Brauchitsch is trying to catch up on a flat tire!"

Despair. Humiliation. Defeat. Nuvolari crossed the finish line a clear winner. Then Stuck, then Caracciola, then Rosemeyer and finally a tearful Brauchitsch. Hühnlein ripped up his speech and grimly walked to the victory stand. Tradition held that the national anthem of the winning driver's nation be played, but the German victory had been so certain—an inevitability—that no recording of the Italian anthem could be found. But Nuvolari apparently had thought of such a contingency and had brought along his own record, which a slightly tipsy Decimo delivered to the press building.

In the midst of his mad game of catch-up, Nuvolari had also spotted a wind-frazzled Italian flag hanging over the main press tribune. It stood in stark, shoddy contrast to the pristine red, black and white Nazi bunting surrounding it. Nello Ugolini recalls that the first words the sweating, exhausted Nuvolari spoke as he crawled from behind the Alfa's wheel were: "Tell the Germans to get a new flag!"

The perfect end to this astounding upset would involve further degradation for a sulking crew of Nazis. But that was not the case. Nuvolari was tremendously popular with German racing enthusiasts, and after the first spasm of nationalistic disappointment was overcome, the Germans gave the tiny Italian an enormous reception. He was especially well liked and respected by the insiders on the German teams and but for the interdiction of Varzi would have had a place with Auto Union, rather than aboard his badly strained Alfa Romeo.

Enzo Ferrari was, of course, nowhere to be seen. He was home in Modena, as usual, and heard of the amazing feat from Ugolini by long-distance telephone. It had to be enormously satisfying for him, to have the Scuderia's ace driver defeat Germany's best on their home turf, but he was too prescient to delude himself. This victory was a fluke in many ways, attributable not only to Nuvolari's supremely inspired driving but also to Brauchitsch's tire failure and to various mechanical ailments that slowed men like Caracciola and Stuck. It was a moment for national celebration, which the Italians engaged in with typical brio, but Ferrari was glumly aware of the overriding reality. Unless enormous funds were forthcoming from the Mussolini government to counter the onslaught of the Germans, there would be no more Italian anthems sung, no more Fascist flags flown over the Grand Prix circuits of Europe. Of that he was certain.

7

Despite Enzo Ferrari's avowed lack of interest in politics, he had to understand that Mussolini's invasion of Abyssinia on October 3, 1935, marked a major change in the fortunes not only of his racing operation but of his entire homeland. It was a foolish, chimerical expedition, mounted more out of spite and misplaced pride than for national gain. The assault on this ancient, arid, flyblown kingdom, now called Ethiopia, had been in the mind of Il Duce since 1928, and was intended purely as an act of vengeance for the Italian defeat and expulsion from Abyssinia in the bloody war of 1895–96.

In his daily affairs, the bombing of barefoot natives in faraway Africa meant almost nothing to Ferrari. Like most Italians, he cared little for government, and as a northerner, he despised Rome as a corrupt sinkhole of highborn poseurs and governmental charlatans. At its strongest, the Fascist Party numbered no more than 800,000 members, and its support was akin to an oil slick on the Bay of Naples—miles wide but only centimeters deep. The Italians had always maintained (and still do) a sub-rosa economy that transcended whatever leftist, centrist, rightist, Communist or Fascist government was in power. Therefore, it was business as usual, save for one critical element: Alfa Romeo was beginning to feel enormous pressure from the government, primarily for increased production of military materials, but also for better performance in Grand Prix competition.

To make matters worse, Vittorio Jano was in a slump. The fabled designer seemed to be losing his way, mainly because he was operating in areas beyond his expertise. Gobbato had assigned Jano to design a radial, air-cooled fighter aircraft engine (design code: AR D2) and it was an abject failure. To Jano's credit, he was an expert in high-revving, small-displacement, water-cooled automobile power plants, which were, in a technical sense, light-years away from the gigantic, slow-turning, air-cooled radial units demanded by the Air Force. The disappointing results of the AR D2 project caused Gobbato to demote Jano from overall technical supervision of Alfa Romeo projects and to limit his activities to cars and trucks.

This cutback of responsibility surely blunted Jano's enthusiasm, if not his long-standing loyalty to the firm, and his further efforts were fitful and uninspired.

He had created a new racing car for the Italian Grand Prix at Monza that had been an improvement on the antiquated P3s, but its potential was limited. The new 8C-35, as the car was labeled, was little more than an update of the old machine, with an improved version of the straight-8 that was directly traceable to his P2 of a decade earlier. To be sure, he had the design for a 4-liter V12 capable of 370 hp almost completed, but his chassis concepts seemed caught in a cul-de-sac that made no sense whatsoever. While the German cars were low and wide, to reduce the center of gravity, Jano's 8C was tall and narrow, a seeming denial of the laws of physics. Yet, according to Scuderia historian Luigi Orsini, Jano was adamant in the contention that a tall car made sense. Arguing from a position that Orsini says was "unworthy of the great engineer," Jano insisted that his high bodywork reduced roll effect, rather than increased it. This was non-sensical (which Jano quietly acknowledged with his later designs for Lancia) and surely affected the performance of the automobile. But Nuvolari used the new independent suspension and the improved hydraulic brakes to advantage and set the fastest lap at the Italian Grand Prix in September before retiring with a burned piston. (He then took over René Dreyfus's sister car and finished second.)

The Italian press was again delirious with expectations. They understood that the Monza cars still carried the old 3.8-liter P3 engines and deluded themselves that when the new V12 was installed the Alfas would take the measure of the Germans. The factory had made an odd arrangement with the Scuderia, choosing to sell six of the 8Cs to Ferrari at a favorable price but stipulating that all prize money would be turned over to Alfa Romeo. Starting money, sponsorship arrangements, and so forth would be Ferrari's, while the team drivers were given discounts on passenger cars they might choose to purchase.

Enzo Ferrari was immersed in the life of a full-time racing *capo*, with all of his energies concentrated on the team's competition schedule. Alfa's passenger car production had almost ceased, and thus his sales from the dealership were negligible. Beyond the sale of a few of the new Alfa 8C-2900 sports cars to selected amateurs and some custom-bodied 6C-2300 "Pescara" sedans built for wealthy gentlemen, his profits were to come solely from motorsports. At this point the Scuderia was grossing perhaps one million dollars a year, but with a staff of over thirty technicians and mechanics and massive costs relating to moving a racing operation all over Europe, Ferrari was still living frugally in the tiny, two-bedroom apartment over the Scuderia. Young Dino remained a frail child, weak in his limbs and often bedridden by illness that prevented him from engaging in normal boyhood activities. Enzo Ferrari was a hard taskmaster and those around him recall that Dino was fearful of his mercurial temper. Because of his father's increasingly limited travel schedule, now almost exclusively restricted to cities along the Po Valley, the little family fell into the daily pattern of lower-middle-class Italian life. Enzo Ferrari was born with simple tastes, and even after he became a rich and prominent Modenese he retained the ways of a simple, uncluttered man. Surely, during the 1930s, when every ounce of his energies and every lira in his pocket were being plowed back into the business, he lived a modest, frugal life.

The team was running a test of the 8Cs at Livorno when Italy invaded Abyssinia, and although the nation was inflamed with jingoistic ardor, with the attack the subject of intense discussion, within the Scuderia, Ferrari makes no mention of the event. Nor does he bother to note that when the team attempted to cross the French border for the February 1936 season opener at Pau, they were turned back at Ponte San Luigi. This was but one of dozens of warning signs that the international situation was rapidly deteriorating, yet Ferrari apparently felt it unworthy of note. (Perhaps he was correct in his judgment. The rebuff of his racing team was one of precious few acts of sanction against Italy and was obviously so trivial as to be meaningless.)

This is not to imply that Ferrari and his operation were completely insulated from the Fascist tides. During this period the team published a periodical called *La Scuderia Ferrari*, which chronicled their activities. It was filled with Fascist bravado and flattery for Mussolini and his associates. While there is no reason to believe that Enzo Ferrari harbored any particular sympathies for the movement, his company journal, which he wrote for and edited, heartily supported the party line. After all, beholden as he was to Alfa Romeo, owned as it was by the government, he clearly had no choice.

As the 1936 season got underway, Didi Trossi was absent. He left, according to the official version, because the now all-pervasive atmosphere of professionalism went against the grain of this grand amateur. But he must have been given a brisk shove by Ferrari, who took over as the chairman and center of power for the Scuderia. This ended the charade of Trossi's presence, for Ferrari had been in complete command from the time the team was formed.

Chiron and Dreyfus were gone as well. Caracciola had arranged a position for Chiron on the Mercedes-Benz team. Whether this was done as an act of simple friendship on behalf of his old teammate or whether it was Caracciola's chance to get closer to Baby Hoffman is unknown, but Chiron left for Stuttgart, justifying his departure by claiming that Ferrari was cutting back on his financial deal. Some have suggested that Mercedes-Benz invited Chiron to join up in order to ease tensions between Germany and France, but there seems little basis for such a theory. Surely there were other French drivers better qualified for the position, including Dreyfus. But the fact that his father was Jewish instantly disqualified Dreyfus, and he returned to France to take up with an ill-fated racing effort by Talbot. Had Dreyfus been given a chance with the Germans, many believe he would have risen to the top rank in the sport. But as in all aspects of the New Order, Jewishness was verboten in motorsports.

The key to the Scuderia's success in generating starting money was of course the enormously popular Nuvolari, who signed on again for 1936. This was perhaps encouraged by the Mussolini government, which simply could not tolerate a defection to the Germans by both Varzi and the beloved Nivola. Brivio was to remain as well, plus a strong addition in the form of twenty-nine-year-old Giuseppe "Nino" Farina. The son of the eldest of the two Farina brothers, of the coach-building firm in Turin, he was something of a Renaissance man, holding a doctorate in political science and also talented in a variety of sports. After flirting with a military career, he took up racing and with occasional guidance from Nuvolari rose quickly in the ranks of Italian amateurs. Noted for his blunt, arrogant driving style— which featured a chin-out Duce-like pose with his arms extended far back from the steering wheel—Farina raced for Maserati with some success in 1935 before joining the Scuderia. Ferrari, who was always attracted to drivers with a faintly suicidal attitude, quickly decided that the young man had a great future. Also joining the team for another term was Carlos Pintacuda, a journeyman who was effective in sports cars (he had won the 1935 Mille Miglia) and showed occasional flashes of brilliance in the Grand Prix cars.

But even with Nuvolari driving with a verve extraordinary for a man half his age, the 1936 season was to be a walkover for Auto Union and a disaster for both Alfa Romeo and Mercedes-Benz. The ever-resourceful Dr. Porsche had increased the displacement of his V16

Auto Union power plants to 6 liters and thereby upped the power output to 520 horsepower. This would be almost 150 more than that produced by Jano's 12-cylinder engine, introduced at Tripoli in May. Moreover, Auto Union featured the extraordinary Bernd Rosemeyer, the Teutonic wunderkind, who in his second full season of racing was to win five of the eleven Grand Prix races he entered. He also led the team to an utter domination of archrival Mercedes-Benz, which had a rare lapse in engineering acumen and fielded short-wheelbase W25s with handling characteristics akin to spooked horses. It was a bad year for Achille Varzi as well; he had become involved with Ilse Pietsch, the ex-wife of sometime Auto Union driver Paul Pietsch, and she, for reasons that are still debated, managed to get the essentially depressive Varzi to share her addiction to morphine. This would ruin his career until the end of the war. But Varzi's decline meant little to the Scuderia Ferrari, or to Nuvolari, who drove with his usual manic style but was able to win only four races, including the American Vanderbilt Cup, for which the Germans did not compete. The V12s—called 12C-36s—were decent automobiles, surely the third-fastest Grand Prix cars in the world, but against the Auto Unions they were powerless. (Rosemeyer even trimmed Nuvolari at his own game of macho, driving past the old master in pea-soup fog to win the Ei-felrennen at the Nürburgring.)

The Spanish Civil War had begun in June and Mussolini had provided massive aid to the Franco-led Republicans. While the Germans and the Russians are remembered as contributing a lion's share of the outside aid, their offerings were minuscule in comparison with the 50,000 troops, 150 tanks, 600 aircraft and 800 first-rate artillery pieces that poured into Spain from Italy. This added war effort (although the Abyssinians had been essentially subdued) increased the work load at Alfa Romeo and led to the arrival of a man whom Enzo Ferrari hated perhaps more than any other in his seventy-year professional life.

His name was Wifredo Pelayo Ricart y Medina, or simply Ricart. He was a formally trained engineer who had known Gobbato for a number of years. His political background is unclear, but in view of the alacrity with which he left Spain and came to Italy, this wellborn Catalan doubtless enjoyed favorable connections with the ruling Fascists. His specialty was aircraft engine design and his initial assignment for the Portello firm was to update and improve their aero power plants—presumably including the notably unsuccessful Jano creation. At thirty-nine, he was a severe, introverted man with the same analytical approach to problems as Gobbato. This instantly placed him at loggerheads with Enzo Ferrari, Jano and the rest of the old guard—a group without formal training who relied on a mixture of racetrack empiricism and raw instinct to create machines. Moreover, Ricart (immediately dubbed *lo spagnolo* by the Ferrari clique) had no abiding passion for motor racing. He was a pure mechanical engineer by trade and tended to approach each project with a kind of aloof methodology that made him more akin to a surgeon than a grease-stained racing-car mechanic.

Of all the individuals that caromed off Ferrari's powerful personality over the decades none seemed to have elicited the same fury as Wifredo Ricart. In his memoirs, Ferrari pettily insisted on referring to him as "Vilfredo Ricard" and recounted truly silly and vicious stories about him.

Surely Ferrari was threatened by Ricart, and with good reason. It must not be forgotten that for all their hard work, the Ferrari-Nuvolari-Jano team was getting unmercifully trounced by the Germans. As an example, on Pescara's long straightaways in the 1936 Coppa Acerbo, Jano's top-heavy 8C-36 driven by Nuvolari was clocked at 152 mph. The Auto Unions ran through the same speed traps at 183 mph! Only Nuvolari's soaring talents and steely courage were able to keep the Alfas in the hunt, and then only on the shorter, twistier circuits. Gobbato, the good Fascist, was no doubt feeling pressure from Ciano and perhaps Mussolini himself. In a nation built primarily on perceived pride and bloated nationalism, the performance of Italy's front-line racing cars was clearly a source of shame and distress.

Griffith Borgeson, the historian who studied Ricart's career and who spent time with his surviving family, totally discredits Ferrari's spiteful reports. There is little question that Ricart sometimes teased the Commendatore with some deadpan humor. If Ferrari took him seriously, he showed a surprising naïveté; more than likely he simply chose to distort the stories in an attempt to portray his rival as an eccentric unworthy of his new role. Ferrari also reported that when Ricart's 3-liter Tipo 162 was first tested in 1938, the tie-rod pickup points were reversed, which caused the front wheels to turn in the opposite direction to what was intended. Again, the implication is clear: to indict Ricart as someone who could not solve the most elementary design problems. The fault surely lay with a draftsman and was easily corrected.

Ferrari's bitterness is explainable, although less easy to forgive. Wifredo Ricart had been assigned the job of getting Alfa Romeo back on track, literally, in the racing wars, and also adding clout to the aircraft side of the business, where lucrative contracts were being let by the IRI. Simply put, the Scuderia's effort was insufficient, as was Jano's. In their defense, the budget constraints were severe and the German teams were capable of outspending them with ease. But Jano had been slow to recognize the benefits of such advances as hydraulic brakes, independent suspensions, low centers of gravity and V-type engines, all of which were adopted in one form or another by the Germans, and in a number of cases by the even more impecunious Maserati brothers. Over the years the sycophants and enthusiasts who have written the histories of Jano's and Ferrari's efforts on Alfa's behalf have artfully managed to obscure the fact that their machines from 1934 onward—the late P3s, the 8C-35s and the 12C-36s—were abject flops. In the face of these consistent failures it is reasonable that Gobbato and the Alfa management would seek radical change.

This they did in March 1937, when Alfa Romeo bought 80 percent of the Scuderia Ferrari and announced to its boss that administration of the racing effort was being returned to the Portello headquarters. Again, Ferrari's recollection of this cataclysm is brief and couched in obscure phrasing ("They also arranged for me to wind up the Scuderia Ferrari and engaged me as their racing manager"), but one must assume that the struggle was long and difficult.

After all, Enzo Ferrari was the Commendatore of Alfa Romeo's Grand Prix racing program. To lose this position had to imply failure, or at least displeasure, in the eyes of his superiors. The period 1935–38 was laden with frustration for a man who wore his pride on his sleeve and who for fifteen years had assiduously—and successfully—curried favor with the company hierarchy. Now all the politicking seemed to be coming to naught.

For a while Ferrari struggled desperately to maintain the integrity of his organization in the face of the changes at Portello. He signed Nuvolari to lead the team again in 1937, although it was an open secret that the great driver was being courted by both Auto Union (now without a drug-numbed Varzi) and by Mercedes-Benz, who, with the aid of a brilliant young designer, Rudolf Uhlenhaut, had debuted improved straight-8s developing over 600 hp in racing trim. Tazio Nuvolari surely responded to pressure from the Mussolini government not to defect until the Alfa Romeo reorganization had been given a chance. He was joined by Farina, Brivio, Pintacuda and Mario Tadini, who was establishing himself as a specialist in hill climbing but little else.

Daring new automobiles were needed and fresh manpower was brought in by Gobbato, who apparently believed that a two-pronged effort centered in both Portello and Modena might work. Bartolomeo Constantini, a longtime engineer for Ettore Bugatti, was placed on staff under Ricart, while a protégé of Jano's named Gioachino Colombo was assigned to Ferrari. A close associate of Jano's since 1924, the thirty-four-year-old Colombo brought impressive credentials to Modena. He had begun his career in Milan at fourteen, starting as a technical draftsman at the noted design school Officine Franco Tosi di Legnano, working on diesel engines and steam turbines. He won a design prize for a supercharger concept he developed, which led to a job with Alfa Romeo. He was on staff in time to aid Jano in the creation of the P2 and continued to work at the side of the master until his assignment to the Ferrari operation in May 1937.

Chaos seemed to reign in both Portello and Modena. Jano was under tremendous pressure to produce a winner with no money. Colombo, by his own admission, may have been a contributor to the trouble. In his memoirs he recalled, "To make poor Jano's life even more complicated, there was the extremely difficult situation which had arisen in the factory—great confusion about roles, too many people all wanting their own way in the matter of planning (and I must include myself in this as well) . . . In fact it's amazing that he managed to build a vehicle at all under those conditions."

The international racing world was in a state of flux. The 750-kilogram formula had produced alarming road speeds of over 200 mph and for 1938 the rules were to be altered to permit only 3-liter supercharged and 4.5-liter unsupercharged engines in yet another attempt to slow the cars down (this too would fail). Moreover, it was being widely reported that the AIACR would further reduce the formula in 1940 to 1.5 liters, supercharged. This meant that Jano, Ricart, Ferrari, Colombo and the others involved in the Alfa Romeo racing department's fortunes had to create a new series of Grand Prix cars for 1938, as well as prepare for even smaller engines for four seasons hence.

As Colombo packed his bags for Modena, Jano's drawing board was stacked with concepts, not only with a stopgap design called the 12C-37 for the immediate 1937 season but also with ideas for a variety of 8-, 12- and even 16-cylinder power plants for the 3-liter formula that would come into effect for 1938. Colombo was to head the Ferrari design team of Bazzi, Federico Giberti, Alberto Massimino and young Angelo Nasi, who was also on loan from Alfa. Massimino was an experienced engineer and designer, having created the chassis and bodywork for the ill-fated Fiat 806, the final racing car to be built by that powerful concern. They were assigned to create a compact 1.5-liter "voiturette" single-seater to race in a class that was beginning to gain popularity among teams and private entrants who simply could not compete against the German might. This type of automobile (called *vetturetta* in Italian) was already being manufactured by Maserati and the English firm of ERA. This *vetturetta* would offer Ferrari and Alfa two benefits: it would permit them to win some races in which the Germans were not entered while offering the chance to develop a full Grand Prix car if and when the proposed 1.5-liter formula came into effect.

Whose idea it was to create such a machine is the subject of debate. Ferrari makes no bones about crediting himself. "It was during this period, in 1937 to be exact, that I had the idea of having a racing car of my own built at Modena. This was the one later to be known as the Alfa 158," he said flatly. However, Colombo recounted the birth differently: "Alfa Romeo's decision to make their *vetturetta* in Modena in the buildings of the former Ferrari team must have been greeted with great relief by Jano. Not only would they not come and interfere with his already precarious plans for the Grand Prix but he would also be rid of me for the time being; I was no doubt a genuinely valued assistant, but perhaps in the midst of all the confusion, I was pressing a bit too hard to be allowed to do some of the designing on my own account."

So Ferrari claimed that he had a car in mind long before Colombo arrived. The designer, on the other hand, made it clear that the *vetturetta* was conceived by Alfa Romeo and was considered by him to be "their" car. There is some historical support for this latter contention, although by no means conclusive. It is known that Alfa Romeo, in the midst of all the Byzantine politics, was at the time considering a number of new cars and engine combinations. No fewer than six new projects were underway at roughly the same time, which appears absurd for a company ostensibly involved in war work and under severe financial constraints. Add to this the warring factions led by Ferrari and Ricart, with Jano and his design staff in the middle, and one has the elements of true chaos.

In the midst of this madness Colombo made his way down the Via Emilia to Modena on a glistening spring day to discuss the new *vetturetta* with Ferrari. Whoever was the father of the concept, there is no question that the decision to fund the car had been made in Portello, although it appears that Ferrari himself had the final say over the configuration of the automobile. Colombo says he brought with him plans for a mid-engine single-seater, based roughly on the concept of the Auto Union car which had dominated Grand Prix competition the season earlier. Ferrari rejected this layout out of hand, remarking, "It has always been that the ox pulls the cart." Therefore the new machine, code-named Tipo 158, would be a conventional front-engine automobile, based purely on the wishes of the Commendatore. (He would again use the same argument in the late 1950s, thereby anchoring his team at least two years behind his new British rivals.)

To Ferrari's credit, he never claimed to be an engineer. He wrote, "I have never considered myself a designer or an inventor, but only one who gets things moving and keeps them running . . . [It's] my innate talent for stirring up men," as he described it. Therefore it was the tiny Modenese team headed by Colombo which created the Tipo 158. However, most historians believe that the engine of the new car, a 1.5-liter, twin-cam straight-8, was essentially one-half of the 16-cylinder Tipo 316 engine block that had been created almost simultaneously by Jano. So one can only conclude that there was enormous cross-pollination of ideas streaming between Portello and Modena during this tumultuous period.

The new 158 was dubbed the Alfetta. Work commenced on a round-the-clock basis, with Ferrari frantically leading his cadre of troops. His days were spent buying and bartering for such crucial items as steel tubing and sheet aluminum, ignition wiring, brake components, fuel and oil pumps, radiators, coolers, steering bits, springs, shock absorbers and a plethora of other pieces being manufactured up and down the Po Valley by small specialist shops. Bazzi was supervising the mechanics, fabricating the cars and operating the dynamometers with which prototypes of the new engines were being tested. Massimino's specialty was the suspension, while Giberti, who was to become one of Ferrari's longtime loyalists, filled in where needed. Nasi assisted Colombo in the little design room of the Scuderia with the masses of complex engineering drawings necessary to create a car from the raw materials being accumulated by the boss.

This is where Enzo Ferrari was at his best—immersed in the genesis of a daring new design that he believed would bring victory on the racetrack. As the summer heat of Modena bore down on them, transforming the workshops of the Scuderia into a fuming hell of simmering, deafening, maddening motion, he seemed to gain strength. Each night the exhausted team would gather in one of the little trattorias around the neighborhood to eat and to muse about their progress, finally to fall into bed for a few hours' sleep, before beginning again at dawn.

In the meantime the official racing season had opened and once again the Germans were gobbling up the competition like angry predators. The new W125 Mercedes-Benzes were even faster than expected and were taking the measure of the Auto Unions, which themselves were leaving the Alfas even farther in the dust. Nuvolari was beside himself and began to drive at a level of zeal that seemed overwrought even by his standards. He lost control and crashed while practicing for the Turin Grand Prix, suffering broken ribs and a concussion. To make matters worse, his father died. Then he left for the Vanderbilt Cup on Long Island aboard the *Normandie* and on the crossing received word that his eldest son, Giorgio, had died of a heart condition. Despondent, he practiced for the race, only to find the American dirt-track star Rex Mays decisively faster, in one of last year's aged 8C-35s, than either him or his teammate Farina.

It was obvious that the great star was ready to bolt, regardless of the endless reminders of patriotic duty issuing from Ciano and others in Rome and Milan. He was driven to the edge of distraction at Pescara in mid-August when Jano and Ferrari brought along a pair of revised 12C-37s—lower, updated versions of the highboy 36s—for himself and Farina. Tazio drove for only four laps before stopping at the pits and disgustedly giving up the car to Farina, who had not started. It was apparent the machine was a disaster. Nuvolari prided himself on being able to master cars that other men found undriveable. Therefore, if he rejected the 37, it had to be beyond hope. "Its inferiority is evident . . . and hurtful," reported one journalist on the scene.

Then Nuvolari appeared at the Swiss Grand Prix at the wheel of an Auto Union. The Italian press was in hysterics, which only subsided when the maestro seemed uncomfortable in the ungainly, overpowered machine with the engine in the back. But the trend was unmistakable; unless somebody, either Jano or Ferrari or Ricart or Colombo, was able to create a racing car that would uphold the flagging honor of Italy—which on the battlefields of Spain and in the company of its new ally Nazi Germany was attempting to represent itself as a superpower—the ax would fall with brutal quickness.

The end came at the Italian Grand Prix. It had been moved from Monza to Livorno because the region was Ciano's home and the political atmosphere was warm and hospitable. Nuvolari had performed well the year before against the Germans, beating them on the 4.48-mile course that ran along the seacoast south of Pisa. Nuvolari chose one of the earlier 12C-36s in hopes of repeating his triumph of 1936, but it was for naught. The awesome W125 Mercedes-Benzes of Rudi Caracciola and Hermann Lang were light-years quicker. Again Nuvolari stopped and in a cold fury gave up his car to Farina. He in turn soldiered on to finish a distant seventh, the best of any Italian in the race.

Two weeks later Vittorio Jano, Italy's most famous racing-car designer, was fired. Rumors swirled that Ferrari would leave as well in order to form a non-Alfa team with Jano. Ferrari had for several years been calling for what he termed a "Scuderia Italia" in which all of the major manufacturers would contribute to a national Grand Prix effort, to be headed by himself, of course. Neither Fiat nor Lancia evidenced any enthusiasm for the idea, although the latter Turinese concern quickly hired Jano, in part to replace its founder, Vincenzo Lancia, who had died earlier in the year.

The Scuderia was under siege. Gobbato, who had total faith in Ricart, was speaking openly of plans to take over the racing operation entirely and move it back to Portello. This only increased the tension in Modena, where Colombo, Giberti, Nasi, Massimino and Bazzi labored like coal miners on the Alfetta. Their work was hampered by a dependence on Portello, where all the castings and forgings were being made before being shipped to Modena. It seemed a mad race against time—one final chance to retain some dignity before the Spanish interloper captured all that was left of the Scuderia. If only the 158 could be completed and then win a few races, a semblance of independence could be preserved.

In order to relieve some of the pressure, Ferrari and Colombo called the press to Modena a few days after Jano's dismissal and exhibited a prototype of the 158. But the situation was obvious even to the most naïve and enthusiastic of the journalists; many months would pass before a 158 would actually roll a wheel in legitimate competition. It was a brave gesture, but hardly enough to mollify the Alfa management.

The final bell rang on New Year's Day 1938. Gobbato announced that the racing operation was being transferred back to Portello, to be run under the aegis of the newly formed Alfa Corse—a company-operated team that would not only supersede the Scuderia but totally absorb it. Ferrari himself saved some face by being appointed as the director of the new organization and certainly benefited financially from the buyout. He was now a comfortable Modenese *borghese*, at least financially, if not socially, but this hardly compensated for the gutting of his beloved Scuderia. Moreover, his position at Alfa Corse meant long commutes to Milan, where a new two-story building had been constructed adjacent to the factory to house the operation.

A fleet of Alfa Romeo trucks arrived to load the machine tools, the racing spares, the sports cars and the four almost-completed Tipo 158s. It was a solemn day as the convoy moved off, leaving the building on the Viale Trento e Trieste vacant save for a few Alfa Romeo passenger cars that were to form the nucleus of a conventional dealership.

With Adolf Hitler once more making increasingly bellicose noises about an *Anschluss* to annex neighboring Austria and Mussolini moving closer to an outright alliance with the Germans, it had to be obvious to most Italians that Europe was veering toward war—a war that could not be avoided by Chamberlain's appeasement policies or the crumbling League of Nations.

Yet, like the last dance before midnight, the racing continued. Ferrari no doubt headed to Milan, his teeth bared, knowing that his hated rival Ricart was at the height of his power. With Jano gone, he was the chief designer and it would be his automobiles that would win or lose on the racetracks, with the exception of the 158s, which, as an act of defense, were pulled ever closer to the collective bosom of the little crew of refugees from Modena.

Then came a second, rather oblique blow. The sparkling career of Bernd Rosemeyer ended in early February on the Frankfurt–Darmstadt autobahn. The irrepressible German was trying to break a high-speed record that had been set earlier in the day by his rival, Rudi Caracciola. Warnings of crosswinds and patches of ice had been ignored ("Don't worry, I'm one of the lucky ones," he said), and he had crashed to an instant death at over 250 mph. That left an opening on the Auto Union team that was surely to be offered to Nuvolari unless Alfa became instantly competitive.

The automobile that would make or break the relationship was the cobbled-together Tipo 308, a machine so badly conceived and executed that no one seemed willing to take credit for its design, although it was welded up in Modena from a Jano design while the team was concentrating on the Alfetta. Regardless of the true source, it was cursed from the start.

During practice for the first race of the season at Pau, the chassis flexed so badly under Nuvolari's spurring that the fuel tank ruptured. The car was engulfed in flames as he veered to a stop and vaulted out with minor burns. That was it. Nuvolari swore on the spot that he would never again drive an Alfa Romeo (he stayed true to his word) and within days announced his retirement. He gathered up his wife and left on a long trip to the United States, but by mid-season he was back with renewed enthusiasm and signed with Auto Union. There he stayed until the outbreak of the war, winning three major races.

A successful test of the Alfetta was finally conducted at Monza on May 5, 1938. With Enrico Nardi driving, the new car rolled out numerous trouble-free laps, its tiny, supercharged engine echoing through the trees. It was a long, low, spidery machine, a three-quarter-scale miniature of the bigger 308 and 312 Grand Prix cars, only much more sophisticated in terms of handling and responsiveness. The Alfetta featured immense, hydraulically actuated brakes, a fully independent suspension and a transaxle, so that the transmission mated directly to the rear differential to maximize weight distribution.

Significantly, the Alfetta's sloping nose carried grillwork resembling an egg crate—a design that would later become a tradition for Ferrari and one that is exclusively associated with his marque.

As the frustrations mounted with the official Alfa Corse Grand Prix campaign, hopes soared in Modena. The marathon to complete the cars ended on August 7 at the Coppa Ciano, to be run, as usual, on the Livorno road course. Three of the 158 Alfettas were loaded up and hauled to the race, where Francesco Severi and Clemente Biondetti would team with young Emilio ("Mimi") Villoresi, whose brother Luigi ("Gigi") was entered in a special Maserati. He was already an established star for Maserati and was imported at the last minute and given a lightweight 6CM-1500 with an aluminum engine block. By then word had filtered out of Modena to the Bologna firm that the Alfettas were formidable new rivals, and Villoresi was assigned the job of spoiling their debut.

The "light car" or *voiturette* race was run as a preliminary to the Grand Prix, but the debut of the Alfettas caused a major stir. Decked out in the traditional Italian racing red, but notably devoid of the Prancing Horse emblem, the 158s were impressive in practice and all started on the front row of the grid. When the flag fell Severi made a lightning-quick start and led the initial stages before Gigi Villoresi's Maserati squirted past. By the sixth lap he appeared headed for an easy victory, but at that point Mimi began to challenge. He moved up, and the two rivals, brother against brother, aboard Italy's two most prestigious racing-car marques, hammered on each other to the delight of the giant crowd. It was the Maserati that broke under the strain, its engine expiring and sending Gigi rolling onto the verge in a cloud of oil smoke and steam. From there Mimi coasted to the checkered flag, giving the Alfetta and the deliriously happy Ferrari crew a major win on its maiden appearance.

Ferrari's stock surely rose in Portello, and there is little doubt that he pressed his advantage with Gobbato. At that point Ricart's Grand Prix efforts were no more successful than the departed Jano's, and surely Enzo Ferrari, the "director" of Alfa Corse, worked endlessly to erode *lo spagnolo's* fragile reputation. The two were barely speaking as the 158s continued to be raced and developed, thanks in large part to Bazzi's singular genius at tuning engines. The cars lost at Pescara, then won at Monza before being humbled by Maserati on their home ground of Modena.

Tensions increased between Ferrari and Ricart. Ferrari, with his ingrained chauvinism, had every reason to be jealous. Here was this outsider, this "foreigner," this Spaniard, who had risen out of nowhere to commandeer the entire racing program of Alfa Romeo, which had been neatly sheltered under the escutcheon of the Scuderia. Now Ferrari was shorn of his team and was operating as a lackey in a giant, contentious organization and answering directly to a man he openly despised. Jealous? That was not the half of it.

"My difference with Gobbato became more acute," he wrote. "I was obliged to tell him that even if I had given up the Scuderia Ferrari, my principles and engineering philosophy remained as before. He replied, 'At Alfa Romeo, I am the manager and I will not get rid of a man who has my trust. Nor, Ferrari, can you expect me to accept your demands without discussion.' I replied that I was sorry that I had made so harsh an answer necessary and added that it was not a question of whether or not my views were to be accepted without discussion, but that I was disturbed by the way in which Ricart's fundamentally unsound ideas were automatically accepted."

Regardless of where the blame actually lay, the annihilation of the Grand Prix Alfas continued through the end of the 1938 season. An added insult came at the hands of the defector Nuvolari, who won the final two races at Monza and at Donnington Park in England at the wheel of one of the hated Auto Unions. Farina managed to salvage some honor by finishing second at the Italian Grand Prix, although his Alfa was three laps down at the end and was the benefactor of high attrition suffered by the Germans.

But there was a plan for 1939. The Italian racing authorities would avenge the trouncing being administered by their German friends. At the Italian Grand Prix in September they announced that in 1939 all major Italian races would be run under the 1.5-liter *voiturette* formula. This seemed a stroke of genius. The Germans had no such machinery and both Alfa Romeo and Maserati, with its just-completed little 4CL, possessed the best there was in the class. Surely the Italians could start winning some races again. What they did not know was that the competitive zeal of Mercedes-Benz was so strong at the time that they would respond with an act of technological bravado unprecedented in the history of motorsports.

In the meantime the political atmosphere at Portello was becoming superheated. Work proceeded through the winter on both the 158 and the larger Grand Prix machines, but the sniping between the Ferrari and Ricart factions was becoming more acute and it was only a matter of time before someone was hit.

This silly, trivial rivalry raged while Italy skidded down the path to war. That Alfa Romeo's management could be distracted by motor racing during a time when the nation was being sapped militarily by Mussolini's adventures borders on the idiotic. Not only had the campaigns in Spain and Ethiopia exacted an enormous toll in manpower, materials and funds, but Il Duce moved in April 1939 to annex little Albania, on the Adriatic. This gave him the bloated title of King of Italy, Ethiopia and Albania but little else. Italy's departure from the League of Nations at the end of 1937 had driven the country ever closer to Hitler, and by May 1939 it was widely known in official circles that Italy and Germany would sign on the twenty-second a major alliance to be called the "Pact of Steel."

Meanwhile, unbelievably, the racing teams were gathering for Marshal Balbo's annual festivities in Tripoli. The desert kingdom had been elevated to full membership in the Italian state and Balbo had added even more frills to his elegant racetrack and to the trappings of glamour that made his Grand Prix the richest and most prestigious on the international schedule. It was a tradition for the teams to arrive early, both to acclimate themselves to the stifling desert heat and to attend the elegant parties organized by Balbo. The event was to be capped by a victory dinner at the white marble palace, the guests surrounded by an honor guard of hooded native Ascari troops. And, thankfully, for the first time since 1934, it appeared that the feted winner would be an Italian who had driven an Italian car.

But at the last minute alarming word arrived. Mercedes-Benz was entering a pair of new cars, specially constructed for the race. They were so untested that one of them was being completed on the ship crossing the Mediterranean. That news brought a sense of relief to the Italian contingent. After all, with the well-developed 158s now generating over 200 hp in racing trim and with Maserati on hand with a special streamliner for Gigi Villoresi, what sort of opposition could the Germans offer up at such short notice?

A great deal, as it turned out. In five months the engineering staff at Daimler-Benz had created a brace of jewel-like V8-powered 1.5-liter cars that were almost perfect miniatures of their larger full-blown Grand Prix machines. In a frantic effort, two Type 165 racing cars, complete with all-new V8 engines, were transformed from working drawings to staggeringly fast rolling stock. Early tests indicated outputs of over 240 hp—or better than 40 horses more than the best Alfas—mounted in chassis that were considerably more advanced in cornering and braking technology.

In the hands of Rudi Caracciola and Hermann Lang the 165s recorded one of the most stunning upsets in motor-racing annals. Running in front of a stunned Balbo and mobs of Alfa brass (excluding Ferrari), the Mercedes-Benzes qualified second and third—behind Villoresi's daringly streamlined Maserati. But once the race started on the superheated, sandswept circuit it was essentially all over for the Italians. Lang leapt off the line and disappeared from everyone, completing the first lap down the long straight before the next car—Caracciola's—even appeared! Villoresi's Maserati retired, along with the other two factory-entered 4CLs, on the first lap. Farina drove with his usual verve to head Caracciola for a few circuits, then fell back, his Alfa unable to maintain the pace. Lang went on to win by almost a lap over his teammate (and openly bitter rival), Caracciola.

This would be the only time the 165s would appear. It was a dazzling display of German racing skill and was surely embarked on as an act of propaganda. There was no reason to expend thousands of man-hours and hundreds of thousands of reichsmarks other than to demonstrate to their Italian allies and to the rest of Europe that the Germans were far ahead in all areas of technology, including military hardware.

A month later the 158s were taken to Monza for further testing. It was assumed that the Germans would make a major assault on the 1.5-liter class, and a number of modifications were made in the search for additional horsepower. Mimi Villoresi was the test driver and he was killed in a high-speed crash. And once again another man swore a lifelong hatred of the marque. Like Nuvolari, Villoresi's brother Gigi pledged that his rump would never darken the seat of an Alfa Romeo, and again like Nuvolari, he was to remain true to his word.

Gigi was also angry about the way Ferrari handled the ensuing insurance issue. The Villoresi family had asked for an insurance settlement from the accident, but Ferrari refused, claiming that Mimi "was ill, which caused him to crash." His accusation infuriated Gigi Villoresi, who insisted that his brother was in perfect health. This quarrel was to affect the relationship between the two men for the rest of their lives—even during the period when Villoresi became a star for the Scuderia.

With young Villoresi dead, the team struggled on against the ever-increasing might of Germany and the weight of its own internal politics. Relief would come only when an infinitely larger agony commenced—albeit slowly—for the Italians and the rest of the world. On September 1, Hitler's panzers plunged across the Polish border, beginning what was to be the bloodiest conflict in the history of mankind.

Within weeks England and France were fighting, although Italy was to remain on the sidelines until the following year. Ciano had long opposed the alliance with Hitler, and he was able to convince Mussolini that his armies were short at least 17,000 tons of crucial war materials and lacked the trains and trucks to transport them even if they existed.

Therefore Italy was to meander along in a kind of quasi-peaceful limbo until June 1940. The Germans had run their final motor race, at Belgrade, Yugoslavia, two days *after* the invasion of Poland. Tazio Nuvolari won for Auto Union, after which the all-conquering German cars were hauled home and placed in storage. It is a commentary on the misplaced priorities of the Italians that Alfa Romeo continued to build and test racing cars at least until 1941, when the nation had been at war for a year.

By then Enzo Ferrari was long gone. His nearly twenty-year string with Alfa Romeo—more appropriately an "umbilical cord," as he called it—was broken late in 1939. He was fired by Gobbato. Various biographers have avoided that word, choosing instead to imply that he left on his own accord, or at the least by mutual consent. Ferrari himself was hardly as timid. He wrote, "The rift thus became unbridgeable and led to my dismissal."

It was over. *Finito.* Enzo Ferrari packed his bags and returned to Modena. Ricart, the winner, would stay, as would Colombo and the others, save for Massimino, who left as well. The bloodletting at Alfa Corse was a major development inside the tiny world of motorsports, but it was, of course, nothing compared with the conflagration that had already begun in Poland and was spreading over Europe. Yet even as the acrid smell of cordite drifted slowly south over the Alpine peaks, Enzo Ferrari had one final gesture of defiance in mind before Italy plunged into the fighting. He would call it Tipo 815. Others would call it the first Ferrari.

8

Enzo Ferrari's dismissal from Alfa Romeo was clearly a greater blow to his ego than to his bank account. By his own admission, the buyout of the Scuderia and severance payments left him in solid financial shape. Moreover, his motorsports ventures had been a success, even when his Alfa cars were uncompetitive. Starting money and lucrative sponsorship contracts had kept the operation solvent and permitted him to create a nest egg that stood him in good stead when he returned to Modena. He maintained a simple life, still living in the small apartment on the second floor of the old Scuderia building on the Viale Trento e Trieste with Laura and Dino, who was now seven years old and beginning to show an interest in the machines that had consumed his father's life.

Enzo Ferrari could look back on the past twenty years with considerable satisfaction. He had begun with nothing, save for a tiny inheritance from his father, and through hard work and chutzpah had become a major figure in international motorsports. At forty-one years, he enjoyed that crucial ingredient in an Italian male's makeup—the respect of his peers. He was known in his native city as a *borghese* of substance and prominence. The Scuderia was a landmark of sorts and the Modenese had become accustomed to seeing not only famed racing drivers like Nuvolari and the late, beloved Campari in their midst but important nobles and Fascistas as well.

Still there was the shame of the Alfa Romeo firing, which had become something of a cause célèbre in the Italian sporting press. There was a certain solace to be gained from the knowledge that his departure had led to a small palace revolt at Alfa Corse and that Enrico Nardi, Federico Giberti and Alberto Massimino soon appeared at his side in Modena. They too had been victims of the bloodletting, although other close associates like Colombo and Bazzi were to remain in Milan for what was to be the duration of the war.

His severance agreement with Alfa Romeo had stipulated that he could not use the old Scuderia Ferrari name or directly engage in motor racing for four years. Gobbato and the rest of the Alfa management were obviously concerned that Ferrari would hook up with another manufacturer and trade on the name that had been so closely associated with Alfa Romeo. Therefore Ferrari set to work establishing a custom machine shop called Auto Avio Costruzione, with the Prancing Horse prominent on all letterheads, sales brochures, and such. By the end of the year Ferrari had arranged to machine parts for a Roman company, Compagnia Nazionale Aeronautica, which was manufacturing small 4-cylinder aircraft engines for lightweight training planes. He invested substantial capital in equipping the old Scuderia building with lathes, milling machines, grinders and shapers. Lacking a foundry, he could not produce his own forgings and castings, but he made arrangements with the well-respected Fonderia Calzoni in Bologna to subcontract such work. With a planned work force of forty, he seemed ready to embark on a new career far removed, both emotionally and financially, from that swamp of intrigue at Portello.

But the Byzantine atmosphere at Alfa Romeo had already been replaced by another troubling situation located but a few blocks from the Scuderia. It had begun in 1937, when the remaining Maserati brothers (the family leader, Alfieri, had died in 1932) decided to sell a controlling interest in their small Bolognese car firm to the powerful Orsi family of Modena. The Orsis, led by Adolfo and his son Omer, had risen from the middle class and had gained enormous wealth through steel mills, agricultural machinery manufacture and operation of the Modenese trolley-car system. The Maseratis were marvelous racing-car builders but poor businessmen. The Orsi sale was intended to increase the firm's business, not only through the continued manufacture of racing cars (never totaling more than seventeen a year) but also through the potentially lucrative sales of Maserati spark plugs, which had been manufactured on a small scale since World War I.

The Maserati operation had been tiny compared with Alfa Romeo's, running only a limited schedule with an official racing team, while depending on its primary income from large, well-financed independents like the Scuderia Ambrosiana from Milan.

Despite their minuscule operation, the Maseratis—with the help of Orsi funds—were now able to produce some superb racing cars. In May 1939 the American champion Wilbur Shaw won the Indianapolis 500 with one of their supercharged 8CFTs (he would repeat again in 1940 and be on his way to an amazing third straight in 1941 when a rear wire wheel, damaged in a pre-race garage fire, collapsed). The Indianapolis race was highly popular in Italy and the one event that Enzo Ferrari openly stated he wanted to win. Surely it was galling for him to read of the lavish praise being heaped on the Maserati brothers, especially during a period when his racing fortunes were at such a low ebb.

But the humiliation was only beginning. In the autumn of 1939 it was announced that the Maserati operation was being moved—cars, tools, spare parts, bag and baggage—to Modena. In a major reorganization Adolfo Orsi was taking over as president and his brother-in-law, Alceste Giacomazzi, was assuming the position of general director. And the impudence of them! Alberto Massimino was being seduced to leave Ferrari and become chief designer. For over a decade Ferrari had been the maestro of local motorsports, the center of attention in the town when it came to automobiles. But now these invaders, these interlopers, were setting up shop on the Via Ciro Menotti, no more than a few long blocks from the Scuderia. The Maserati racing operation would continue with Bindo Maserati ostensibly a key figure, although all of the brothers were now reduced to highly paid employees, with little say over general policy. But the Orsis had serious expansion in mind. With war seemingly imminent, the company was planning to increase business by mass-producing spark plugs, batteries, small electric three-wheeled delivery trucks and a wide variety of lathes, grinders and milling machines. A lucrative sideline overhauling military trucks was also planned.

Compared with this massive crosstown effort, Enzo Ferrari's job prospects seemed positively minor-league. Then in December 1939 he was approached by young Alberto Ascari, the son of the late Antonio, and his friend Marchese Lotario Rangoni Macchiavelli di Modena, a wealthy and respected Modenese nobleman. Ascari, a chubby, easygoing youth of twenty-one who was known to his friends as "Ciccio" ("Tubby"), was a motorcycle racer of growing repute. He was riding for the Bianchi factory team while running a fuel transporting operation between Italy and North Africa in partnership with his friend and mentor, Luigi Villoresi, which provided him with a deferment from the Army. Rangoni was an amateur enthusiast who had competed in a number of minor Emilian events and maintained a small shop for his cars near the Scuderia.

These two came to Ferrari with an idea to build a pair of sports cars for the upcoming Mille Miglia—which had, in artful Italian style, been reorganized as the Gran Premio di Brescia and would be run on a shortened course between Brescia, Cremona and Mantua. The old route had been abandoned after a 1938 crash in Bologna that killed ten spectators and injured twenty-three more. But it was still the Mille Miglia, except that the circuit involved 102 miles of open roads, as opposed to 1,000. The route would simply be lapped ten times, ironically exposing the spectators to ten times as much danger as on the old circuit.

Ferrari agreed to build the cars, reaching a decision at a dinner on Christmas Eve 1939. When most families were together, either feasting or praying, on one of the holiest nights of the year, Enzo Ferrari was conducting business. For most Italians, even those who, in Ferrari's words, "have not the gift of faith," Christmas Eve was a sacred time reserved for family gatherings and obligatory attendance at church services. That Enzo Ferrari would spend such a period with racing associates underscores not only the singular intensity with which he engaged in business but also the hypocrisy of his testimonials of devotion to his son and wife. This behavior was hardly unusual for him, and others also found him boring ahead with his work on various Easter Sundays, Christmases or any one of the multitude of holidays and feast days that dot the Italian calendar.

The big race was to be run on April 28, 1940. Thus they had only four months to build the two racing cars. Because of the time constraints, Massimino, by far the most experienced designer on the staff, chose to "hot-rod" a pair of Fiat 508C Ballilas, small-displacement sedans, named for the Fascist youth organization, which had been manufactured since 1932. The Ballila featured a four-speed transmission, hydraulic brakes and a nicely engineered independent front suspension. It had become popular as the basis for small, home-built racing specials (and would also see service in the Italian Army as a light truck). But the engine was minuscule, and could not be bored out to meet the specifications of the 1.5-liter sports class that Ascari and Rangoni planned to enter.

Massimino's solution was to make his own engine block— a straight-8 that would employ a pair of specially modified Ballila cylinder heads. It was a brilliant solution, and one celebrated by Rangoni, who already had a cut-down 508 sports car fabricated by a local modifier and engine tuner, Vittorio Stanguellini. The new block was cast in Bologna, but the Ferrari shop did the rest, including the machining of the new crankshaft and the myriad of small bits and pieces needed to put together a new car—a car that would be called simply the "815" (for 8 cylinders and 1.5 liters. Others would note, however, that it was "158" reordered. A Ferrari irony?).

The bodywork would be done by Carrozzeria Touring Superleggera, the Milanese shop that had been started in 1926 by a group that included Ascari's uncle, Vittorio. In those days the world of high-performance cars in Italy was restricted to a tiny fraternity involving perhaps no more than a thousand prominent builders, designers, mechanics and drivers. All were related, either by blood or by business, and lived for the most part along a crescent formed between Turin and Modena. The 815 styling would be called "Torpedino Brescia" and featured full-fendered bodywork with a wide-mouthed, ovoid-shaped grille and a rather elongated tail. The second car, Rangoni's, was somewhat better detailed than Ascari's and had a more graceful, slender rear end. The cars rode on Borrani wire wheels and were fully open roadsters, with the driver protected only by a full-width windshield made of Plexiglas, a material that was just beginning to come into vogue.

The automobiles were finished ahead of schedule—a prodigious feat in itself. A perfect ending to the story would have them being driven to a grand victory, but it wasn't to be. Ascari and his friend and co-driver, Spoldi, led the 1,500-cc class on the first lap, which began and ended in Brescia. Rangoni and his partner, Enrico Nardi, were close behind. But trouble in the valve train, probably a rocker arm, sidelined Ascari on the next round. Rangoni then took the lead and extended his advantage to over half an hour on second place before bearing failure took him out.

The 815s had shown good speed and potential, but the entry of Italy into the war was less than two months away and the plans Massimino had created to update and improve the design were to be canceled by the fighting.

The Gran Premio di Brescia was won by an all-German team of BMW sports cars led by Count Huschke von Hanstein and Walter Bäumer, who averaged 105 mph over the route in their 2-liter 328 coupe. This pair, plus the remainder of the team, took the politicization of motorsports to perhaps unprecedented depths by wearing starched white coveralls emblazoned with the twin lightning bolts of the dreaded SS, or Schutzstaffel. Once Hanstein assumed his postwar position as the charming and widely respected press relations chief of Porsche, photographs of him in his Mille Miglia regalia were understandably seldom seen outside Germany.

On June 3, 1940, Italy blundered into the war at the wrong time, on the wrong side, with the wrong equipment and with the wrong army. Its best troops had been worn out in the Ethiopian and Spanish campaigns, and morale, which had been excellent in World War I, was low. Worse yet, Benito Mussolini had been fascinated with flashy armaments—quick, maneuverable tanks, planes and warships—all lightly armored and technically sophisticated but unable to withstand the brutal onslaughts of firepower that were about to descend on them. Not only did Italy lack crucial raw materials; its industrial power was sapped by inefficient management and a churlish labor force that never responded to the urgings of the Fascist government.

Hitler publicly flattered Mussolini by claiming they were equals, but in fact the Germans never revealed much of their military plans, forcing Il Duce and his staff to improvise on their own.

In the autumn of 1940, buoyed by the Wehrmacht's seemingly easy victories, Mussolini attempted to flex his muscles. With 200,000 of his troops already tied down in Ethiopia, where the British and the South Africans were attacking, he took on the little kingdom of Greece. It seemed an easy mark, just over the frontier from Italy's Albanian possessions. But the Greeks proved to be more than able to defend themselves on the rugged terrain of the Epirus Mountains and the campaign immediately bogged down. Then squadrons of British aircraft from the carrier HMS *Illustrious* blasted the Italian fleet at Taranto in the boot of Italy, sinking three battleships and disabling half the remaining ships. The fleet scurried back to Naples, never to be seen at sea in force again. Worse yet, the British rose out of Egypt and surged westward across the Sahara, capturing Tobruk and taking on ten Italian divisions with a pair of armored units. The toll: 130,000 Italians captured and thousands killed while the British lost a total of 438 troops. Il Duce was learning that this business of fighting was considerably more difficult than causing the trains to run on time.

Ironically, Mussolini's failure in Greece would keep him in the war for a while longer. Hitler was prompted to move against Yugoslavia and Greece to bail out his floundering ally. This in turn caused the British to divert part of their African army back to Egypt to protect what appeared to be a Nazi move toward the Middle East. Hitler countered by sending the Desert Fox, Erwin Rommel, to Africa to assault the unprotected British flank in Libya. Throughout this super-power maneuvering the Italians were merely lightly armed specta-tors, being shoved hither and yon by the dictates of stronger, more resolute allies and enemies alike.

While Italy blundered into the first year of the war, one of its citi-zens found himself suddenly isolated on a faraway shore in unlikely circumstances. Luigi Chinetti was happily ensconced in France when, in early 1940, he received a telegram demanding that he re-turn to Italy and join the Army. Chinetti defiantly wired a reply that he had already served in World War I. However, Chinetti came to America in 1941 on a chimerical expedition to the Indianapolis 500 and decided to stay. He was to remain in New York—as an enemy alien working on exotic cars—for the duration.

Enzo Ferrari was too old at forty-two to be eligible for military service. While some in the Alfa Romeo management were still fid-dling with racing cars, he had given up his chosen profession and had diverted his energies to the war effort—or, more correctly, to profiting from the war effort (as would millions of noncombatants on both sides). To repeat, there is no evidence, aside from some strident slo-ganeering in the Scuderia newsletter, that Enzo Ferrari had a particu-lar predilection for Fascism—or any other political movement, for that matter. He seemed to view the war as little more than an inter-ruption of his racing operation and therefore stoically set about mak-ing the best of it on a personal basis. This was hardly unique among the Italian populace.

Mussolini never enjoyed the goose-stepping enthusiasm accorded Hitler by the Germans. He faced a solid anti-Fascist coalition of old royalists, socialists, academics and a strong Communist Party cen-tered in the industrial north. To his credit, and directly opposed to the inhuman actions of Hitler and Stalin, Mussolini did not incite po-groms against his opposition and precious few of his opponents were killed or incarcerated. As with so much of his regime, even his brutal-ity was more show than action. It is estimated that between 1927 and 1943 about five thousand people were condemned to death by the

Fascists, but only twenty-nine were actually executed. A number of dissidents simply left the country, including the brilliant atomic scientist Enrico Fermi, who became a powerful and valued presence in the development of America's atomic bomb.

Ferrari's war efforts are, like so much of his life, rife with contradictions. What is known is that he began producing exotic German machine tools—sophisticated hydraulic grinders that he says were used in the manufacture of ball bearings. Others claim differently. Ferrari's version states that he was introduced to Turin businessman Corrado Gatti by his friend and associate Enrico Nardi. Gatti is said to have suggested that Auto Avio Costruzione get into the business of copying German-made Jung grinding machines under patent. This Ferrari says he tried, but was refused. The Germans claimed that they never granted such patent rights, and besides, they airily said, they doubted whether anyone but themselves had the technical expertise to manufacture such sophisticated devices.

But Ferrari discovered that German patents were not enforced under Italian law and decided to copy the grinders without permission. The effort was a great success.

However, there is another, slightly more detailed version of the story, outlined by Franco Cortese in 1984 to Italian historian Angelo Tito Anselmi. Cortese was a thirty-eight-year-old gentleman racing driver of some note who was instrumental in putting Ferrari in touch with the Brescia firm of Ernesto Breda, a powerful manufacturer of armaments and military vehicles for the government. The chairman was a relative of Count Giovanni Lurani, a prominent motorsports journalist, sometime driver and a widely respected enthusiast who was well connected inside the sport. It was Cortese, through Lurani, who introduced Ferrari to Breda as a potential supplier of grinders the company needed, not for ball bearings, but for components of its Type 37 machine guns.

In his memoirs Ferrari chooses to ignore the fact that he did work for Breda, or that his grinders were employed for anything but the manufacture of ball bearings. But by 1942 his firm was capable of undertaking highly sophisticated machining; Ferrari's name was trumpeted up and down the Po Valley by Cortese, who had become a highly effective "road man." This was hardly an easy task, considering the increasing reluctance of his employer to leave Modena. For example, Cortese had arranged a deal with Breda for Ferrari to manufacture a complex gear-reduction unit for a special landing craft being built for the Italian Army. Breda was converting their big Tipo D17 6-cylinder engines from their Littorina self-propelled rail cars for use in special landing craft intended for the invasion of Malta. Ferrari was to fabricate the gearboxes which would permit the Breda engines to be modified for marine use. It appeared to be a lucrative contract for the small firm, and the Breda officials were eager to get on with business when they summoned Ferrari to Milan to sign the contract.

Ferrari imperiously notified Cortese that he would not drive to Milan, and that if the Army and Breda wanted to close the deal, they would have to come to Modena. Phone calls were exchanged between an apoplectic Cortese and an intransigent Ferrari. The "miracle man" was adamant; he would not travel the seventy-five miles up the ancient Via Emilia to Piacenza, there to cross the Po and continue the final twenty-five miles to Milan. Grumbling in disbelief, Cortese and his entourage of executives and Army brass gathered up eight automobiles and motored off to Modena to complete the deal. The mountain came to Mohammed! It would not be the last time.

By early 1943 it was obvious that the Axis chances for victory were about to disappear entirely. They had crested early in 1942, and now, backed by the massive industrial power of the United States, the increasingly effective Allied armies were driving like runaway freight trains. The invasion of Sicily had begun in July 1943 and it seemed that Montgomery and Patton meant to push straight up the boot of Italy. Italy was on the edge of anarchy. Strikes swept through Milan and Turin. Within a few weeks Mussolini was removed from office in a bloodless coup and replaced by General Pietro Badoglio, who artfully assured the Germans he would continue to fight while secretly trying to sue for peace with the Allies. Hitler responded with typical ferocity, heaving Badoglio from power, rescuing Mussolini from exile in the Apennines in a daring raid, executing those who deposed Il Duce, including Ciano, and turning Italy into a military fiefdom—no longer an ally, but rugged real estate to be held by Field Marshal Albert Kesselring's hardened troops at all costs.

To the north, Ferrari was ordered to move his factory to safer ground. Since late 1942 the Allies had extended their bombing raids to the major industrial centers of the Po Valley, and an order for decentralizing industry was issued. Ferrari still employed about forty workers—a number of them now women—at the old Scuderia on the Viale Trento e Trieste and business was booming. He was ready for expansion, convinced that within months the German defense in the south would collapse and peace would bring new opportunities to resume building racing cars.

He selected a piece of property in the village of Formigine, a few miles south of the city. But after considerable negotiation the owner declined to sell. He then asked Mino Amarotti, a friend in the small community of Maranello, to help him find additional land. The town was a fair distance from Modena, about ten miles, but Ferrari already owned a small plot there, an aged stone farmhouse in the middle of a cherry orchard. Amarotti helped persuade the local miller to sell his adjacent house and a parcel of land across the Abetone–Brennero highway. These three pieces would form sufficient property on which to construct a factory. A triangular complex of wood-and-concrete sheds was quickly thrown up and another hundred workers were hired for the thriving Auto Avio Costruzione. But the property would not be his. Ferrari apparently put the land in Laura's name to preclude any future legal or governmental complications that might arise. The times were uncertain, and while it appeared that the Germans would be slowly driven out of Italy, it was impossible to foresee what sort of regime would replace them. Whatever might happen, Ferrari would be in a better position if the buildings and the property were divided between him and his wife. Their relationship was little more than a legal arrangement bonded by the presence of their son and the ironbound Italian traditions against divorce. But both under-stood the value of land in those dangerous days and the Maranello split was surely endorsed by both parties.

By the time Montgomery and his troops leapt across the Strait of Messina on September 3, 1943, Ferrari's little operation was thriving. The contract with Alfa Romeo forbidding the use of his name had run out and the old Auto Avio Costruzione name now included "Scu-deria Ferrari."

The Germans proved to be both tough fighters and brutal admin-istrators. They managed to slow the Allied advance up the boot at Anzio and along the Gustav Line at the base of the Apennines, where the ancient monastery at Monte Cassino buttressed their defenses. To the north they lashed Italian industry to increase production, which had been sluggish at best. Their major supplier was Fiat, which spent most of its energies in work slowdowns and in trying to prevent its workers from being deported to German factories. The giant Turin complex was supposed to produce 180 aircraft a month. But due to endless acts of sabotage, strikes, absenteeism and overt malingering, the maximum output per month from September 1943 to the end of the war was 18! The Germans demanded 1,500 aircraft engines a month. They got between 90 and 300. Truck production was equally paltry, and half of the vehicles disappeared into the black market. Furious with the Italians, the Germans announced in early 1944 that all Fiat tooling and machinery was to be shipped to Germany. That triggered a general strike which spread across northern Italy and im-mobilized what feeble war effort remained. Fiat was not moved.

Even Ferrari's tiny operation did not escape the heavy hand of the invaders. In September 1944 a contingent of German officers arrived at the factory in Maranello to check inventories and production lev-els. They must have been pleased with the condition of the facility, for Ferrari's workshops would have met even the most compulsive German's standards for cleanliness. However, the Jung copies imme-diately caught their eye and it was icily announced to a thoroughly uncomfortable Ferrari that the entire inventory of grinding equipment would be requisitioned by the German government. They drove away, leaving him—in theory at least—possessing only an empty factory which sat on land owned by his wife.

In reality, nothing changed. The Germans were now totally occupied with holding off the Allies, who pounded north through the Caesar Line at Pescara. Masses of Italian guerrillas swarmed over the countryside. They were so strong and determined that eight German divisions had to be diverted to hold them at bay. On November 4, 1944, the Americans launched one of their daily bombing sorties, using the workhorses of the theater, four-engine B-24 Liberator bombers. Their target was the concentration of German troops along what was presumed to be the final line of defense south of the Po River. Called the Gothic Line, this loose linkage of bunkers, revetments and supply depots ran well south of Maranello, but the Ferrari works still qualified as a legitimate if not high-priority target. A few bombs thudded into the factory's porous concrete, caving in part of the roof and damaging some of the "German" machinery, but within a few weeks the operation was again running at full tilt.

At this point Enzo Ferrari was deeply immersed in facets of his life that were to have a much greater long-term impact than a few hundred pounds of explosives. By far the most volatile was a lovely, fairhaired woman named Lina Lardi. She was one of dozens of women who had come to work at the factory during the middle years of the war, and her riveting good looks no doubt quickly triggered the attention of the boss, whose sexual appetite was in full force at age forty-six. Lina Lardi was from Castelvetro, a rural village about eight miles from Maranello, dominated by a craggy hill surmounted by a medieval keep and an ancient church. Little is known of her background. What is known is that she possessed serenity and a gentle nature that were to provide an emotional sanctuary for Enzo Ferrari for the remainder of his life. She was also to provide him with the sole heir to his empire.

As Lina Lardi entered his life, so did new visions for the future of his company. The Germans were reeling everywhere from the onslaughts of the Allies. "Festung Europa" was being breached from the east, west and south and it appeared that peace would soon come to Italy. Men were already resuming their dreams, and certainly Ferrari's thoughts were turning to the days when raucous racing automobiles carrying the Prancing Horse would spill out of the gate onto the Abetone road. His reputation was solid within the Italian automotive community and it was only a matter of time before a major manufacturer would appear on his doorstep with a proposal.

That was to be the venerable firm of Isotta Fraschini, one of whose aged Tipo 1Ms Ferrari had driven three times in 1920. The company was an old and honored marque, having been started in 1900. Isotta Fraschini created some of the most advanced luxury cars in the world during the 1920s, but like so many limited-production manufacturers, the firm was ravaged by the Depression, and by 1936 it had been absorbed by Caproni (after a buyout attempt by Henry Ford had failed). Production was diverted to aircraft engines and diesel trucks. But as the fighting appeared to be ending, the Isotta management harbored plans to reenter the car business. They were referred to Ferrari by the same Count Lurani who had helped with the Breda connection.

Once again a meeting was set in Modena, with a small contingent of Isotta management making the slow and dangerous trek down from Milan. They spoke of building a sports car when peace came, and their ideas centered on a V12 engine. Alberto Massimino was present at the meetings, recalled Count Lurani, and it was Massimino who was to make the initial drawings.

The meeting is significant because the inflated Ferrari legend says that plans for a V12 did not arise until his alliance with Gioachino Colombo the following year. Most biographers support the notion that it was Ferrari himself who created the V12 out of thin air, although this is nonsense; Mercedes-Benz, Auto Union and Alfa Romeo, to mention but three, had long since proven the value of the design. But Lurani insists that the 1944 discussions with the Isotta Fraschini management influenced Ferrari's decision to employ the

V12 configuration himself. In any case, the story dead-ends. Shortly after the meeting the Isotta Fraschini works were leveled by American bombs and much of the management team was dispersed, either to go into hiding or to be captured by the Italian *Partigiani*. But if Lurani is correct, and there is little reason to doubt him, the famed Ferrari V12 may have had its origins, not in the mind of Enzo Ferrari, but in the boardroom of Isotta Fraschini.

As the days shortened and a perpetual grayness fell over Emilia, 1944 dribbled to an end in an atmosphere of dread and uncertainty. The nation was in chaos. Who was to know what form of government—or series of governments—might end up ruling a shattered and fragmented Italy? Business prospects looked bleak for Ferrari, what with the Isotta Fraschini discussions coming to naught and the umbilical cord to Alfa Romeo severed forever.

Worse yet, the temptations of the vast Orsi coffers proved too great for Alberto Massimino. The fine engineer, who had been such a valued addition to the 158 development team and in the creation of the 815 sports cars, was finally persuaded to leave Ferrari and move across town to the rival Maserati works. The defection was a blow because Ferrari was counting on Massimino to design new engines—and perhaps complete automobiles—once the fighting ceased.

In February 1945, American bombers returned and hit the Maranello works even harder. Damage was heavy and months would pass before repairs could be completed. Ferrari's world was literally crumbling around him. Worse yet, dramas loomed in Castelvetro and in Modena.

Lina Lardi was five months pregnant. Ferrari understood that he would be the father of an illegitimate child by the middle of the year. An entirely new and shadowy second life was about to open up for him. Moreover, the situation in Modena was equally depressing. Not only was the relationship with Laura in its normal state of simmering hatred, but Dino was ill. The boy, now twelve, was evidencing symptoms of a disease that, according to the doctors who examined him, would ultimately take his life. The irony of how he might simultaneously gain and lose a child could not have escaped him.

As the war pounded down around him, Enzo Ferrari, for the first time since those bleak days in 1918, began to sense defeat. Anarchy, brutality and death were everywhere, in the shell-pocked cities, on his once-serene Padana plain, in the now-bloodied Apennines above him, even in his own household. Peace was slowly coming to the country, but would it be any better than the ravages of war?

9

The war was surging across Italy like a series of tornadoes, touching down with insane fury in one region while blithely skipping past others and leaving them untouched. Most of the bombing damage had occurred in Turin and Milan to the west, while the south had suffered most of the artillery damage as the Germans and Allies slugged it out along the spine of the Apennines. Aside from a few sporadic bombing raids like the pair that thudded down on the Maranello factory, Modena had not suffered seriously. Electric power was at a premium and most essentials were in short supply, but the fertile acreage around the city had kept the residents' bellies reasonably full for much of the duration.

For Enzo Ferrari the early months of 1945 immersed him in a personal drama that transcended the brutality of the war. On May 22, Lina Lardi bore him his second son. Unlike Dino, now a frail and rather uncertain thirteen-year-old, the new child appeared healthy, with his mother's strong genes. He was named Piero and was immediately sequestered in the Lardi household in Castelvetro. Regular visits would be made to mother and child by Ferrari, but the existence of Piero Lardi, illegitimate son of the renowned local motorsports power, was to be hidden for years and known only to a few of his closest confidants.

Piero Lardi entered the world a month after the death of the brash countryman whose hubris had carried Italy on such a bizarre roller-coaster ride for the past two decades. In April, Benito Mussolini had been captured by *Partigiani* on the shore of Lake Garda while attempting to skulk out of the country disguised as a German soldier. The brutalizing of his body—as well as that of his mistress—which was hung like a side of beef from a Milanese lamppost, had marked the Italian people's final, symbolic act of self-cleansing from the Fascist farce.

By June 1945 most of the 150,000 irregular *Partigiani* guerrillas, for the most part controlled by the powerful Communist Party, had turned in their arms and returned home. Ahead lay a massive reconstruction project for Italy, politically, spiritually and physically. The northern industrial region had received a fair pounding, and the nation's bridges, highways, electrical power plants, telephone lines, water purification systems and sewage plants required considerable rebuilding. This was a bleak period of cynicism and corruption for Italy. The people were turning on the Fascists with a fury reserved for those who raise hopes to absurd, hysterical heights. Thousands of party members and the 800,000 civil servants were hunted down. Many were lynched or imprisoned by rump courts. Others were hectored by "purity committees" backed by the labor unions, a coalition of leftist parties and the Communists. Still, the Italians were relatively quick to forgive. By 1946 an amnesty had been declared and only 3,000 to 4,000 former Fascists and known war criminals remained in jail. In fact, the "punishment" meted out by the *Partigiani* had much of the comic-opera quality that had characterized Il Duce's regime. The conquerors would often take a northern Italian village and immediately declare the town hall the *casa di popolo*. A loud and theatrical hunt would then be undertaken for the local Fascists and Blackshirts. Much shouting and posturing would take place as the guilty slipped out of town and those who remained simply pleaded innocent. Within days, life returned to normal.

Ugo Gobbato wasn't so lucky. He was assassinated by a lone gun-man as he left the Portello works. The gunman was never found, nor was the motive clear. Because of his efforts in behalf of the Fascist government, it is probable that the killing was politically motivated. But it is possible that a disgruntled worker, harboring an unknown grievance, simply used the confusion of the times as a convenient cover for a murder that had no social or political overtones whatso-ever. Regardless, Gobbato was a victim—deserved or otherwise—of the madness that gripped Italy.

Another old friend had already disappeared. Eduardo Weber, the carburation genius—whose units had become exclusive to all Alfa Romeos in 1937, and who had been associated with Ferrari for twenty years—vanished just prior to the end of the fighting. It was speculated by some that he had been taken by the Communist partisans, to be held for ransom. But this was never confirmed and Weber's body was never located.

Initial assessments of the nation's war damage turned out to be somewhat overblown. The infrastructure of bridges, highways and tunnels had been heavily damaged, and initially industrial output had plunged to about one-fifth that of 1938 (although much of the drop was attributable to inflation, labor disputes, weak foreign ex-change and strikes rather than physical wreckage). In the south, the essentially agrarian economy was operating at half strength. But once repair crews cleared away the rubble in cities like Milan, Turin and Modena, it was discovered that factory production capacity was actu-ally reduced by no more than 10 percent.

The major threat to Italy was political and economic. The old right wing and the new leftists—led by a strong Communist Party—were vying for power, and in the chaos the financial stability of the nation teetered on the brink. The man most responsible for pulling Italy out of this bleak situation was a severe, humorless, deeply reli-gious man from the north, Alcide De Gaspari. A longtime opponent of Mussolini, he had spent most of the last decade as a political refu-gee inside the Vatican. It was De Gaspari who would marshal the centrist forces under the banner of the Christian Democrats, neutralize the Communists and lead Italy into what is recalled as the "economic miracle."

This "miracle" was in many ways attributable to the Marshall Plan and its flood of Yankee dollars during the late 1940s. Whether or not De Gaspari—or anyone else, for that matter—could have lifted Italy off the mat without American largesse is a matter of conjecture. Regardless, it aided Enzo Ferrari by quickly creating a class of wealthy sportsmen who were eager to spend their newfound lire on the latest, flashiest sports car from Maranello.

Through a combination of good fortune and artful political dodging, Enzo Ferrari appeared to dance through the rubble relatively unscathed. The defeated Germans and their patents were now irrelevant, which meant that his machine tool business would be nicely positioned to take advantage of the industrial reconstruction that was bound to come. The factory had been quickly repaired, and Cortese was ready to solicit business, although electric power was still scarce and chronic labor unrest plagued the entire Po Valley. Still, compared to the complications with Dino and the unexpected presence of baby Piero, the commercial aspects of his life seemed rather positive. His holdings in real estate, coupled with a business that dovetailed neatly with Italy's peacetime needs, presaged a stable and prosperous future for a man rapidly closing in on his fiftieth birthday.

For Enzo Ferrari, 1945 was to mark a beginning, not an end. The idea of backing down, of slipping into the gray obscurity of the Modenese gentry, probably never entered his mind. The arrival of peace brought a welcome respite to his countrymen, but for him it was a call to arms. Motor racing would soon resume and he meant to be part of it—this time not as the stalking horse of Alfa Romeo, or the contract builder for rich sportsmen, but as the manufacturer of his own automobiles. There has been some debate over when and how Ferrari chose to discontinue the machine tool business and begin the production of automobiles, but according to Cortese's and Colombo's recollections, it seems clear that the shift was evolutionary, a slow transition that spanned perhaps five years.

His reason for returning to the automobile business was simple: it was all he really knew. The machine tools had been transitional, merely stopgaps for the duration of the war. But his reputation, his contacts, his passion, centered on cars and there was no logical reason that this would not be his area of concentration once the fighting stopped. The brief fling with Isotta Fraschini had only reinforced his resolve to build automobiles in a factory that had now expanded to 40,000 square feet, or about four times the size of the old Scuderia. He still lacked a foundry, but otherwise his equipment was sufficiently broad-based to begin the arduous and complex task of creating an automobile from the ground up.

But where to begin? He first contacted his old friends. Bazzi would come. Alfa Romeo, crippled and in political disarray, was months away from resuming any projects relating to automobiles, so the veteran magician with engines agreed to rejoin Ferrari and move back to Modena. That would help to compensate for the loss of Massimino and the skilled Vittorio Bellentani, both of whom had been lured to Maserati. Quite obviously the man Ferrari wanted and needed most of all was Vittorio Jano, the one designer in Italy whom he totally trusted, both professionally and personally. But Jano was not about to move. He had labored through the war with Lancia and was hard at work on a new, advanced V6 coupe to be called the Aurelia. While Lancia had no immediate plans to engage in racing, the operation was stable and devoid of the political intrigue that had cursed his last years at Alfa Romeo. Moreover, Lancia was based in his beloved Turin, and at the age of fifty-four, a move to the dreary, dusty outback that was Modena seemed a major gamble considering the tenuousness of Ferrari's plans. If Enzo Ferrari was to find a designer he would have to look elsewhere. The place would be Milan. Gioachino Colombo was looking for work. Ostensibly he remained on staff with Alfa Romeo, but his situation was worsening by the day. The company's trade union had formed a "purity committee" and was investigating the role of various managers in the now-defunct Fascist Party. Colombo had been a member, although the level of his enthusiasm was unclear. He was therefore placed on a leave of absence by the rump court. These were uncertain, impassioned times, and although it turned out that most of the charges were ultimately swept away in a wave of rhetoric, the assassination of Gobbato was surely vividly recalled by Colombo. This made a move to the quieter precincts of Modena seem attractive.

He responded to Ferrari's call in July 1945, less than three months following the establishment of peace. He and Enrico Nardi made the difficult drive over shell-pocked roads and through bombed-out villages in a prewar car fueled by black-market gasoline. The day was steamy and humid. The pair crossed the Po at Piacenza via a barge that had replaced the shattered Via Emilia bridge. Work was proceeding on a temporary pontoon bridge downriver while repairs to the main span were made, but Colombo and Nardi were forced to sail inelegantly across the grand old river. It had been eight years since he had made his first journey to Modena to help create the Tipo 158. But that project would be a lark compared to what lay ahead. In 1938 the ample resources of Alfa Romeo and the government had been at their disposal. Now they would be alone, underfunded and with little prospects for immediate success.

Colombo met Ferrari in the little office of the Scuderia, where so much discussion of the 158 had taken place. This time there was initial small talk of old friends, now scattered to the winds by the fighting, and of the situation at Alfa, where the remnants of the company were turning out such modest items as cheap cooking utensils while plans were slowly forming to manufacture low-priced cars for the masses. Before his suspension, Colombo had been transferred to the design staff for diesel engines, which held no interest for him. He told Ferrari that there were no current plans for Alfa Romeo to resume any sort of motorsports activity, which seemed understandable considering the shape of the European economy and the fact that half the world was still immersed in the Pacific war. In fact, it must have seemed a bit bizarre to him that his host would even be considering automobile racing at such a time.

But that was exactly what Ferrari had in mind. "Colombo, I want to go back to making racing cars," he announced. "What do you say, how would you propose to make a fifteen hundred?" He was referring to a 1,500-cc engine—a power plant of the same size as the 158's and the displacement that he assumed would be the new Grand Prix formula once racing resumed. Colombo understood instantly. After all, a number of prewar cars from the *vetturetta* formula were known to exist: the 6-cylinder British ERAs, the powerful 4-CLs from Maserati and the remaining Alfa Romeo 158s. This 1.5-liter class would be the new battleground, and Enzo Ferrari meant to be a combatant. But what would be the configuration for the new power plant? "In my view you should be making a twelve-cylinder," said Colombo.

"My dear Colombo, you read my mind!" Ferrari responded. Colombo later recalled that the meeting was brief and to the point. Long-range planning, time schedules, finances and the like were discussed in general terms over a meal in a nearby trattoria, after which Colombo and Nardi drove back to Milan. This version, recounted by Colombo in his recollections published in 1985, once again places in doubt the oft-told story that the V12 engine rose full-blown out of Enzo Ferrari's imagination. As has been said, the Germans had already proved from 1938 onward that such a layout offered great performance potential. And of course the hated Ricart and his associate Bruno Trevisan had also produced a number of powerful V12 variations before and during the war. Therefore the design was hardly revolutionary and, to his credit, Ferrari never made such a claim. But his later suggestion that his inspiration for the V12 came from the Packard staff cars he had seen in World War I borders on the nonsensical. Those power plants were immense, slow-revving, flathead crudities compared with the exotic versions of Mercedes-Benz, Auto Union and Alfa Romeo, but it is likely that Ferrari's pride prevented him from crediting those contemporary rivals with the inspiration.

To make such a small-displacement 12-cylinder engine would be difficult, in terms of both design and manufacture, but if it worked the dividends would be tremendous. At that point no one in the world was making such an engine, and if it could produce sufficient power its uniqueness alone would bring notoriety to the new Ferrari car. That such an audacious new product might one day exact revenge against the Milanese giant that had dishonored Ferrari was probably reason enough for him to launch Colombo on the project.

Colombo was operating on shaky ground. He remained theoretically in Alfa's employ, although on de facto suspension. But with bills to pay he didn't hesitate to moonlight on the Ferrari assignment. It was mid-August, during the summer holiday of *ferragosto* (a national siesta held sacrosanct by Italians even while their country was in shambles—in retrospect, no less rational than dreaming up racing cars in the midst of the chaos), when Colombo traveled to his sister's house at Castellanza, near Varese, for a family gathering. There, in the shadow of the Alps, he drifted into the courtyard after lunch and began doodling on a piece of wrapping paper. The engine being the most critical component, he roughed out the cylinder head—that supremely sensitive area where the valves, lifters and spark plugs create the theme of an engine, whether it be a high-revving racing powerhouse or an efficient passenger-car motor or a low-output but reliable power plant suitable for trucks and tractors.

Colombo's V12 had a single overhead camshaft operating the valves on each bank. The design was similar to that of several larger units he had worked on at Alfa. He decided on a daring new plan to duplicate a practice being used on competition motorcycles—that of making the stroke of the pistons slightly shorter than the diameter of the cylinders. This "oversquare" design would become standard in years to come, but in the precarious summer of 1945 it offered a key to Ferrari's future. Back in Milan, Colombo converted the small bedroom of his apartment into a studio. After borrowing a drafting board from Carrozzeria Touring Superleggera, the coach builders, he set to work in earnest. It was not easy, laboring amidst the noisy clutter of the family, among the bedding and the laundry.

By mutual agreement with Ferrari the new car would be called the 125. The reasoning seemed simple enough. Each cylinder displaced 125 cubic centimeters, which multiplied by 12 equaled 1,500 cubic centimeters (or 1.5 liters, or 91 cubic inches). This practice of naming his cars on the basis of a single cylinder's displacement was to become a relatively standard practice for Ferrari, although in later years he would capriciously vary the nomenclature, based sometimes on the total engine displacement plus the number of cylinders. (Even then some of the designations defied logic and are of interest only to serious students of Ferrari arcana.)

But was there another reason for the designation? The numbers were a rearrangement of 512, the ill-fated mid-engine design of Ricart's. Colombo's engine was surely derivative of that bold effort, and if there is any validity to the theory that the prewar 815 was a numerological play on the Alfa Romeo 158, surely it is possible that the 125 arose out of Ferrari's same subtle urge to turn the tables on the opposition.

Ferrari began to apply the pressure. He needed a design immediately in order to return to the real wars on the racetrack. He had burned his bridges with Alfa Romeo and was committed to building his own marque. He had, with his consummate skills as a promoter, developed sponsorship support from Marelli, the ignition manufacturers, Pirelli, the tire people, and Weber, the makers of carburetors—although they were now half-owned by Fiat (a complete takeover would ensue). Time was wasting, and Ferrari, using his finely honed skills to alternately cajole and encourage, had Colombo whipped into a frenzy of activity. In September, Colombo called in Angelo Nasi, the old 158 design-team compatriot, who was now unemployed. Colombo set "Lino" to work designing a five-speed gearbox for the car, which was a welcome relief from the mundane job of working on industrial diesel engines while he was still at the Portello works.

Lest one forget, motor racing was a monstrously popular sport in Europe during the first half of the century. Even to this day, it ranks with soccer and bicycle racing as an obsession of the masses. As the war ended, a car-crazy public could not wait to rush to the streets for the first race. Less than four months after the shooting had died down, the engines revved and hooted on the Bois de Boulogne in Paris on September 9, 1945. There French ace Jean-Pierre Wimille drove his prewar Bugatti single-seater to victory (one of the last wins the famed marque would enjoy) in a race nicknamed the "Prisoners Cup" (formal title: Grand Prix de la Liberation). Competition would start on a large scale in the spring of 1946, although the cars would initially remain a patchwork of prewar machines that had been hidden and preserved during the fighting.

The national automobile clubs of the various nations (excluding Germany, which had been banned) had gathered in Paris and organized under the banner of the Fédération Internationale de l'Automobile (FIA) and agreed that Grand Prix racing would be run under a formula that permitted engine sizes of 4.5 liters (271 cubic inches) unsupercharged. Ferrari had been sure of this development, thanks to his broad political connections within the sport. He considered himself on the right track with his new 1.5-liter V12, although to be competitive it would have to be fitted with a supercharger and radically increased in terms of power output. This he and Colombo were planning on as a second phase of development.

Slowly the car took shape in the early autumn. With the roads between Milan and Maranello improving, communication between Colombo's bedroom studio and the factory became easier. Ferrari continued to hedge his bets by manufacturing machine tools, but the operation was steadily being converted to building cars. A young man named Luciano Fochi had been hired, almost as an afterthought, and the twenty-one-year-old was to become a key element in the creation of the early automobiles. He had been taken on by Alfa Romeo as an apprentice draftsman at sixteen and had worked there until entering the Army. Now out of the service and adrift, the slight, sad-eyed man had returned to Modena to live with his parents. He had gone to Maranello seeking work but had been turned away because of his youth. But by coincidence he had arrived on a day when Colombo was visiting. The old Alfa hand had recalled Fochi's skill at the drafting table. He was hired and assigned the job of transforming Colombo's often rough, sometimes unworkable sketches into formal blueprints and schematics from which Bazzi and his artisans would build the actual automobiles.

By October 1945 the Colombo version of the 125 was on paper. Ferrari liked what he saw, although the nuances of the layouts escaped him. The engine, with its multiple cylinders and small dis-placement, was interesting, but hardly original. A number of makers, dating back to the 1920s, had long since exhibited similar downsized sophistication. It promised good power, especially when the super-charged version was completed in the second phase of the program. The suspension, independent at the front and with a solid axle at the rear, bordered on the antique. Single cantilever leaf springs were em-ployed, rather than the lighter, more compact torsion bars and coil springs that had been used so effectively by the Mercedes-Benz and Auto Union engineers.

The frame was light, stubby and quite rigid, based on design themes developed at Alfa Romeo by Jano, Colombo and Ricart. Co-lombo had laid down a setup similar to that employed on the 158 and had contracted with the specialty-steel-fabricating house of GILCO Autotelai in Milan to make the oval-shaped frame tubes that formed the structure. Maserati was also using GILCO tubing for its ever-expanding arsenal of single-seaters and sports cars. At this point the crosstown rivals were far ahead of Ferrari in all phases of the business. One of their cars had run in a small sports-car race at Nice in April 1946, further adding to their lead over Ferrari. It was already known that a new 2-liter straight-6 design by Massimino was almost com-pleted.

In final form the 125 weighed about 1,430 pounds, as opposed to the 1,580 pounds of the similar-sized Maserati A6s that were being built almost simultaneously. This was good news to Ferrari, having hectored Colombo to keep chassis weight at a minimum. Still, he was obsessed with horsepower, reasoning that on any given racetrack a greater percentage of the surface was devoted to straights than to cor-ners. Therefore outright acceleration and top speed were much more important to him than the nuances of road holding and braking. This was a hangover from his early days, when suspensions were in fact little more than those found on horse carts and raw power was the critical factor. Over the years he would proudly repeat, "I build en-gines and attach wheels to them," which puts his priorities in per-spective and helps to explain why, until his death, Ferraris often enjoyed a reputation for superior power, but were never in the van-guard of sophisticated chassis design.

In fact, no Ferrari ever built was a glittering example of daring technology. The 125, with its short-stroke, small-displacement V12, was one of very few examples in the history of the company that could be described as remotely revolutionary. A myth has grown up around the cars relating to their advanced designs, but actually Enzo Ferrari was extremely conservative and was often left at the starting gate by more creative builders (his reluctance in the future to adopt such obviously superior components as mid-engine layouts, coil spring sus-pensions, disc brakes, monocoque chassis, magnesium wheels and fuel injection exemplifies his crude approach to design). Mike Law-rence, the British racing historian, put it this way: "Apologists have suggested that Ferrari was resistant to any idea which his firm did not originate, and the late use of disc brakes is cited as an instance. If Fer-rari was so insistent on original ideas, the marque would not have got off the ground, for it has *never, ever* introduced a new idea to motor racing." Although the Commendatore would surely have been infuri-ated by such an accusation (regardless of its obvious truth), he admit-ted that the 125 was purely derivative. "The first Ferrari . . . came into being as an orthodox car; it was not the result of any experimentation. All we wanted to do was build a conventional engine, but one that was outstanding."

As the days began to grow shorter in the autumn of 1945, Co-lombo and Nasi seemed to have the engineering details for the most part solved, and it was expected that Bazzi and his craftsmen could actually begin construction before the winter set in. Life on the home front was getting more complicated, and Ferrari was sharing his con-cerns over Dino's declining health with his closer associates. It is also probable that some of them, possibly including Colombo, knew of Lina and little Piero, although when and how Laura made the discov-ery is not known. Whatever the case, Ferrari's life was consumed not with domestic problems but with his elemental, ego-bound project of creating the first automobile to bear his own name.

In November, Ferrari received a frantic call from Colombo. He was being summoned back to Alfa Romeo. The directors and the workers' committee, who were operating as a kind of revolutionary guard, had gotten wind of Colombo's back-door work for Ferrari. The bad blood that had led to his suspension had hardly been forgotten. At the same time the administration had been considering a return to motor racing, feeling that it was an effective way to publicize their passenger-car line and to showcase Italy's technical expertise to a doubting world. The 158s, although dismantled and dusty, were still potent racing cars and their revival seemed an obvious first step. And who better to resuscitate them than their creator? The five remaining cars had been hidden during the war, in pieces, inside a dairy-products factory near Lake Orta. And with the Auto Unions isolated behind Russian lines and the state-of-the-art Mercedes-Benzes moldering at their war-shattered company in Stuttgart-Untertürkheim, the 158s were by default the fastest racing cars in the world. All it would take was the ministrations of Colombo to return them to the tracks.

A series of typically noisy and contentious negotiations were held during which the Alfa Romeo trade unions threatened to strike if Colombo did not return to the Milanese fold. He had little choice, considering that his contract was still in effect, and recalling that not too many months earlier the same workers had considered backing him up against the nearest stone wall with other accused Fascists and Nazi collaborators. Gioachino Colombo gave one final, futile shrug, waved goodbye to his old friends in Modena and headed northward once again on the Via Emilia.

The loss of Colombo was a double blow. Not only was he needed to carry the 125 project to fruition, but his departure meant that he would be on the hated opposing side, developing the 158s into Ferrari's strongest opposition. And to add to the irony, the Alfas had been constructed in his own Modena shop! He would be beaten by his own cars! Colombo was an engineer of the old school that had produced Ferrari and Bazzi. He was used to long hours, improvising and changing components as needed. He cared little for theory and mathematical formulas favored by his younger, university-trained counterparts and was tuned to the traditional ways of the Ferrari artisans. His departure left Bazzi, Attilio Galetto, the machine shop foreman, and Federico Giberti, who was now the company's purchasing agent, to translate his incomplete legacy of blueprints into an actual prototype engine and automobile. Without the engineer at their side to direct the modifications as they went along, the project was stalled.

Then a familiar face appeared on the day before Christmas. It was Luigi Chinetti. He had driven down from Paris with his new wife and baby son in an ancient front-drive Citroën with tires so bald that a companion had to ride on the hood to help provide traction as they assaulted the snow-clogged Alpine passes. Chinetti found Modena battered and monochromatic, half-lit, its senses still dulled by the bombing and the fighting between the Fascists and the *Partigiani*. Modena was packed with old friends, racing companions from the 1930s, drivers, mechanics, metal shapers, machinists, pattern makers, team owners, small manufacturers, car traders, hustlers, deal makers, hangers-on, pretenders, dreamers: all trying to sort through the rubble and begin again. The place reeked of high-performance automobiles. Small shops specializing in fabricating transmissions, engine blocks, custom car bodies, steering wheels and accessories of all kinds were scattered around the sprawling, morose, table-flat city dominated by the Romanesque campanile of La Ghirlandina and the hulking Cathedral of St. Geminiano. But Modena's reputation as a place where fast cars screeched up and down its broad avenues was attributable to one man and to the Scuderia he had created there fifteen years earlier. And it was this man that Luigi Chinetti first sought out upon arriving on that snowy day in December.

He found Enzo Ferrari without difficulty, holding forth, as expected, at the same dingy, umber-hued, two-story building on the Viale Trento e Trieste. Ferrari was in a bad way. He looked haggard beyond his forty-eight years. His shock of hair, combed back in pompadour style, was peppered with gray. His eyes, always puffy and languid, were red-rimmed with fatigue. His hulking frame was slumped behind a desk in the large, unheated office. In the semidarkness Chinetti could see the row of trophies and the photographs of drivers who had won at the wheel of Scuderia cars, all of which had been unmoved and undusted since before the war. The place seemed like a musty tomb reopened to a crack of light after centuries of isolation.

Out back, the shop, which Chinetti had remembered as being filled with cheerful, chattering mechanics and glistening Alfa Romeo racing cars, was chill and empty—a dowdy museum of outdated frivolity. In those postwar times, in a broken, mean-spirited place like Modena, a racing car of the Scuderia Ferrari had seemed as absurd and obsolete as the comic pomposity of the defunct Mussolini regime.

The greeting between them was joyless. Perfunctory. Chinetti called him simply Ferrari, in the lower-class fashion. Others had referred to him as Cavaliere in the early days, then as Commendatore after the honor was bestowed upon him by Mussolini's puppet king, but Chinetti refused to use such a Fascist-sponsored appellation. To him, this once-powerful figure in Italian motorsports had been Ferrari, nothing else, and it would remain so until their final parting.

On that day, in the dingy icebox that was Modena, they spoke of the future. Each tried to conjure up some optimism, but it came in brief spurts, only to fade into a morass of broken dreams. Ferrari talked about the war, about how he had bought land to build a small factory in the nearby village of Maranello. Machine tools had been fabricated for the Italian and German armies until the Americans had bombed it twice. He seemed to revel in the tragedy of the devastation, of the lost business, but Chinetti suspected otherwise. He knew him as the consummate actor, as a man who dwelled in the pathos of the moment, draining from each word, each grandly constructed phrase, the last drop of drama. He knew him as a man who never revealed his hand, who never trumpeted the full measure of his own success. He had heard from others that Enzo Ferrari had done well in the war and that the factory in Maranello was not seriously damaged and that a few automobiles were already being made there. Hence his visit.

Yes, said Ferrari, he had commissioned Colombo, his old friend and one of Alfa's best engineers, to lay down the plans for a small V12 racing car. But Colombo was in trouble with the Communists in

Milan, who harassed him about his Fascist sympathies. Yes, Bazzi, faithful old Bazzi, and a few others, for the most part youngsters, were beginning to dyno-test two of the prototype engines and would soon start fabricating a car, and then the Scuderia would try to return to the racing wars, such as they were in the rubble-strewn streets of Europe. Yes, the Scuderia would race again, but part of the future lay in machine tools. Ferrari seemed to pledge that never again would he commit everything to the tenuous world of motorsports, where the clash of brittle egos could wreck a business as quickly as a car spinning off an Apennine hairpin. After all, was not Alfa Romeo a shell of its former self? Had not the Allied bombers transformed the once-great maker of bullet-fast cars into a maker of kitchen utensils, windows and doorframes? No, said Ferrari, he would not take the risk again. There would always be a demand for the German-designed Jung grinders that he and his toolmakers had copied so masterfully during the war. Now that the Nazis were defeated and their nation in ruins, there would be no more danger of patent infringements. Surely a rebuilding Italy would need such machinery.

Chinetti spoke of his years in America, where a war effort had been mounted that by comparison made Italian industrialists look like Etruscan potters. In the rush to spit out millions of airplanes, tanks, trucks, rifles, artillery pieces, landing craft, prefabricated buildings, portable bridges, helmets, canteens, crates, uniform buttons, pistols, binoculars, radios, radars and God knows what else, the American colossus had created miraculous tools of mass production assembly-line techniques so ingenious that full-size cargo ships could be built by a man named Kaiser in less than a week! How, asked Chinetti, could Ferrari hope to compete against this Western industrial might?

Chinetti described the miracle that was America. The aggressiveness of its people, their childlike acquisitiveness and their bourgeois crudities. Yet there was a rich upper class whose tastes were European. He had seen them, met them, gained their trust as he kept their aging automobiles from Crewe and Stuttgart and Milan operating through the war years. They would pay a king's ransom for such elite machines and he, Luigi Chinetti, meant to exploit that naïve colonial lust.

America, the man-child of a nation, could be either a blessing or a curse, said Chinetti. If Ferrari chose to remain in the machine tool business, he would be crushed by the tidal wave of industrial power from the West. But by exploiting the new wealth that same industry was bound to produce, both could prosper. Chinetti was firm: build five cars a year for America and he would sell them for unseemly amounts of money; give him the opportunity and he would sell twenty a year—a number that on that dank day in Modena sounded as astronomical as if they had been discussing General Motors' annual output.

Ferrari agreed. Yes, if the little V12 of Colombo's worked and if the 125 was a decent automobile . . . yes, he might build a few cars for the provincials.

This encounter was to produce much confusion in later years. Chinetti was to claim that it was he who convinced Ferrari to build automobiles after the war during that Christmas 1946 meeting. But Ferrari had the Tipo 125 project underway and there is no reason to believe that he was not considering production and sale to private customers. After all, his role models were Alfa Romeo and Bugatti, both of whom had successfully built racing cars and limited-production road cars. Logic demands that he would follow their example. If Chinetti deserves credit, it is for identifying the lucrative American market, which for the isolated, increasingly provincial Ferrari might as well have been on the dark side of the moon.

Regardless of Chinetti's optimistic predictions about sales potential, the immediate problems were critical. With Colombo gone the development of the new car was at a standstill. Worse yet, the Maserati interlopers were moving ahead with a new Massimino-designed A6 sports car that would rival the 125 in size and performance. Moreover, the Orsis were still gloating over their cars' performance in the previous May's Indianapolis 500. Dirt-track star Ted Horn had come in third in the venerable 8CTF Maserati formerly driven by Wilbur Shaw, while Gigi Villoresi finished seventh after having lost many laps while changing no fewer than three flawed magnetos. These honors only increased Ferrari's frustrations, which on occasion rose to open fury when the Maserati brothers chose to use the Abetone road for testing and the impudent yowl of their engines rattled the windows of Ferrari's factory office.

Production of high-performance cars was beginning all over the world, even in Spain, where Ricart had fled at the end of the war. There was talk that he was planning a sports car to be financed by the Spanish government (this would happen, in 1951, with the introduction of the advanced but poor-selling Pegaso). Ferrari was apoplectic— he needed a replacement for Colombo. Finally his old associate managed a solution. Colombo recommended an out-of-work Alfa Romeo engineer named Giuseppe Busso, although he was unable to join Ferrari until June. By then the 125 project was hopelessly behind schedule and Ferrari practically chained the inexperienced Busso to his drafting table. The swarthy Turin native, whose laconic behavior masked a surprisingly stubborn streak, manfully waded into the strange waters.

He moved from Turin with his family and settled into a small apartment in Modena. At first the going was easier than he expected, with Ferrari and Laura hosting him at long, Lambrusco-lubricated dinners at their quarters over the Scuderia. In an interview given in 1986, Busso recalled those days with a certain fondness. He noted, however, that Ferrari was openly concerned about the health of Dino—and of Laura—and "the anguish of seeing him cry many a time when talking about Dino may have contributed to my affection for Ferrari." Busso also noted that others found his sympathy for Ferrari "unusual among those who have had dealings with the Commendatore." Busso's mention of Dino's illness, which obviously preoccupied Ferrari as early as the spring of 1946, is among the first known acknowledgments by any of Ferrari's associates that the boy was already seriously ill.

Busso's training had been in aircraft power plants, and his knowledge of cars was purely theoretical. His only advantage lay in the area of supercharging, where he had gained experience with Alfa Romeo during the war. But he was bright and resourceful and set to work immediately to carry on Colombo's creation while also setting out to develop a supercharged version of the 1.5-liter engine for Grand Prix competition. Head down, slide rule in hand, Busso plunged ahead. The pressure on the young engineer was enormous. Colombo's designs were loose and incomplete and his plans for a supercharged *gran premio* single-seater made almost no sense at all. But Busso labored on, badgered on a daily basis by his boss and haunted by the progress of Maserati. Their factory was but a few blocks from the old Scuderia and there was much intermingling between the mechanics and the workmen. There were no secrets. (A 1940 letter from Adolfo Orsi to Nuvolari notes that Modena is "a small town where rumors could spread too easily.") No doubt the agonizingly slow pace of the 125 project was a source of considerable amusement at Maserati and in Milan, where the Alfa management had returned to competition in 1946 with a four-car team headed by Jean-Pierre Wimille, Didi Trossi, Consalvo Sanesi and the drug-free Achille Varzi. The updated 158s, featuring two-stage-supercharged 260-hp engines (improved by Colombo), were winning every race they entered by wide margins. In the faraway hinterlands of Maranello, there seemed no possibility of beating them, much less even fielding a car, in the foreseeable future.

Considering his lack of hard-core experience with automobiles, Busso was surprisingly effective in translating Colombo's hazy concepts into reality. His principal order of business was to get the engine into shape. The first example was mounted on Bazzi's engine dynamometer on September 26, 1946, and fired up with modest results. Unsupercharged and undeveloped, the little V12 produced about 60 hp at a modest 5,600 rpm, mainly because of a crude ignition system and bearing layout, both of which dated back to the late 1930s. Busso and Bazzi worked hard to improve the situation, but Ferrari was steaming with impatience. More than a year had passed since the project had begun and now, with the machine tool business slowly being phased out, he had no choice but to convert to motor racing as his sole means of support. Desperate, he sought more help.

Thirty-year-old Aurelio Lampredi arrived in early October and took over as deputy manager of the tiny technical office under Busso. He too was an aircraft engine expert, having worked in the advanced design department of Reggiane-Caproni of Reggio-Emilia during the war. He was shocked to find Busso laboring alone in the company of what he called "little boys"—four draftsmen led by Fochi who were in their late teens and early twenties. Lampredi was a tall, confident engineer who immediately decided that Busso lacked the authority and practical knowledge needed to butt heads with the willful Ferrari and to gain the confidence of race-hardened shop hands like Bazzi and Galetto. Lampredi was more attuned to the Ferrari way of doing business. Although four years younger than Busso, he seemed surer of himself and quickly established a better rapport with Bazzi and the men in the shop. He had a solid academic background, with a degree in mechanical engineering from the High Technical Institute at Freiburg. Thus he had, if anything, better credentials than Busso, as well as broader experience in both the aircraft and steel industries.

The pair clashed immediately over two knotty problems left behind by Colombo. He had created a unique type of floating connecting-rod bearings that were unworkable. Busso and Lampredi labored side by side seeking a solution, growing more contentious by the hour in the pressure-packed environment. Worse yet, Ferrari was now demanding action on the supercharged Grand Prix engine, and Busso had done a total redesign on Colombo's confused, incomplete plans for the twin-overhead-camshaft cylinder-head layout. Lampredi agreed that the Colombo creation was beyond hope, but said the same about Busso's solution. Built his way, the heads would be exceedingly complex and difficult to fabricate and maintain. There was a practical side to Lampredi, perhaps developed while working on military aircraft engines. He placed a strong emphasis on simplicity and reliability in his designs—qualities that were to become Ferrari hallmarks.

By the end of 1946 the communication between Busso and Lampredi had been reduced to a series of muted grunts. They had somehow managed to get more power and reliability from the prototype engine, while Bazzi and company were beginning to weld together the lightweight GILCO thin-walled frame tubing for the first three 125s. Scuderia Ferrari was months, perhaps years, away from reaching the racetracks and the autostradas of Italy, but this in no way stopped Enzo Ferrari from taking a brash gamble. In December he called a press conference at Maranello and grandly announced his new line of Ferrari cars—the 125 sport, a 125 *competizione* and a 125 *gran premio*. These annual press gatherings were to become a Ferrari tradition that would last another forty years, but on this occasion, during the bleak final days of 1946, the affair was a desperate bluff. In the back shop, behind the façade and the flurry of brochures, Ferrari knew he had an engineering group that still faced substantial obstacles in transforming Colombo's designs into workable automobiles. A month before, Ferrari had told Cortese to stop taking orders for machine tools. That financial security blanket had been tossed away. Bazzi's group was building only 125 sports, with no plans in hand to start a single-seat *gran premio*. The announced *competizione* grand touring coupe was virtually unthought of.

Surely the conversation with Luigi Chinetti during his visit to Modena remained in Ferrari's thoughts. If Chinetti was right, the North American market was a potential cornucopia. Providing all went according to plan, a limited-production street version of the racing cars would be undertaken. But Ferrari knew the game that had been played so well by Bugatti and Alfa Romeo and even now was being played in Modena by the Maserati-Orsi combine. The rules were simple: build successful racing cars that improve the image of the company and rich men will flock to the door to obtain similar cars with which to play out their fantasies. (In later years Ferrari would divide his customers into three groups: the "gentleman sportsman" who is "convinced he knows how to handle a car almost like a racing driver"; the "fifty-year-olds" who are "making a dream come true" and trying to "snatch back a little of their youth"; and "the exhibitionist" who "buys a Ferrari merely because it is, as it were, the chinchilla among automobiles.")

By announcing the new line of cars he could expect a flurry of orders from wealthy sportsmen and perhaps an infusion of deposit funds to carry on the work. The publicity might also bring in more sponsorship money from accessory manufacturers. In all, it was a virtuoso use of publicity—and a practice he would develop to a high art later in his career.

The Italian automotive media received the news of Ferrari's entry into manufacturing as a major story. For the moment the situation improved and the little staff enjoyed a brief regeneration of spirit. But reality quickly intervened. A long, gray winter lay ahead with massive loads of work to be completed before even one prototype automobile could be rolled out of the shop. Ferrari quickly returned to his badgering managerial style, and after three months Lampredi, fed up with the politics, the acrimony and the slow progress, resigned. Ferrari, at his theatrical best, implored him to stay on. The 125 must be run in the spring and raced in the summer! He tried to convince Lampredi that the labor expended would pay great dividends down the road. The argument was long and explosive, but in the end Lampredi relented. Work continued through the winter of 1947, which old-timers around Modena recalled as one of the worst in memory. Ferrari himself said it reminded him of the stories his family had told about the awesome storms that swept the Po Valley during the winter he was born, now nearly fifty years past.

Finally, nearly two years from the time that he and Colombo had first discussed creating the 125, Enzo Ferrari was able to climb aboard a bare-boned prototype and take an initial test drive. It came on March 12, a late-winter day in 1947 that he had surely often suspected would never arrive. The machine, little more than a collection of frame tubes and four steel disc wheels supporting the sculptured aluminum-alloy engine, had been rolled out of the factory by the small retinue of workers, which included Bazzi, Giuseppe Peiretti, the headmaster of the coachbuilding department (who was unhappy with the situation and about to quit), Galetto and of course Lampredi and Busso, who were barely speaking. While the moment was charged with anticipation, a leaden tension gripped the little knot of men as they stepped into the flinty sunlight that bathed the cobbles of the courtyard.

Bundled against the wind, Ferrari planted his considerable bulk on the bench seat of the spindly automobile. He brushed back a few locks of his graying hair and fiddled briefly with the controls, fingering the collection of switches on the instrument panel and punching the brake and throttle pedals before moving the gear lever through all five forward positions. Then he reached for the ignition key.

Luigi Bazzi stepped forward. Loyal Bazzi, now fifty-five and six years older than his longtime colleague, fixed his hard gaze on the engine that had been the source of so much hard work and acrimony for nearly two years. He nodded to Ferrari, and the engine crackled, its twelve tiny pistons stepping into a rhythm of combustion like perfect mechanical soldiers. Bazzi smiled, perhaps recalling a time over twenty years before when he had climbed into the seat of Antonio Ascari's P2 Alfa Romeo and had demanded quietly, "Let me hear the heartbeat of the creature." The workers stepped back, gasping at the impudent whine and rumble of their joint creation. Those ahead of the bare snout of the radiator cleared a path across the cobbles.

Ferrari slipped into first gear and eased out the clutch. Its single plate grabbed and the skinny Pirellis spun against the chilled surface, seeking traction. Accustomed to such cranky behavior from powerful, high-performance automobiles, he quickly corrected and guided the machine through the factory gates and out onto the Abetone road.

Ferrari headed north, away from Maranello and onto the flat, poplar-lined stretch toward Formigine. The choice of routes is revealing. Had he turned south, the road beyond Maranello would have led into the foothills of the Apennines, where the going was curved and hilly. Yet Ferrari chose the long straight, where only horsepower counted. The handling of the car was secondary, hence the rather primitive chassis and suspension systems that were to become associated with his products for decades to come.

Free of the factory gates, he accelerated through the gears, the wind gnawing at his face and drawing tears from his ungoggled eyes.

Behind him the whine of the twin exhausts echoed across the still-brown pastures and orchards that lined the narrow highway. The V12, fresh off the test bed, revved with remarkable ease and smoothness, more like an electric motor than a fiendishly complex small-displacement engine meant for motor racing and a new line of fast grand touring road cars. At 1.5 liters, the power plant was only fractionally larger than those propelling the mundane Fiats that had infested the streets of Modena before the war. The power output was still no more than 65 hp, a low total for an alleged high-performance automobile, but understandable for an engine at the initial stage of evolution. Hopes for prodigious 200-plus numbers lay with Busso's plans to add a supercharger for Grand Prix competition.

Ferrari braked hard, and the 125 skittered to a stop on the outskirts of Formigine. He turned into a farmer's lane and stopped. He scanned the broad flatlands, still dappled with pockets of snow. As he stood on the threshold of the fiftieth year of his life, he was about to create an automobile bearing his own name—built in this harsh Emilian backwater that increasingly shackled him to its brooding, hostile landscape.

Ferrari refired the engine and turned back toward Maranello. He shifted up through the five gears, watching the tachometer edge toward its redline of 6,800 rpm. He felt a spatter of oil against his cheek and lifted the throttle, coasting the rest of the way to the factory. He rolled to a stop, his large nose aglow from the lashing of the wind. Bazzi came up, looking dour as usual. He had already spotted the tiny streak of oil on the cowling and began probing the engine for its source. He quickly found a loose bolt on the right-side cam cover. He nodded to Ferrari. It was nothing. Bazzi was like that. Always there. Always able to make it right. He was an alchemist with engines and a man possessed of a steadiness and quiet resolve that transcended the constant explosions of temper and ego that buffeted the factory. Bazzi was the unbreakable weld that held the operation together and there was perhaps no man Ferrari had ever known—or would ever know—whom he trusted more. He was already one of a tiny coterie of close associates who knew of Lina and Piero and who often accompanied

Ferrari on his nocturnal visits to Castelvetro. If there was an exception to Luigi Chinetti's trenchant observation that Enzo Ferrari "didn't like anyone," it was Luigi Bazzi.

Ferrari lifted himself from the seat of the 125 and circled the automobile in silence, like a skeptical customer in a showroom. Lampredi came up as Bazzi slipped behind the wheel. The little crowd stepped back as the engine was refired, filling the courtyard with its sassy bark. Bazzi tossed away his ever-present beret and squirted out the gate. He was not a good driver, although like all Italians he would sooner have denied his faith in the Holy Mother than admit it. Ferrari listened as his old friend wound out the little V12 and a broad smile crossed his face. Yes, he too had felt the heartbeat of the creature and it made him ache for the grand confrontation with Alfa Romeo that was now sure to come.

It would begin at Piacenza. Two months had passed since the skeletal 125 had made its first run down the Abetone road, and now it was time to test its mettle in competition. Much had changed. Bazzi and his crew, working with Peiretti and his small band of metal shapers, had built two complete 125s. This had been accomplished in company with a desperate Busso, who stood alone in the design department. Lampredi was gone. He had endured the hectoring of Ferrari and the daily arguments with Busso until the end of March, then had resigned in a noisy confrontation with the boss and had gone off to work for Isotta Fraschini.

The loss of Lampredi had been a setback for Ferrari, but he refused to let it impede his plan to debut the 125s at Piacenza on May 11, 1947. The entire factory labored long hours to complete the two cars. One was a full-bodied, "thick wing" version with bulbous fenders and a rounded, rakish tail. The other was called the "cigar" by the mechanics, owing to its spare tubular shape. Small motorcycle-type fenders covered the wheels. Each automobile was devoid of the vivid identification marks that would brand most future Ferrari sports cars—the Prancing Horse escutcheon, the "egg crate" grillwork and the distinctive "Ferrari" logo on the engine valve covers. The 125s' snouts carried notably ugly square grilles with simple horizontal bars.

In the rush to get these machines to the racetrack Enzo Ferrari had forgone aesthetic niceties. He was, after all, a year behind, and a return to competition, no matter how modest, was imperative.

And Piacenza would indeed be a modest debut. This was no Grand Prix race. This would be no mortal confrontation with the hated Alfa Romeo 158s. Piacenza would be a simple contest through the streets of the *città centro* organized by the local auto club. There would be ample starting money, because Ferrari remained a master of negotiation. Surely the presence of the famed Scuderia would bring out additional spectators to gawk from behind the flimsy hay-bale barriers that lined the course. Therefore the new Ferraris were lavishly covered in the local sporting press. The design staff at Portello, with Colombo back in their fold, were watching, albeit from a distance, with considerable interest. They knew full well that this foray at Piacenza was merely a probe by Ferrari, with Colombo's plans for a supercharged Grand Prix version still in the Maranello grand strategy. What they did not know was that Busso and Lampredi had found them to be essentially unworkable and that as last-hour preparations for Piacenza were being completed, Busso had already begun a major push to modify Colombo's rough concepts into a workable engine.

What was known by Colombo and his associates at Alfa Romeo was that even Bazzi, with his tuning genius, had managed to tweak no more than 115 hp from the little V12. Busso now faced enormous challenges if he was to transform the 125s into anything except interesting minor-league sports cars suitable for competition in the streets of small Italian cities.

The *ne plus ultra* of motorsports, the international Grand Prix schedule for Formula One, was starting to take shape. The first race on the five-event schedule would be the Marne Grand Prix at Rheims in June, followed by the Belgian Grand Prix at Spa, the Italian Grand Prix at Turin, the Swiss Grand Prix at Berne and the French Grand Prix at Lyon. Alfa Romeo was the overwhelming favorite, possessing not only the best cars but the best team of drivers. Nino Farina, now only thirty-eight and at his peak, was number one, backed by the magnificent Achille Varzi, who at forty-two had kicked his morphine habit and seemed ready to race seriously once again. Didi Trossi, formerly of the Scuderia Ferrari, was on board as well, and when the planets were in the proper orbits he could be expected to drive brilliantly. The only non-Italian was the courtly former Bugatti driver Jean-Pierre Wimille, rounding out an Alfa Corse possessed of great skill and maturity.

But the first Grand Prix at Rheims was nearly a month away and this permitted Ferrari to score a coup. He summoned old friend Farina to the colors and the star responded. He would handle a 125S (as the car was now designated) at Piacenza. Ferrari wanted the driver he admired above all others to be in the seat of the other 125S (for "sport"), but Tazio Nuvolari could not answer the call. He was contracted to the upstart Cisitalia team formed by Turin industrialist Piero Dusio for the Mille Miglia in June, and his failing health, ravaged as he was by emphysema, precluded a drive for his old boss. His lungs were giving up after decades of ingesting exhaust fumes, and after but a few laps the bloody coughing would begin. He was trying a new type of face mask to filter out the filthy air, but the benefits were limited. At fifty-five, the fabled Nuvolari seemed at the end of the line. But the fires were hardly banked and he ached for the day when his body would permit him more lashing, sideways drives at the wheel of any car—be it a Maserati, Cisitalia or the new Ferrari. If his health improved, Nuvolari assured Ferrari, he was ready to drive later in the summer.

To fill the second seat of the 125S, Ferrari turned to Franco Cortese, now winding down his career as the representative of Ferrari's expiring machine tool business. Cortese had already resumed racing and in March had been a member of the so-called Dusio Circus— a collection of five expert drivers including Taruffi, Chiron and the brilliant twenty-eight-year-old Alberto Ascari. The cars were Fiat-based, 1,100-cc single-seaters. He had won at Cairo and that victory, coupled with some fine drives before the war for Maserati and the now defunct Scuderia Ambrosiana, had gained him Ferrari's respect. He was a smooth and mannerly driver, but hardly the prototypical *pilota* for the Scuderia. Ferrari openly admired the harum-scarum style of Nuvolari and his banzai attacks on the racecourse. Ferrari loved Guy Moll's wild tactics, for as long as they lasted, and he still looked upon Farina as one who might assume the mantle of Nuvolari—although as he approached forty Nino's potential had no doubt been reached. Cortese was an above-average gentleman driver and a good man to shake the bugs out of a new car like the 125S, but he could never be expected to dominate a race like the Nivola of yore or rising stars like Ascari. Still, Cortese was available and was selected to run at Piacenza as second banana to Farina.

The weekend at Piacenza, as would be typical for the Scuderia, was laced with high emotion. The *Auto Italiana* motoring correspondent circulated through the pits with a rumor that a contingent of Alfa Romeo racing executives—including Colombo—had come down from Portello to see the new Ferrari. That was not the case (Colombo decided it would be notably unwise, considering his delicate position with the workers, to go anywhere near a Ferrari). But amazingly, the most important player in the cast was absent as well. Enzo Ferrari, whose first racing car was about to debut, was not present at Piacenza. After he and his tiny staff had labored for the better part of two years to complete the first automobile that would bear his name, Ferrari failed to appear for its initial race! Much was to be made of his penchant for not attending races in his later life, but why he chose not to travel the seventy-five miles to Piacenza to oversee the introduction of perhaps the single most important automobile to bear his name remains a mystery.

Practice on May 10 was a mixed bag. Cortese took an immediate liking to the smooth-running, full-bodied "thick wing" 125S and set the fastest lap time. This gave him the pole position for the race. Farina did not perform as well. Squirting around the narrow, roughly rectangular street course, he spun and whacked a curbstone. The car was easily repairable, but following practice Farina imperiously informed Ferrari by telephone that he would not race the car. Nino's volatile temper was well known, as was Ferrari's, and while no record of this confrontation exists, one must assume it was monumental in terms of both rhetoric and decibel level. Ferrari would no doubt have considered this an act of unforgivable treachery had Farina not been the best driver in Italy. Why did Farina defect at the last moment? It is possible that he felt considerable pressure from Portello and decided that a race for rival Ferrari—no matter how minor—might place his Grand Prix ride with Alfa in jeopardy. Perhaps it was a moment of pique with either the new car or its owner, but the upshot was simple: Franco Cortese, a man never to be confused with any of the great stars of motorsports, would do lonely battle for the burgundy-wine colors of the Scuderia in the first-ever race with a car bearing the Ferrari name.

There was heartening news from Modena and Turin. The Maserati team, which had developed potent new A6GCS-2000 sports cars, had withdrawn its entry. The cars were not yet ready for competition. Cisitalia was also expected to enter their new P46 1,100-cc cars, but chose to sit out the race.

These no-shows reduced Cortese's competition to a ragged collection of prewar sports cars driven for the most part by wealthy amateurs and second-rate semiprofessionals. The cars lined up on a cloudy afternoon along the tree-lined Via Farnese straightaway. Crowds packed the curbs, no more concerned about the danger than if a procession of prams was about to pass by. Beside Cortese in the front row were an old Maserati driven by Mario Angiolini and a modified BMW 328 with Giovanni "Nino" Rovelli at the wheel.

The cars, nineteen of them, were gridded three by three for the start. Bazzi and a cluster of Ferrari mechanics watched nervously as Cortese climbed into the open cockpit of the 125S and pulled on his linen helmet. He was about to engage in the world's most dangerous sport no better protected from injury than the competitors in what is considered to be the first true motor race, the Paris–Bordeaux run of June 1895. Over half a century had passed with no appreciable advances in safety. Cortese, like his fellow competitors, was dressed in light street clothes, although some wore thin coveralls as protection from dust and grease. His head was covered with a linen aviator's helmet to prevent undue mussing of the hair. Leather crash helmets, first used in motorcycle competition, had become mandatory in American racing in the late 1930s, but it was considered weak and unmanly to don such gear in Europe. Moreover, Cortese followed the aged practice of eschewing a seat belt, believing, like his peers, that it was better to be tossed from the crashing car than to be trapped inside. Considering that the machines carried fuel in containers no more fire-resistant than milk cartons and offered no rollover protection whatsoever, there was a certain macabre logic to viewing ejection from a speeding automobile as a viable safety option.

The race would be contested over a distance of 99 kilometers (approximately 60 miles), or thirty laps around the rough two-mile circuit laid out through the narrow streets of Piacenza. (War-damaged Monza still had the only permanent racetrack in Italy, with no plans to build another; all other races would be run on public thoroughfares closed for the occasion.)

The flag dropped and Rovelli moved away with Angiolini on his tail. Cortese's engine bogged down and he stuttered off a poor third. Young, headstrong Rovelli skated through the corners and had a long lead as he completed the first lap. But the BMW was running much too quickly at the end of the Via Farnese straight and, despite his panicked braking, thudded into a pile of hay bales, barely missing a mob of scattering spectators. He was out, giving the lead to Angiolini's Maserati. Meanwhile Cortese had managed to get the Ferrari running on all twelve cylinders. The lead cars had distanced themselves from the rest of the field, composed for the most part of aged Lancia Aprilias and assorted home-built specials. On the twentieth lap, Cortese, cheered on by the crew, snatched the lead from a helpless Angiolini. The Ferrari was singing its high-pitched siren song all around the Piacenza track and with three laps left had opened up an insurmountable lead of twenty-five seconds. But the little engine began to cough, gasping for fuel. Cortese struggled on for a few kilometers, then coasted to a stop with a broken fuel pump.

Angiolini went on to win with the Maserati, but the Ferrari had proven to be the quickest car in the race by a wide margin. The competition had been thin and no serious manufacturers or top drivers had been bested. Nor was there a victory wreath mounted on the cowl of the grimy, oil-smeared 125S as it was loaded into the Fiat transporter for the drive back to Maranello. But it had proven to be an automobile packed with potential. It was, as Enzo Ferrari later mused to Bazzi, "a promising failure."

In retrospect, it was more a failure than a promising debut. The race has been celebrated by many Ferrari enthusiasts simply because the Tipo 125 was leading when it retired. But the stark fact remains that the competition was trivial at best. With a solid semiprofessional driver on board a freshly designed automobile, the race ought to have been a walkover. Considering the feeble, second-rate competition being fielded in the regional sports-car events that Ferrari entered during 1947, the 125 was a bad joke. Understandable teething problems notwithstanding, the Tipo 125 was an ill-handling, meekly powered tub compared with the Maseratis, tiny Cisitalias and Fiat-powered hot rods it faced. Considering the time, money and talent employed to create it, the 125 ought to have been overwhelming. But not only was its short-wheelbase, stiffly sprung chassis prone to oversteer, but its engine, owing in part to Colombo's complex, outdated bearing design, would rev little more than 6,000 rpm, or about 800 rpm below its intended redline.

Still, there were few serious rivals to oppose it, and two weeks following Piacenza, Cortese took the car to Rome for the "Roman Spring Season" races on a street circuit around the Baths of Caracalla. This time he won, the first victory in history for a Ferrari, but one that is little celebrated even by Ferrari acolytes, considering the impotence of Cortese's rivals. The little team raced five more times in June, gathering two class victories, breaking twice and winning overall once. One of Cortese's breakdowns came during the Mille Miglia, a commercial disaster since success in that famed event would have meant major sales. But a week earlier he had won at Pavia in a car shod in a body built by local coach builder Ansaloni that was so bloated and ugly that Ferrari's workmen nicknamed it *l'autobotte* ("the tank truck").

At the end of May, Nuvolari told Ferrari that he was ready to drive again, but his services would cost a king's ransom. Despite his age and infirmity, Nuvolari remained the best-known and most beloved sports figure in Italy and he would not come cheap. In fact, Ferrari agreed to pay him 145,000 lire for two races in July. But the outlay was a bargain, in terms of both publicity and on-track success. Nuvolari debuted at Forli and won both the 1,500-cc class and the hearts of the adoring throngs who lined the track. He drove two weeks later at Parma, where he and Cortese ran first and second, with Tazio displaying some of his old flair by snatching up the "race queen" at the finish line and scurrying off with her in the passenger seat of the Tipo 125. He was not seen again until the awards banquet later that evening!

Nuvolari's presence had been a great tonic for the operation, but the cure was only temporary. He had contracts with the Cisitalia team that would allow him only one more race for Ferrari (at Montenero in August), and he would not be available again until the following year—assuming that his deteriorating health would permit him to carry on. With no other major drivers available, Ferrari had to content himself with the steady, but unspectacular Franco Cortese and his backup, Ferdinando Righetti, who had impressed Ferrari earlier in the season by competing against the Tipo 125 in an underpowered 1,100-cc Fiat-Stanguellini.

Despite the blush of success and publicity generated by the colorful Nuvolari, it was obvious that he and Ferrari had beaten up on notably bush-league competition. One thing was clear: after the races the Tipo 125 was still barely competitive with prewar home-built specials, and was likely to be blown off the track when the new Alberto Massimino–designed Maserati A6GCS appeared on the grid. Although the Maserati brothers were about to complete their ten-year servitude under the Orsis, choosing to remain in Bologna (they never moved to Modena) and start their own operation under the new OSCA banner, they, Massimino and Gino Bertocchi had completed a superb 6-cylinder two-seater that Ferrari knew through the grapevine was generating prodigious power.

Busso was laboring hard on two fronts: to spur the performance of the 125 and to push forward the *primo* project—the supercharged

Formula One version of the V12 engine. Help arrived from an unexpected quarter. Giulio Ramponi, the old Scuderia colleague, who had long since departed for the more stable realms of England, returned to Maranello bearing a much-coveted gift. Having heard that the Achilles' heel of the engine was Colombo's overly complex and unreliable crankshaft needle bearings, he brought along a set of "thin wall" bearings being made by a rich associate, Guy Anthony "Tony" Vandervell. Vandervell was a tough former car and motorcycle racer who had made a fortune in the war after gaining the British licensing rights to the American Clevite bearing, a revolutionary design that was first applied to World War II aircraft and would later become a universal component of modern internal-combustion engines. It was immediately obvious to Busso and Ferrari that the "thin wall" bearing was a solution to the engine's anemic revving.

The addition of the Vandervell bearings to Colombo's engine was done without his knowledge, as he was once again embroiled in a contretemps back in Milan. While at Alfa Romeo he had embarked on yet another moonlighting project similar to his Ferrari assignment. This involved not a racing car but a tiny economy two-seater called the Volpe ("fox"). Colombo's design for the midget was neatly conceived, but the parent company, ALCA, had been overzealous in accepting orders and by the summer of 1947 over 1,500 Italians were demanding their money back for nondelivery of automobiles. Colombo had managed to avoid the crossfire between the ALCA management and their enraged customers, but the scandal had seriously damaged his position at Alfa Romeo and he let it be known that he was available for more work in Maranello. This triggered some stubborn maneuvering between Ferrari and Colombo, although the *ingegnere* agreed to some part-time consulting and announced that he would attend some of the final races of the season.

Back in Maranello, Busso had completed a modest redesign of the V12 engine, increasing its displacement to 1.9 liters and changing the car's designation to Tipo 159. Horsepower was increased somewhat, but the handling of the machine remained as treacherous as ever. Bazzi, who like all Italians tended to drive beyond his skills, lost control of one of the 159s on the Abetone road and crashed hard, breaking a leg. This would remove his steady, talented hand from the project of fine-tuning the 159 for the most important race of the season, the late-September confrontation between Ferrari and Maserati on their home turf of Modena.

Maserati was preparing to debut a pair of its lithe A6GCS sports cars—featuring radical Fantuzzi bodies with an odd single headlight in the grillwork. The drivers were a formidable pair. Gigi Villoresi had signed on with his protege, Alberto "Ciccio" Ascari—a youth who was already being recognized as a major talent. Villoresi was still angry with Ferrari over the death of his brother, which only added to the intensity of the confrontation between the crosstown rivals. While there was considerable camaraderie (and the exchange of rumors) between the two teams' mechanics, the managers of the firms were locked in a struggle for bragging rights in Modena. Ferrari considered the Orsis to be interlopers on his turf and below his station. Moreover, their products were aimed at the same clientele of gentlemen drivers, which qualified them as commercial enemies of the first order. As for the Maserati brothers, he viewed them as legitimate competitors on the racetrack, but was never close to them personally. His writings, as always, are as revealing for what they do not say as for what is put in print. The Maserati name is barely mentioned, and aside from a few references to the late Alfieri, none of the brothers are worthy of acknowledgment in any of the memoirs.

The Modena race was sufficiently serious to bring Colombo down from Milan to have a look at the cars entered for Cortese and Righetti. Sparks flew immediately when he examined the front suspension under the watchful eyes of Ferrari and Busso. The man whom Ferrari respected as an engineer like no other, with the exception of Jano, studied the layout for a few silent, tension-filled moments. He then pronounced imperiously, "It's all wrong. I'll deal with it myself."

The affront was unbearable for Busso, who had been laboring like a monk for months on a machine which had been badly conceived in the first place by the very man who was insulting him. He suffered in silence, but immediately pledged to leave the operation and return to

Milan—a prospect that seemed to be encouraged by Ferrari, who was now paying full attention to Colombo's pronouncements. (Busso was to leave by the end of the year, to return to Alfa Romeo, where he remained until 1977, rising in prominence with the design of numerous Alfa passenger cars.)

Although the two 159s were clearly inferior to the Maseratis in practice, Ferrari's two drivers soldiered on gallantly during the early stages of the race. Ascari and Villoresi quickly took command and began motoring around in a smug, all-conquering nose-to-tail formation while Ferrari fumed in the pits. Cortese's car stalled on the course and after it was abandoned was hit by an old Delage driven by Giovanni Bracco (perhaps in an alcoholic daze, as was his habit). The two machines tumbled into the crowd lining the track, killing two and seriously injuring a score of others. The race was stopped after twenty-four laps. Ciccio won easily, with his friend Villoresi in second, while Righetti struggled on to finish a distant, ignored and ignominious fifth. The day at Modena had been a triumph for the Orsis and a disaster of the first magnitude for Enzo Ferrari. Clearly, something had to be done to exact revenge in the next big race, scheduled for two weeks hence at the Circuito del Valentino in the lovely, shaded riverside park of Turin.

Colombo spent a few hectic days prior to the race trying to sort out the handling quirks of the 159—its tail-lashing oversteer and rock-hard brakes that had given Cortese and even Nivola some bad moments. One car would be entered at Turin and it would carry a new driver. Frenchman Raymond Sommer, the forty-one-year-old son of a wealthy Parisian felt and textile manufacturer, was a longtime acquaintance of both Ferrari and Luigi Chinetti. He had teamed with Chinetti to win the 1932 Le Mans 24 Hour race and had repeated the following year in concert with Tazio Nuvolari. Both victories had come aboard Alfa Romeos (although not the Scuderia's), and Ferrari must have been familiar with the sturdy, pleasant Frenchman's tenacious, if rather unflowery style behind the wheel. Colombo mentions in his recollections of the event that Cortese "was missing" and Farina's part in the race "had remained in the planning stage."

Whatever the real meaning of these cryptic explanations, Sommer was the appointed driver. What is known is that Farina had been fired from the Alfa Romeo team, in part owing to a dispute with Achille Varzi. It is possible that Chinetti had a hand in the Sommer connection, having been at the time working in Paris to generate some sales for Ferrari and having known Sommer for years as an enthusiastic and well-heeled potential customer. Whether or not Sommer paid for the ride at Turin is unknown, but Colombo recalled that following the race Sommer attempted to purchase the 159.

And well he should have, because he drove it to a convincing victory—the first relatively significant win for a Ferrari automobile, although the competition still remained no more than the brace of potent Orsi A6GCS Maseratis and a few antique Alfas that had been dredged up for major Italian sports-car races. The vaunted Alfa Romeo Grand Prix team, the true target of the Ferrari vendetta, was quite obviously not entered in such a minor event.

Sommer won easily after Ascari and Villoresi retired early with similar transmission failures. This prompted a moment of quasi-operatic drama, according to Enzo Ferrari. The Sommer triumph had meant so much to him, he wrote, that he sought out the same park bench upon which he had wept in the frozen winter of 1918 and sobbed again—only this time with joy over his team's win. This may or may not be apocryphal. Ferrari was known to cry (some claim he was such a consummate actor that he could well up tears on cue) and certainly the win was important for the fledgling Scuderia. But was it more important than the very first victory by Cortese in Rome or the stirring win by Nuvolari in Forli? It probably was, simply because in Ferrari's mind, which retained slights as if they were locked in amber, the win was, in a small, very personal way, an act of vengeance against the immense Fiat operation that had so ignominiously turned him away twenty-eight years earlier. In fact, a few of those close to him believe that he did not consider that debt of honor fully repaid until Agnelli completed the buyout of his company in 1969, fifty years later!

The victory at Turin apparently added a final weld to the bond between Ferrari and Colombo. It was agreed by both that a formal employment contract should be consummated. Yes, Gioachino Colombo, formerly of Alfa Romeo SpA, Alfa Corse, and one of the eminences in European engineering circles, would become a member of the Ferrari engineering staff on a formal basis. It would take only a brief friendly meeting over a hearty lunch to finalize the details. Ferrari naturally summoned his new engineer to Maranello. Colombo, sensing an advantage, insisted that the meeting take place in Milan. Ferrari, becoming more sedentary by the day and assuming his doge-like propensity for having parleys take place in his own throne room or not at all, refused. The meeting would be at Maranello, he declared. But Colombo would not relent. When an impasse seemed certain, Franco Cortese intervened and arranged a compromise—the Croce Bianca Hotel in Piacenza, halfway between Milan and Modena. This same neutral ground was the site of the 1935 meeting between Nuvolari and Ferrari. This was a significant capitulation—which only underlines how badly Ferrari wanted Colombo back in the fold. Surely fuming at the indignity of it all, he crammed his large body into his miniature 903-cc, 29-hp Lancia Ardea and chugged up the Via Emilia for the conference. The number of times he would again give way in what he surely considered to be an act of shameful subordination could be counted on the fingers of one hand. The only compensation for such a gesture was Colombo's agreement to formally join the firm at the end of the year. But before a move could be made, the great *ingegnere* had a number of complex legal matters to clear up in Milan with the management of both Alfa Romeo and the harried leaders of the soon to be defunct ALCA operation. However, Colombo agreed to begin work on the so-called 125GPC (Grand Prix Compressor) immediately and by the end of November had submitted detailed plans for a neat, short-wheelbase single-seater.

Sommer's victory at Turin had generated some bonuses. The car's performance in front of the wealthy northern Italian elite was a small breakthrough for business. These fine-tailored sportsmen were the core clientele at whom Ferrari was aiming, and sure enough, nobles such as Count Soave Besana and Count Bruno Sterzi of Milan, as well as the exiled Russian prince Igor Troubetzkoy (then married to Woolworth heiress Barbara Hutton), later appeared in Modena to examine the possibilities of purchasing new cars. By December several were being built—to join the three that had been created during 1946–47. One would be the first coupe—a sturdy little machine topped by a squarish Allemano body that would bring the Scuderia its first Mille Miglia win the following spring.

Better yet, Aurelio Lampredi was on the hook again. The Isotta Fraschini Monterosa project had failed and the man who had shown so much potential before clashing with both Busso and Ferrari was beginning to do some part-time consulting work. Ferrari must have savored the possibility of having both Colombo and Lampredi on his design team, although during preliminary discussions he indicated to Lampredi that Colombo would act merely as a consultant to the firm. This seemed a satisfactory option to Lampredi, who agreed to rejoin Ferrari at the end of the year. He had never met the man whom at the time he still considered a "magician." This esteem would soon deteriorate into open hostility, but as 1947 came to an end, the union of Ferrari, Colombo and Lampredi sounded appealing to all those involved. Certainly Ferrari had to be delighted at the prospect of being able to use his openly acknowledged skills as "an agitator of men" to prod and poke the two designers into a rivalry that would unleash a flood of creativity. In the years to come he would raise this technique of generating *mano a mano* contests among his staff to a high art. Already the pitting of Busso against Lampredi and Colombo had produced a better machine. Surely the coming contest between Lampredi and Colombo would be even more fruitful.

10

Enzo Ferrari's life was becoming locked into a mundane ritual. Each morning he arose in the apartment above the old Scuderia and, after seeing young Dino off for class at the nearby Corni Technical Institute, he trooped off for a shave by his favorite barber. Mornings were generally spent in the cluttered office of the Scuderia, where meetings were held, important customers greeted and pressing correspondence—usually in his distinctive, purple-penned longhand— disposed of. After lunch he made the ten-mile drive south along the Abetone road to the factory, where Lampredi, Bazzi, Giberti and Company would be hectored to press on with the forgings, castings and bits of raw metal—still scarce in postwar Italy—that would make up the first Ferraris.

There he held sway in a large office on the ground floor, its walls painted a soft blue. At the center was an immense desk that forced visitors to practically prostrate themselves when shaking hands across it. (This was the final indignity for many after having endured a silent, chilly, frustrating wait in an unheated room near the gate before being allowed entrance.) Unlike the office he still kept at the Scuderia in Modena, which was stacked with trophies and photos, the Maranello space was stark and unadorned. It did not need to be otherwise, because Ferrari was seldom there for long periods. Most of his day was spent in the factory workshops, where Bazzi and the mechanics were completing work on new chassis and a cadre of expert machinists were doing the final finish work on engine and transmission castings.

Nearby, in the drafting rooms, Lampredi and Colombo—who was commuting from Milan on a part-time basis—were completing drawings for the new designs that were critical to the next phase of the operation: winning races and creating a line of salable sports cars that would in turn create sufficient capital to—what else?—win more races. Colombo's efforts were concentrated on pushing forward with the 125GPC, or *gran premio* compressor, the supercharged Formula One car that remained Ferrari's supreme goal. Lampredi in the meantime was about to finalize the Tipo 166, the bored-out, 2-liter version of the 125/159 normally aspirated sports-car engine that was to form the bedrock of the consumer lineup.

Ferrari's evening routine had two forms. On most nights it was back to the Scuderia for a family dinner and a night of reading on a variety of subjects ranging from politics and philosophy to the sporting gazettes, in which the Scuderia Ferrari was still relegated to small type compared with the more visible efforts of Alfa Romeo and even Maserati. But on other evenings he would say to Bazzi, Pepino Verdelli, perhaps Colombo, "Let's take the long way home." This would mean a diversion to Castelvetro and into the arms of the serene Lina and little Piero. There, say eyewitnesses, his entire persona would soften. His posturing, his gruffness and his explosive temper would disappear once inside the shuttered little home, and he would tolerate from the rambunctious child all manner of teasing and ridiculous horseplay. This shedding of pretense was not unusual for an important man around his mistress, but for a man like Ferrari, who was developing an overblown, magisterial aloofness in public, this reversion to the ways of an old *padrone* was a shocking transformation.

He was also playing the dutiful son. As with all Italian men, his mother remained a strong element in his life, although he constantly served as intermediary—and occasional referee—between her and Laura. Both were willful, vocal women and their encounters, while mercifully few, were always contentious.

As he was about to celebrate his fiftieth birthday in February 1948, his business future looked relatively bright. Lampredi and Colombo were, somewhat surprisingly given their different engineering philosophies, getting on well, often having long dinners together after work to carry on their discussions. Colombo was from the Jano school, a hands-on innovator and experimenter accustomed to fiddling and modifying a design until it worked. His plans were often little more than inspired sketches of high-revving, highly stressed, supercharged racing engines. Lampredi, on the other hand, was a trained aircraft engineer who believed that carefully conceived, mathematically correct plans must precede any actual fabrication and that a minimum of changes should be necessary. Automobile engines, according to him, ought to be like those in airplanes—reliable, relatively simple and easy to maintain.

Far away in New York, Luigi Chinetti had set up shop in a dingy, one-bay garage on Manhattan's West Forty-ninth Street. The new shop caused few heads to turn, even among the core of wealthy sports-car aficionados who lived in the area. To most of them the name Ferrari meant nothing, although the involvement of Chinetti in the tiny dealership had a certain impact. Their plan was to promote a series of amateur road races, and if the new Italian marque performed anywhere near Chinetti's hyperbolic descriptions, sales would be assured. In the short term Chinetti planned to manage his operation in both Paris and New York, but ultimately to emigrate to America and exploit what he considered to be the mother lode of new Ferrari aficionados. All he needed was automobiles—which Ferrari had promised, provided he could generate sufficient funds to begin a small production effort. Chinetti knew that thousands of Americans had returned home from Europe with an interest in high-performance cars. A new national organization, the Sports Car Club of America, was formed and a race through the streets of Watkins Glen, New York, was being planned for the autumn of 1948. The format would ape the clubby, totally nonprofessional events of the British Racing Drivers Club, a stuffy collection of English sportsmen who raced only as lily-white amateurs and whose unspoken motto was "the right crowd with no crowding." It was this newly formed clique of rich dilettantes, located for the most part around New York's and Philadelphia's fancy suburbs and in Palm Beach, that Chinetti cannily targeted as his customer body. He had already become acquainted with the core group during his exile in Queens during the war, and now he began to use his considerable powers of persuasion to sell the most daring and aggressive of the lot on the glories of the Ferrari—although it would be nearly a year before one would actually reach the American shores.

Ferrari himself understood the potential of the same group in Italy, and had long since endeared himself to the brothers Gabriele and Soave Besana, Milanese nobles who ordered a Tipo 166SC (Spider Corsa) at the end of 1947. They thereby are credited with becoming the first customers of Ferrari (although there is some debate over whether or not Sommer purchased the car he drove at Turin in September 1947). Two more aristocrats, Bruno Sterzi, also of Milan, and Prince Troubetzkoy, also ordered similar 166 Spiders for their team, Gruppa Inter. Like the A6GCS Maseratis also being built in Modena, these early cars were triple-purpose machines. They could be driven on the public roads, they could be run in sports-car races like the Tour of Sicily or the Mille Miglia and they could, shorn of fenders and other road equipment, be entered in international Formula Two events (for open-wheel machines with engines of 2 liters or less). But for the most part these early customers, being skilled amateur competitors, intended to use their new Ferraris on the posh boulevards of Rome and Milan and in selected sports-car races. It would be up to Ferrari's hired hands, including Sommer, Cortese, Righetti and Nuvolari (when and if he was healthy and available—he had been bedridden most of the winter), to do the serious driving and to win races.

The first major event of the season was the Tour of Sicily in April, and the factory was able to complete three cars in time for entry. Count Soave Besana was teamed with Bruno Sterzi; Franco Cortese had another factory-entered 166SC; and Troubetzkoy was entered with Mille Miglia specialist Clemente Biondetti in a new, bloated, full-bodied 166 roadster with unknown bodywork (but probably by Allemano). Biondetti was a wealthy Tuscan who had won the 1938 and 1947 Mille Miglias in Alfa Romeo 8C-2900s and who was acknowledged to be one of the toughest, most tenacious long-distance drivers in the world. Troubetzkoy would ride with him as navigator and part-time relief driver under the name Igor.

Biondetti and Igor won in Sicily, which, in terms of commerce, was as important a victory as Sommer's at Turin the year before. It brought a flurry of orders—including one from Troubetzkoy—and set the stage for a strong showing in the upcoming Mille Miglia. It would also mark the last great drive of Tazio Nuvolari.

Accounts differ wildly of how the immortal driver, now fifty-six years old, in terrible health and still heartsick over the loss of his second son, came to drive in the Mille Miglia. Some say Ferrari heard that Alfa Romeo intended to sign the little man for one last drive to defend their Mille Miglia winning record (one loss—in 1931—since 1928). He then drove to Nuvolari's home in Gardone and persuaded him to race. Others claim that Nuvolari came to Maranello and, after being shown the cars in the shop, decided to race on the spur of the moment. Both these stories are probably apocryphal.

It is documented in Cesare De Agostini's excellent biography of Nuvolari that he had intended to run in the Mille Miglia for several months prior to the May race and had a ride contracted with the Dusio-owned Cisitalia team. But on April 27, a week prior to the start, his car was wrecked by another driver in a road accident and Nuvolari was notified by telegram from Dusio that no Cisitalia would be available. It was only then that contact was made with Ferrari.

Regardless of the lurid romanticizing that followed the teaming of the two men, there was probably little sentimentality associated with the union. Recall that Nuvolari had left Alfa Romeo and Ferrari amidst notable acrimony and that their relationship through the 1930s had been stormy at best. But Ferrari surely wanted Nuvolari in one of his cars, as he had in 1947—for the sheer publicity value. While the two men were never particularly close friends, they certainly respected one another.

A noted Italian motoring journalist of the day, Corrado Filippini, reported in his journal *Auto Italiana* that Ferrari escorted Nuvolari through the workshops and gave him the choice of two cars, a 166SC that was to be sold to Prince Troubetzkoy, who graciously offered it to Nuvolari, or a boxy, Allemano-bodied "Berlinetta" coupe that had been in preparation since late 1947. Nuvolari chose the open Spider over the coupe. He reasoned that the closed car would be too restrictive for his damaged lungs. This was a wise decision. Clemente Biondetti, who then got the coupe by default, was to suffer. He and his co-driver, Giuseppe Navone, would choke from engine fumes, be deafened by noise and be blinded by the chronically fogged-up windows over the entire route in the poorly designed machine.

Still, the notion of aging, heroic, underdog Nuvolari and his mechanic, Scapinelli, facing the rain, snow and ice of the Apennine passes in an open car while his principal adversary, the patrician Biondetti, cruised along in presumed luxury prompted a national orgy of sympathy. Nuvolari responded in grand style. Alberto Ascari led early in his Maserati A6GCS but retired at Padua. Cortese then moved to the front before his gearbox—the perpetual Achilles' heel of early Ferraris—snapped and he was out. A special 6C-2500 Alfa Romeo, which was an updated prewar car with fresh bodywork, was driven hard by Consalvo Sanesi, but it too failed while lying second. No fewer than ten Cisitalias started, but all were powered by small, overstressed engines and were not competitive. By the time the field had streamed down the Adriatic coast and had begun to climb the Apennine passes of Furlo and Scheggia, the crafty Nuvolari began to use the numerous curves to his advantage (he called the twisty sections "my resources"). He was driving with a surgical discipline that belied his reputation for mad behavior behind the wheel. By Ravenna, Nuvolari had the lead and was opening the gap on his rivals. But the car was shredding beneath him. When he reached the Rome checkpoint his left front fender had vibrated off and the entire hood had blown away. (The crowds would attribute this to Nuvolari's bronco-busting driving style, but it was in fact a flaw in the bodywork.)

Chin out, his left elbow pummeling Scapinelli's ribs as he sawed through the corners, Nuvolari was driving like a man reborn. Then he overcooked it on a bend and spun into a ditch, damaging the rear suspension and breaking Scapinelli's seat loose from its mountings. Undeterred, he navigated the fearsome Futa and Raticosa passes in a series of perfectly executed schusses and by Bologna had a twenty-nine-minute lead on Biondetti, who was struggling in the fetid tin-drum coupe.

Nuvolari surged onto the Po Valley flatlands seemingly headed for a brilliant victory. Then a damaged spring shackle gave way and the little Ferrari slewed to a stop at Villa Ospizio, near Reggio Emilia. Furious, Tazio Nuvolari climbed out, ending the last titanic drive of his incomparable career. He would race a few times more before his death in 1953 (ironically, in bed) but never again with the brio he exhibited for the last time in the 1948 Mille Miglia.

Clemente Biondetti drove on to easily win his third Mille Miglia, finishing more than an hour and a half ahead of the second-place team aboard a minuscule Fiat 1100S. In fact, among all the Ferraris, Maseratis and Alfa Romeos entered, Biondetti's was the only con-tender to finish in the top ten. A Ferrari had won Italy's most presti-gious sports-car race and, as expected, more orders arrived.

While the Ferrari team was involved in an active schedule of rela-tively minor sports-car and Formula Two competitions during the summer of 1948, Enzo Ferrari turned his attention to a pair of major goals: the completion of the Colombo-designed 125GPC single-seater, which had been on the drawing boards for nearly three years, and a small production run of elegant 166 series customer cars that could be employed both as sports-racing cars and as *gran turismo* road cars.

The first order of business was the racing car. Alfa Romeo and Maserati already had excellent prewar machines in the field, and rumors were rife that a group of wealthy British sportsmen were backing a project by British Racing Motors (BRM) to produce a supercharged 16-cylinder monster. Piero Dusio had commissioned the Porsche design studio—at the time operating out of a converted sawmill in Austria—to create a revolutionary mid-engine, flat-12-cylinder, four-wheel-drive super-car in the spirit of the old Auto Union. Because their boss, Ferdinand Porsche, was at the time being held for ransom by the French (after they had duped him into designing the Renault 4CV for them), the firm took on the project in hopes that the Dusio funds would help pay for his release. This car, the Cisitalia 360, was doomed to failure because of Dusio's financial problems, but it was surely the most advanced car of the postwar period.

Alfa Romeo was still master of all it surveyed. The old 158 engine had been improved (by Colombo) to produce 310 hp and, with revised suspension, had been updated to the Tipo 158D, a machine considerably advanced over the one that had rolled out of the Scuderia ten years earlier. Maserati had not been idle either. Massimino had radically improved the old 4CL by giving the chassis an advanced coil-spring front suspension and the engine a two-stage supercharger that boosted power to 260 hp. The new cars, to be driven by Villoresi and the rapidly improving Ascari, were labeled the 4CLT/48 San Remos, after their debut victory in that minor event.

The French government was also funding a project by the Centre d'Etudes Techniques de l'Automobile et du Cycle to produce an exotic, two-stage-supercharged, V8-powered Formula One car. The engine was powerful enough, but the chassis was unsatisfactory and the ambitious enterprise was dropped in 1947 after two cars were constructed.

Against all of this Colombo labored with the reality that his supercharged V12 was pumping out a modest 225 hp on the dynamometer and would rev no more than 7,500 rpm, although it was designed to operate best at 10,000 rpm. Moreover, the chassis into which the feeble engine would be placed was hardly state-of-the-art, mounted as it was on the proven but uninspired sports-car leaf-spring front suspension and a treacherous swing-axle rear setup with torsion bars (soon to be replaced by a single leaf spring).

As work progressed on the 125GPC, Alfa Romeo confidently entered the 1948 Grand Prix season with four of the finest drivers in the world: Achille Varzi, the titular team leader who was rapidly returning to prewar form; the brilliant, pug-nosed Jean-Pierre Wimille; Didi Trossi, who had matured into a disciplined backup driver; and factory regular Consalvo Sanesi, a dependable fourth who had taken the place of the fiery Farina, who had been sacked the year before for arguing over the allotment of the Number One slot to Varzi. Their first race would be the Swiss Grand Prix on the tricky Bremgarten circuit outside Berne. On the chilly, rainy weekend of July 4, Varzi took the new 158D out for a late-afternoon practice run on the tricky 4.52-mile natural-road circuit. In a momentary lapse of concentration, this impeccably controlled driver, who seldom ever placed a wheel wrong, clipped a curbstone and the big Alfa flipped over. Varzi tumbled from the cockpit and was killed instantly. Stunned, the Alfa team announced its withdrawal from the race in mourning for their forty-four-year-old leader, but Varzi's wife, Norma, insisted that they carry on in his name. This they did, with Trossi winning and Wimille finishing second—probably restrained by team orders. Varzi was buried in his home city of Galliate after a funeral that attracted thousands of distraught followers.

This was but the beginning of a series of disasters that would strike the Alfa Romeo team during the season. Shortly after Varzi's death, Didi Trossi was diagnosed as suffering from lung cancer. Although he would manfully carry on in declining health, he was to die in a Milan clinic early the following year. Moreover, the Alfa factory, still strapped for cash during its recovery period after the war, was hard at work trying to create the new 1900 series passenger cars—their first original-production automobile since the mid-1930s. The design was being handled by the brilliant Orazio Satta Puglia, known simply as Satta, who was largely responsible for the continued development of the 158 after the departure of Colombo to Ferrari. But there was some question how long the rather impecunious company could finance both the creation of an extensive new line of passenger cars and a major effort in Grand Prix competition.

No matter, Alfa Romeo won again in the French Grand Prix at Rheims, led by Wimille, who was now being spoken of as the best in the world. While Maserati entered numerous races across the continent, Alfa restricted its activity to the major international Grand Prix events, of which there were but four scheduled in 1948. The next was set for September 5: the Italian Grand Prix on the 4.8-kilometer Parco Valentino circuit in Turin. Satta and Company knew that this would mark the first of a series of encounters with their old ally and current rival, Enzo Ferrari.

Colombo, Bazzi and the Ferrari crew worked through the summer to complete three of the 125GPC cars in time for the Turin race. With Nuvolari again too ill to drive, Ferrari selected Farina, who was available, as his top hired gun. Raymond Sommer, who had won with such élan the year before on the same track, was brought aboard, as was a newcomer, a man the crowds would know simply as B. Bira. He was in fact Birabongse Bhanudej Bhaanubandh, a royal prince of Siam. While never the member of a first-line Grand Prix team, the Cambridge- and Eton-educated Bira was a respected and often plucky competitor who put in some excellent races until his retirement in 1954. The precise reason for Ferrari's retention of Bira is unknown, but it is possible that the prince "rented" the ride, contributing to the Ferrari coffers for the privilege of participating in a race from which he otherwise would have been excluded.

Because Farina was a native of Turin, he was selected to do the initial testing on the Parco Valentino circuit. This was done in the early morning, before normal traffic would take over the roads. One can imagine the citizens of Turin awakening to the snarl of an unmuffled, supercharged V12 engine echoing off the banks of the Po as Farina hammered around the track in the predawn gloom.

The Grand Prix was run in pouring rain, with Sommer having qualified a surprising third fastest. He ran with the leaders, while Farina whacked a curbstone in a moment of customary overexuberance and holed his radiator. Bira went out, once again with the chronic

Ferrari affliction of transmission failure. Sommer carried on with surprising gusto and lost second place to Villoresi's fully extended Maserati by less than a car length. Wimille won easily for Alfa Romeo.

The debut of the Ferrari Grand Prix car had hardly been impressive. Not only did it put out nearly 100 hp less than the all-conquering Alfas, but its short wheelbase and crude suspension made it an unruly, oversteering brute to drive. Worse yet, Colombo's cylinder-head design was the source of severe coolant leaks, and it would take the patient genius of Bazzi to find a cure.

Again, the usual celebration of the Ferrari myth obscures the fact that the 125GPC was a weak and disappointing contender. After all, it was the only postwar car in the field in 1948 and was competing against Alfas and Maseratis that were at least a decade old (albeit updated). Despite nearly three years of development, the car must be judged as a mediocre effort. Enzo Ferrari needed a stroke of luck to make the car a winner. And as would happen a number of times in his long career, he was to receive it the following year.

Happily, the concurrent project of developing the 166 sports car and a touring car was bearing more fruit.

In March 1948, Ferrari had contacted his old associate Felice Bianchi Anderloni, the proprietor of Carrozzeria Touring Superleggera, a man he had known since their racing days in the early 1920s. As was his custom, Ferrari chose to deal with known quantities in the business whenever possible, and he had worked with Touring both during his years at Alfa Romeo and when Anderloni had fabricated the bodies of the two prewar 815s. Touring (which had quickly shed its Fascist-mandated name change when peace had come) was but one of a dozen or more elite custom coach builders located in Milan and Turin. It had been through Busso's connections with Allemano that the Mille Miglia–winning 166 had been created, but its style was elongated and boxy and the firm was not chosen to produce the first legitimate *gran turismos* for Ferrari.

That honor went to Anderloni and his craftsmen. By May 1948 they were fabricating two distinct body styles at their small Milanese shop located on the Via Ludovico de Breme. One would be a four-seat Berlinetta specified by Ferrari, the other a small, full-bodied open roadster. The specifications were, according to Ferrari, to be based on the tiny Lancia Ardea coupe that he used as his personal car—meaning that the two automobiles would be near-miniatures as compared with the immense, long-wheelbase Alfas that Touring had turned out on the old 8C-2900 chassis before the war.

Enzo Ferrari's aesthetic contributions to this project were limited, if nonexistent. While his cars were to be celebrated for their splendid lines, there is no evidence that he influenced their external shapes in any way. If he was able to sketch, no examples of his work have ever been revealed. If he ever used his considerable powers of persuasion to force men like Anderloni to create shapes or styling themes that were vivid expressions of his own artistic sensibilities, no such credit has ever been given by the stylists themselves. Ferrari's contribution to the external beauty of his cars was his ability to select the best *carrozzerias* in Italy, whose proprietors in turn created beautiful shapes for their client. However, Ferrari often placed his imprimatur on designs by approving the seating position. Because he was a large man of substantial girth, the steering wheel was placed well forward of the seat and angled sharply to accommodate his preferred driving position. This made driving a Ferrari difficult for small men and women, who discovered that the preferences of the Commendatore had made it impossible for them to reach either the pedals or the steering wheel.

The plan was to complete the two Anderloni automobiles in time for the Salon dell' Automobile in Turin on September 15–25, a few days following the running of the Grand Prix. Luigi Chinetti had already indicated that he would buy one of the cars, making it critical that the completion deadline be met. Whether or not the pressure to complete the brace of Ferraris affected his health is not known, but in June, Anderloni died suddenly and the project briefly came to a halt. Then his son, Carlo Felice, took over the business and work resumed.

By midsummer a sizable number of the cycle-fendered 166SC sports-racing cars had been completed, but these represented a distinct line of development. The Touring cars would be civilized machines, complete with tops, windows, heaters, full road equipment and lush leather upholstery. An intense rivalry existed between the various Italian *carrozzerias*, and each tried to outdo the others in terms of original styling and exquisite craftsmanship. Each body was hand-formed by artisans who beat the aluminum panels into shape over wooden forms, or "bucks," then lovingly laid them onto the distinctive Touring network of light tubular framing—a technique that produced their feathery, but very strong *superleggera* (super-light) coachwork.

Ferrari had legitimate reason to believe that he would be successful in selling a limited quantity of *gran turismo* automobiles and had worked out a plan with Anderloni whereby he would order batches of from three to ten automobiles at a time. His staff would deliver the bare chassis to Touring, at which point the customer would be able to choose certain details, including paint, upholstery and in some cases special external trim. Until the Fiat takeover nearly a quarter century later, all Ferrari road cars would be produced in this way and therefore almost none would be identical.

As Sommer was splashing across the finish line at Turin, Anderloni's craftsmen were applying the final touches of bright red paint to the two 166s. Both were coated in a distinctive vermilion that had a faint metallic tint—a vivid contrast to the early rather murky burgundy hue that covered the flanks of the first 125s.

The Turin show was one of the largest of its kind in the world, and all the elite manufacturers and *carrozzerias* were represented. On September 14, 1948, the gallery was opened to the press for a special preview and the journalists found the two Ferraris separated. The little roadster was the single representative of the new Ferrari firm while the neat four-seater occupied a central position at the Touring booth.

The roadster stunned the crowd. It was a lovely symphony of rounded, sensuous curves that belied the short wheelbase of the machine. The grille was wide-mouthed and brash and featured, not the egg-crate configuration (which would become a trademark a year later), but rather a series of horizontal polished aluminum bars. The windshield was no more than a contoured plate of Plexiglas and the soft leather upholstery, in contrast to the flame-red paint, was fawn-colored.

Someone immediately labeled the car *barchetta*, or "little boat." Journalist Giovanni Canestrini wrote that he found the styling "disconcerting" but the overall judgment of the two cars was positive and they were quickly sold. Old Ferrari regular Count Bruno Sterzi purchased the compact four-seater Berlinetta coupe. Luigi Chinetti immediately resold the Barchetta to wealthy Los Angeles Cadillac dealer and enthusiast Tommy Lee. This was an unbelievably hectic month for Chinetti. Not only was he arranging for the sale of the first Ferrari to an American customer, but he was racing at Montlhéry with British sportsman Lord Peter Selsdon in the 12 Hours of Paris. They drove one of the early 166SC racing machines and won easily—the first long-distance victory for a Ferrari outside Italy and a significant breakthrough in terms of notoriety. Moreover, the car would once again be sold to an American.

Briggs Cunningham was an extravagantly wealthy heir to a Cincinnati Procter & Gamble fortune who had become hooked on exotic cars and amateur sports-car racing before the war. He had honeymooned in Europe aboard his 6-meter yacht, which he shipped from the United States, and had purchased an Alfa Romeo 6C and a Mercedes-Benz SS for onshore transportation. At that time, Ferrari was a prominent figure in the Alfa milieu and surely Cunningham, as an aficionado of European-style motorsports, was familiar with Ferrari's accomplishments. He was an obvious sales candidate for an early Ferrari, and Chinetti arranged for the rich American to purchase his Montlhéry-winning 166SC. The price, delivered in New York, was $9,000—a princely sum indeed for a car that had been flogged through at least one twelve-hour race and was in the truest sense a used car. Nevertheless, when it finally reached America in 1949—beating the Lee car to these shores by several weeks—it was received like a papal benediction.

But in September 1948 Ferrari had considerably more to worry about than the potential of the North American market. He had embarked on an ambitious project to convert the old mule stable across the Abetone road into a small trade school. Young men were being taken in from surrounding communities to be trained in machine work, pattern making and metal fabrication, thereby producing a steady reserve of workers as the factory's activities increased. In addition to the start-up costs of the school, the racing program was becoming a financial drain. The Formula One car was clearly inferior and needed radical updating, in terms of both the engine and the ill-handling chassis. But by employing what he referred to as "immoral techniques," Ferrari came up with a quick fix to keep his cars in the public eye. He suggested that the normally aspirated (called "atmospheric" at the time) 166 sports-car engine be loaded into the gran premio chassis, thereby creating a single-seat Formula Two car (engine displacement: 2 liters, unsupercharged). This was undertaken on Sommer's Turin car and he was entered in the fifth Grand Prix of Florence at Cascina on September 26. He won easily, in a car that was lighter and much more responsive under cornering and braking. Despite the fact that the event was dampened when Ermini spun into the crowd and killed five spectators (not in a Ferrari), the team returned to Modena full of short-term hope.

The full team reappeared at Monza on October 17 for the celebrated postwar opening of the track. Enormous damage had been inflicted on the track surface by legions of Allied tanks, half-tracks and weapons carriers that clanked down the main straightaway during an April 1945 victory parade. For the reopening the full might of Alfa Romeo was brought to bear and once again the Ferraris flagged in the face of it.

Wimille was the winner, followed by his three Alfa Romeo teammates running in military formation. Sommer struggled manfully in third place for a time, then entered the pits, coughing, and retired with a massive asthma attack. Farina motored helplessly in fifth, miles behind the Alfas, before his transmission broke and he too was finished for the day.

The team's next outing was at a minor race in Garda, where Farina won against weak amateur opposition. Count Bruno Sterzi was given the Formula Two car that Sommer drove at Florence and finished second. The final race of the season came at the Penya Rhin Grand Prix, run at Barcelona, where Bira led briefly before he joined his colleagues, Farina and local driver Julio Pola (a potential customer), on the sidelines.

In November, Chinetti salvaged some honor for the marque by returning to Montlhéry and gathering up some international records in the 2-liter class, but it hardly erased the disappointment of the Grand Prix effort.

Then, in December, Ferrari suddenly concluded a deal with the organizers of the South American Temporada series in Argentina and Brazil. Farina was to drive the single Ferrari in the six races. The reason for the campaign was purely commercial; the Perón government was offering lavish starting money and Ferrari recognized that if the car was successful, sales would follow in the rich, untapped market. Lampredi and Colombo quickly cobbled together a supercharged 2-liter engine and placed it in one of the single-seaters for the Formula Libre (anything goes) events. Farina was to be gone for three months and it was hoped that he would be successful against his major opposition, Villoresi and Ascari in official team Maseratis.

The Temporada was to have two major if oblique effects on the fortunes of Enzo Ferrari. The first occurred at the first race, run on the Palmero Park circuit outside Buenos Aires. Jean-Pierre Wimille, who had come to South America to drive a small Simca-Gordini after Alfa declined to enter, lost control in practice and was killed. The accident has been attributed to the blocking of the course by a spectator, or a mounted policeman, or both. But the tragic result removed the third—and surely the best—driver on the Alfa team. Varzi was dead, Trossi was dying and now Wimille was gone as well. The decision was made in Portello to withdraw from competition for the 1949 season. While the loss of their drivers was surely a factor, there is no question that uncertain economic conditions, labor unrest and the costs involved in the creation of the new 1900 series passenger cars were equal factors in the withdrawal. The second development in South America which was to have a long-range influence on Ferrari was the debut in international competition of a balding, rather portly, bowlegged thirty-seven-year-old named Juan Manuel Fangio. He had been a star in Argentinian competition for some time and was on the verge of beginning a European campaign that would raise him to near-immortality in the sport. As it was, this almost mystically skilled driver, lovingly called "El Chueco," won two of the six races and dazzled the Europeans with his poise and daring. Farina won one Temporada race, took second once and retired with mechanical failures four times. He returned to Italy to find the motor-racing scene in chaos.

Alfa was out. The field was clear, save for Maserati, and they too suddenly dropped out. In February the Orsis shut the doors of the factory, announcing that they would stay closed until June while a major regrouping was undertaken. The factory team would withdraw from competition, although a small quantity of San Remos would be built and maintained for a few private teams. But a major emphasis would be placed on the development of a grand touring car based on the A6 chassis with bodywork by the rising firm of Pinin Farina (soon to be called Pininfarina). Maserati was also to continue the manufacture of machine tools and small electric trucks, thereby removing itself from direct competition with Ferrari. This alteration of direction by the Orsis was the first of a series of tactical errors that would ultimately take their company out of the ranks of the elite automakers. While Ferrari was zeroing in on the tiny world of exotic cars, the Orsis were doing just the opposite, expanding in a variety of directions and thereby blurring the focus of the old, much-honored firm. They would make several massive and impressive comebacks, but the decision of February 1949 was to reap both long- and short-term benefits for their old Modenese rival. Of immediate interest to Ferrari was the news that both young Ascari and his mentor, Gigi Villoresi, would be unemployed.

The sudden disappearance of his two major tormentors, Alfa and Maserati, was a stunning victory for Ferrari. Moreover, there were no real contenders to take their place. The Cisitalia-Porsche project was stillborn for lack of funds, and the vaunted BRM effort from England was hopelessly behind schedule. One of its sponsors, Tony Vandervell, had become sufficiently impatient to begin inquiring about the possibility of Ferrari's building a new 125GPC for him.

How things had changed! Ettore Bugatti was dead, and the remnant of his company was engaged in military contract work. Mercedes-Benz was still banned from competition. The Auto Union works had been dismembered and taken captive in East Germany. The Maserati brothers had formed OSCA in Bologna and were said to be designing a 4.5-liter V12, but the operation was underfunded and going nowhere. That left only Talbot-Lago, a French team that fielded a series of ancient prewar machines that could be run as both Grand Prix cars and—with fenders—as sports cars. Their low-revving, underpowered, 6-cylinder, unsupercharged engines had but one advantage: on occasion their superior fuel mileage (nine to ten miles per gallon) permitted them to outlast the thirsty supercharged Italian machines. Ironically, these cars would serve as an inspiration for the first truly successful Formula One cars from Ferrari.

Two major challenges lay ahead for Enzo Ferrari as 1949 arrived. The first was to improve the power and reliability of the Grand Prix car. The second was to round up a team of first-class drivers. Sommer was a fine man, a sportsman and an unyielding competitor, but he lacked the fire and flair of a Nuvolari or a Farina. Now that Maserati was out, Ferrari immediately set to work trying to lure Ascari and Villoresi into the fold. But this would have to wait until the pair returned from Argentina.

Colombo remained as a commuter from Milan, arriving every week or two with new plans and ideas for updating the Formula One car. This involved creating a two-stage supercharging layout similar to the Alfa setup and adopting the twin-overhead-camshaft cylinder heads that had been originally designed by Giuseppe Busso. This seemed the obvious solution, based on Colombo's long philosophical involvement with the Alfa Romeo school. But his counterpart, Lampredi, who was on full-time duty at Maranello, had different ideas. He remained a devotee of simplicity and reliability, and was therefore intrigued with the crude but effective way that Talbot-Lago did business. It was reaching a point where the Alfa Romeos were getting no more than two miles per gallon on alcohol fuel, and in a 300-mile race, that could mean two pit stops, even with immense tanks mounted in their tails. But with a 4.5-liter engine, reasoned Lampredi, that could reliably pump out over 300 hp, mileage might be doubled, thereby offering a simple key to victory.

Colombo was immediately skeptical. Lampredi's plan was the antithesis of all he stood for in terms of engine design, and the argument, while friendly, began to gain a hard edge as the spring came on. This sparring was surely no surprise to Ferrari; nor did he discourage it. In fact, he seemed to side with Lampredi during the discussions, which only forced his old friend and associate Colombo to work even harder. It was a perfect situation: two talented and highly motivated men racing on two different paths toward the same goal. This was a key to Ferrari's managerial style, and a situation that he would actively create numerous times in his career.

The season started well for the sports cars, with the Tour of Sicily once again falling to Biondetti, as well as the Mille Miglia, which Ferrari won—his unprecedented fourth victory in that most difficult of the traditional over-the-road races. But it was the upcoming Grand Prix season that obsessed Ferrari, and he began to set his hook to capture the most talented brace of drivers in Italy—Villoresi and Ascari. Villoresi was the key. He remained the patron for his younger—and probably more talented—sidekick, and Ferrari understood perfectly well that if Villoresi could be hired, he would get Ascari as well.

Villoresi recalled the first postwar meeting with relish. He had been loaned a Ferrari for two races, one in Belgium, the other in Luxembourg, and upon returning to Italy had driven down from Milan for a meeting with the vaunted Commendatore (Ferrari still preferred that title at the time). "I found him in bed, in his apartment above the Scuderia," he remembers. "Ferrari was in a darkened room, his hands folded across his chest. His eyes were closed. He looked as if he was lying in state. I waited two, three minutes. He simply lay there. Still. Saying nothing. Finally I said, 'Enough of this nonsense. I am leaving.' At that point Ferrari 'woke up' and we made a deal."

Clearly, Villoresi would not fall for Ferrari's bluff. Lesser men could have been intimidated by such a charade, but Villoresi was a wellborn star whose driving talents were much in demand. Ferrari needed him more than Villoresi needed Ferrari—especially with the prize of young Ascari in his pocket. As noted previously, Enzo Ferrari possessed an unerring sense of whom to bully and whom to accommodate, which may have been his most valuable business skill. But a handful of drivers—Nuvolari, Villoresi, Fangio, Lauda—held such strong hands that Ferrari simply could not force them to bend to his lash.

With Villoresi and Ascari on the team, as well as the talented, hard-driving Felice Bonetto, and his major rivals sidelined, it might have been expected that 1949 would have been a walkover for the Ferrari Grand Prix cars. It was not, despite the fact that Ferrari had considerable help from a pair of rich, highly competitive Englishmen. Tony Vandervell had tried to obtain a 158 Alfa Romeo and, after being turned down, arranged with Ferrari to purchase one of the early 125GPCs. Peter Whitehead, a gentleman driver of considerable talent, also obtained one of the machines, thereby becoming one of a handful of men ever able to privately purchase a contemporary Grand Prix car from the Scuderia. (On occasion Ferrari sold obsolete cars to collectors, but seldom did he let go of cars that could be raced against him.)

As for the official team, the results were mixed at best. In their first race, at the Belgian Grand Prix, the 125s were quick, but Louis Rosier, a grizzled veteran better known for his skills in sports cars, beat them in one of the ancient 4.5-liter Talbots. This victory, in which Rosier motored on without a pit stop while the fuel-thirsty Ferraris were repeatedly visiting the pits, may or may not have been an epiphany for the Commendatore. There is no question that Lampredi was already advocating a 4.5-liter unblown V12, but the loss to Talbot at Spa may well have been the final blow to Colombo's argument that more and more supercharging was the answer.

The team rebounded a month later by running one-two at Berne in the Swiss Grand Prix, with Ascari and Ferrari both winning their first international Grand Prix. Ascari triumphed again in a minor race at Rheims, beating the newcomer Fangio (driving an independent yellow-and-blue Maserati backed in part by the Perón government), then moved on to the Dutch Grand Prix, run on the sand-swept circuit at Zandvoort. There Ascari crashed when a front suspension bit broke and a wheel came loose. He barely missed a brick barrier and ended up against a sand dune, unhurt. Villoresi went on to win, beating Prince Bira, who nonetheless set the fastest lap with his San Remo Maserati.

In the meantime Luigi Chinetti had won what was perhaps the most important race of them all in terms of the company's long-term success. He and Lord Selsdon had taken a privately entered 166MM Barchetta sports car to Le Mans, where the fabled 24 Hour race was being run for the first time since the war. Facing off against a fleet of well-prepared French Delahayes and Talbots were two Ferraris, the Chinetti-Selsdon 166 and the sister car driven by Parisian banker Pierre Louis Dreyfus (alias "Ferret") and Jean Lucas. The little cars, with half the displacement of the bigger Gallic machinery, ran with the leaders until Lucas lost control at the White House kink and crashed. That left Chinetti to drive on alone, with Selsdon ailing from some unknown malady (rumors swirled that he was hung over; Chinetti denies this). With the hard-bitten forty-three-year-old Milanese doing the driving for over twenty-two of the twenty-four hours, and using a jury-rigged funnel in the cockpit with which to hand-feed oil to the leaking V12 while on the run, Chinetti drove on to a widely heralded victory. American audiences were often myopic in their knowledge of racing, unaware of competitions outside their borders. Le Mans was one of the few recognizable European races that had any cachet—so the fact that Ferrari's North American representative was the driver of record meant a potential bonanza in sales. (Later that season, in September, Briggs Cunningham debuted his 166SC at Watkins Glen, which added further luster to the new Ferrari marque on the far side of the Atlantic.)

As if Chinetti needed further credentials to establish himself as one of the greatest endurance drivers of all time, he teamed with Jean

Lucas later that summer to win another 24 Hour race—this one on the difficult Spa-Francorchamps circuit in Belgium. Leading by a wide margin, Chinetti spun on a patch of oil with only minutes left in the event and struggled home a winner with the bodywork on Lucas's Barchetta (presumably the one he had also crashed at Le Mans) in a shambles.

The Grand Prix team's next outing during the steamy summer of 1949 was in England at a converted Midlands airport circuit called Silverstone. Ascari and Villoresi were on hand with their single-stage 125s, facing no fewer than six San Remo Maseratis—one of which was driven by Farina—and four of the heavy, anvil-strong Talbot-Lagos. Joining the official Scuderia team was Vandervell, who entered his new 125 with top amateur Raymond Mays, a Lincolnshire woolens heir and principal in the ERA and BRM projects, at the wheel. However, the new car behaved so badly, even on Silverstone's table-flat straights, that Mays politely stepped out and offered it to tyro Ken Richardson, who promptly stuffed it into a trackside ditch. Far from pleased, Vandervell shipped the car back to Maranello—a gesture which marked the beginning of a series of setbacks in the relations between Vandervell and Ferrari.

Ascari won the race following a wild, wheel-rubbing duel with Farina, who recovered from a late spin to keep Villoresi out of second place. A week later Farina turned the tables on Ascari at Lausanne, Switzerland, where he won easily, beating the Ferraris by over a minute. Only one car was ready for the French Grand Prix on the triangular, 4.86-mile open-road circuit outside the ancient cathedral city of Rheims. Again a tortoiselike Talbot—this one in the hands of veteran Louis Chiron—won easily after Villoresi's transmission broke. Some honor was salvaged by privateer Whitehead, who finished second and set the fastest lap at 105.1 mph.

By August, Colombo and Lampredi had completed work on the two-stage-supercharged, double-overhead-camshaft engine. Extensive testing was carried out both at Monza and on the Abetone road. In those days a racing car could be licensed for road use simply by painting the letters "Prova" on the tail, followed by "Mo" (for the city of Modena) and a two-digit number. Thus equipped and with one of the factory testers or perhaps a race driver like Bonetto or Ascari at the wheel, an unpainted single-seater would squirt through the factory gates and lope up the road, its engine revving. It was inevitably followed by a wheezing Fiat 509 touring car with its bodywork converted to a pickup. On board would be a shirt-sleeved Ferrari, Bazzi, a few mechanics and a load of tools to support the test run. Even after the Modena autodrome was built around the perimeter of the city park and most of the factory test program was moved there, it was not uncommon to see a racing car being tested on the Abetone road. (To this day each production Ferrari is driven on a test loop over a network of rural Maranello roads before being cleared for sale. But unlike the old days, sensible velocities are maintained.)

So went the tests of the new car, which was showing a creditable 310 to 320 hp on the dynamometer (as opposed to about 260 hp for the single-cam version that had been used so far that season). However, the engine was overheating, and the chassis, while its wheelbase had been lengthened, was far from a handling dream. Still, work rushed on, it being the custom for major Italian racing-car builders to showcase their new wares at the Monza Grand Prix each autumn. There Ferrari would face two serious adversaries. While considerably less formidable than the Alfa Romeo entry, Farina and Piero Taruffi were aboard independently entered San Remo Maseratis, and Talbots were in the hands of the likes of Rosier and the veteran Philippe Etancelin.

The day was insufferably warm, so that the job of driving one of the Ferraris—with its furnacelike engine mounted inches in front of the driver's feet—bordered on torture. Still, Ascari, showing the brilliance that was in evidence in race after race, season after season, dominated the contest after the Maseratis and his teammate Villoresi fell by the wayside. (Transmission failure, again, for Gigi.) Ascari developed such a lead over Etancelin's Talbot that he was able to make a leisurely stop to add oil, change a spark plug and down a slug of *acqua minerale* before getting back underway and winning by over a lap.

This victory assured Ascari and Ferrari the Italian championship and gave them bragging rights over their rivals, such as they were. Nevertheless, the achievements of 1949 were debatable at best. While celebrants of the Ferrari myth recall that the cars won several major races in 1949, including Le Mans, the Mille Miglia and numerous Formula Two races, the fact remains that Ferrari entered the Grand Prix arena with what was ostensibly a new design and could barely fend off the independent opposition comprised of modestly funded Maseratis and Talbot-Lagos. It was hardly a devastating assault, and it was bound to be blunted even more in 1950 if rumors floating out of Milan were true: Alfa Romeo, after regrouping for a year, was said to be ready to enter the fray once again with the wondrous 158s. Their driver lineup would be impressive. The youngster would be thirty-eight-year-old Juan Manuel Fangio, teamed with forty-three-year-old Farina and the grizzled old "Abruzzi Robber," Luigi Fagioli, who was making a comeback at age fifty-two.

Ferrari would counter with Ascari and Villoresi, plus occasional help from Bonetto and Sommer. But would it be enough? The Alfas were still producing more horsepower, with considerably more reliability, and while Colombo labored valiantly to get more power from the so-called double-stage V12, Lampredi's arguments for an "atmospheric" counterpart became more persuasive with each passing hour. Surely something would have to be done, because fielding the Colombo cars against the improved 158s raised the distinct possibility of another thrashing by their Milanese antagonist.

11

For a month following the 1949 Italian Grand Prix, Enzo Ferrari pondered the dim prospects offered by Colombo's Formula One car. The situation was becoming intolerable. Clearly the super-charged engine was not up to the task of dominating the world's most elite racing class, and with Alfa Romeo poised to return in 1950 with a powerful team, action had to be taken. The Fédération Internationale de l'Automobile had announced that a World Champion would be crowned at the end of the season, making the stakes for success all the higher.

Late in October he gathered his staff in his Modena office. Colombo and Lampredi, who were beginning to feud openly, were seated in opposite corners of the room. The Commendatore curtly announced a major shift of assignments. From now on, he said, Lampredi would be responsible for developing the Grand Prix car, in line with his plans for a 4.5-liter unsupercharged or "atmospheric" engine. Colombo's new assignment was to develop the sports and touring automobiles for the firm—clearly a demotion.

In the flowery, circumspect prose style practiced by so many Italian historians, this incident is dismissed as a polite, almost routine reassignment of duties. It was in fact a major upheaval. Colombo was

a volatile and prideful man who understood that he was being replaced by an engineer he considered to be his junior not only in years
but in ability. One can imagine the grand crescendo of egos as Ferrari
tried to dragoon Colombo into accepting his new role. Shrieks of
outrage echoed through the Scuderia. Insults were hurled. Threats
darkly repeated. Colombo vowed to quit and return to Milan. Ferrari
countered with soaring pronouncements about contractual and
moral obligations to the firm. When Colombo rose to walk toward
the door, Ferrari grandly picked up the telephone and called the
Modena prefect of police. He demanded that Colombo's apartment
be searched to ensure that none of Ferrari's plans might be secreted
back to Milan, where they might be used against him. Outraged at
Ferrari's behavior, Colombo remained to continue the argument. It
was classic Italian theater, with two strong-willed, tough-talking men
slugging it out verbally across a desk in a smoky Modena office.

Enzo Ferrari held the high cards. Lampredi's concept for the new
engine held real promise, and he had made up his mind to change
direction. The sports cars—the Mille Miglia contingent and the rich
men's passenger cars—were at best afterthoughts. He knew full well
that if Colombo departed, a dozen eager young designers would be
rattling the factory gates to replace him. Colombo remained something of a pariah among the noisy, contentious members of the Communist Party, which controlled the labor unions in the big automobile
firms, and therefore he hardly enjoyed carte blanche in Milan or
Turin. The failed Ferrari Grand Prix car had damaged his reputation
in the Italian motorsports community, although there was no doubt
that he was a talented engineer still capable of producing excellent
designs. The soaring rhetoric and the gutter insults of the Ferrari office subsided; Colombo would stay on for the time being. However,
there can be no doubt that Enzo Ferrari understood that the demotion and the loss of face by Colombo meant that his tenure at Maranello would be short-lived.

Despite the prospect of losing Colombo, Ferrari faced 1950 with
considerable hope. Lampredi was confident that his "atmospheric"
engine could take the measure of the Alfas, especially with the bril-

liant young Ascari and his old pal Villoresi mounted in their seats. Luigi Chinetti was operating out of a loft on Nineteenth Street in Manhattan and still splitting his duties between Paris and New York, but the intense, dark-eyed little man was beginning to report increasing American sales and enthusiasm for the new marque. Each day more and more important men would appear on the doorstep of the Modena office. Others would go directly to the Maranello works to order either high-powered road cars or special sports machines for big races like the Mille Miglia or Le Mans. Some, like Prince Bernhard of the Netherlands or King Leopold of Belgium, would be waltzed around the factory with pomp and circumstance. (Smoking was forbidden in work areas, and whenever Ferrari and other dignitaries approached, dozens of lighted cigarettes were thrown into toolboxes. This would produce an absurd scene wherein the Commendatore and his guests toured shops where each worker's toolbox appeared to be on fire.) Others, including eager customers like the Belgian Olivier Gendebien and Eugenio Castellotti, both of whom would become star drivers for the firm after each bought half a dozen cars, were forced to wait for hours in the crude little reception shack at the Maranello gate before being allowed admittance. This indignity sometimes backfired. Tony Vandervell had purchased a second, updated 1.5-liter Colombo Grand Prix car and, like the first, found it wanting and returned it. Modifications were again completed and finally a third, much-improved car was purchased, but the relationship between the two willful men was to deteriorate into open warfare. Ferrari, for reasons unexplained, became paranoid that Vandervell was using his relationship to supply technology to the flagging BRM Grand Prix project, despite the Englishman's assurances that he had long since severed any connection with that organization. Nevertheless, Ferrari remained skeptical and the stormy relationship ended when he left Vandervell to sweat and fume for three hours in the gatehouse "waiting room" on a fetid Maranello afternoon. The Englishman stalked out, never to return. His revenge would come five years later when his own Vanwall Grand Prix cars took the measure of what he had come to deride as "those bloody red cars."

Such setbacks were hardly troubling to the increasingly imperious Ferrari. He was on track with the eager Lampredi, who pushed
ahead to prove his point with his new engine. This left the sullen
Colombo to transform his little V12 into a reliable engine for road
cars. (In company parlance, Colombo's version would be the "short"
V12, Lampredi's the "long" version.) With the hated Maserati firm
still trying to market machine tools, its midget delivery trucks not selling and its race cars immobilized, Enzo Ferrari was for the moment
the king of motorsports in Modena.

While the Orsis had labor troubles during this period, the Ferrari
workers remained quiet. An employee from that era remembers the
boss as being "cheap but fair." While money was surely tight—
although perhaps not as tight as the ever-complaining Ferrari
claimed—the staff was paid regularly on the eleventh and twenty-
second day of each month. Being from the working class himself,
Ferrari maintained a strong affinity for his mechanics and artisans,
and the longevity of those employees was often greater than that of his
drivers, team managers and senior staffers, many of whom came from
the upper classes.

"I think he always cared more for the mechanics than for the drivers," observed a veteran of the Modena facility. "He spent each morning from eight to eleven o'clock in Modena. Then he'd go to
Maranello for the afternoon. He'd be back by five like clockwork to
check the workshops. He had a fetish for cleanliness. Every tool had
to be in place before we'd climb on our bicycles and leave for the day.
It was strange. When the cars were winning, he'd be a madman,
screaming at everybody. But when the cars were losing, he was quiet.
Very quiet. Like a humble man.

"He knew us all like family. He knew the names of our children,
our wives. He knew when someone was sick. He seemed to know
everything that went on in Modena."

It was at this point that a lanky youth named Romolo Tavoni came
to work as a secretary for Ferrari. He was to remain with the firm for
perhaps the most tempestuous decade of Ferrari's life and was to become immersed in the increasingly Byzantine politics of the place,

the politics that seemed to energize the way Ferrari engaged in business.

In the beginning Tavoni took care of routine correspondence and helped to compose painstakingly crafted press releases. Ferrari had long been obsessed with the vocal Italian sporting press, and part of his mornings were spent planted on the toilet, searching through the major newspapers for comments about the firm. Any items he might find would be red-penciled and filed away for future use—mainly as retaliation against the writer for any presumed slanders. Responses would come as veiled references in the aforementioned statements to the press and during the annual press conferences that were held to introduce new products. As the years passed, his struggles with the Italian press would occupy more and more time as Ferrari increasingly portrayed himself as the unjustly hectored David fighting gallantly against the journalistic Philistines.

At home, life was still dully routine. Laura was a distant, cynical woman who maintained a connection with daily operations through her financial interest in the company, although she came no nearer the workshops than to putter in her garden behind the Scuderia or to sweep up litter on the street out front. Dino was a dutiful schoolboy when his health permitted. Involvement with his father was inconsistent at best. Some associates recall that Dino appeared in the factory on a regular basis. Others remember him being alone for the most part and on the sidelines of the operation. The truth probably lies in the middle ground. Ferrari was a tough taskmaster when it came to his son. "Dino was afraid of his father," recalls one colleague. "Ferrari was hard on the boy, although he obviously took pride in him. One day Dino drove a car from the Scuderia in Modena out to the Maranello factory. Ferrari was furious. Dino wasn't healthy and the idea of the boy driving a car without his permission gave him fits." Despite his outbursts of temper—which were daily occurrences—Ferrari intended for his legitimate heir to take over the business and there is reason to believe that Dino cared deeply about automobiles. But Enzo Ferrari was driven by ambition, not by family obligations, and it is logical to assume that the child entered his life only when it was convenient. Moreover, Lina Lardi and little Piero were constant distractions in nearby Castelvetro, leading Ferrari to remark to Tavoni shortly after his arrival that "a man should always have two wives."

He was obsessed by sex. Long conversations at dinner with close associates centered on women. Ferrari prided himself on his conquests. "Women were simply objects," recalls one who worked closely with him for years. "He didn't really care for them. They were symbols to be carted off to bed—notches in his belt, that's all." (Years later, when Ferrari was in his eighties, he hosted a small birthday lunch at the Cavallino restaurant across from the factory. His guest of honor was an old colleague and former member of the Scuderia. The man fancied himself something of a Casanova, and during dessert Ferrari asked him in his customary point-blank manner, "How many women have you had in your lifetime? Be truthful." The guest thought for a moment, then answered proudly, "At least three thousand." Ferrari pulled back in mock amazement. "*Only* three thousand?!" he sneered.)

Clearly, Enzo Ferrari was never mistaken for a liberated male. Italy is hardly a hotbed of the women's liberation movement, and Ferrari, as a man of the early twentieth century, carried a simplistic, hopelessly chauvinistic view of women to his grave. He wrote that "feminine superiority is apparent above all in the matter of marriage; it is the woman who chooses her mate, not vice versa. In fact, any woman who is passably good-looking can count on at least three potential suitors. We men are taken into consideration as potential husbands, observed carefully, weighed up and perhaps chosen. We think we have wooed and won, whereas in reality we are merely the slaves of our desire, on which the woman has played with consummate skill." Ferrari also noted that men are vulnerable to women because of elemental hormonal passions. Men "are capable of anything under the urge of reasons that stem from desire," he observed.

"I am convinced," he wrote, "that when a man tells a woman he loves her, he only means that he desires her and that the only perfect love in this world is that of a father for his son." (That was written in 1961, five years after the death of Dino and while Ferrari was juggling three women in his life.) Even as he drifted into his eighties, he remained a "slave of desire."

Ferrari's friend and affectionate biographer, Gino Rancati, recounts the adventures of a priest from Bari who was a pal and fellow carouser of Ferrari's. Padre Don Giulio came to Modena periodically for what was referred to as "his practice." This was a polite reference to the priest's visitations with his mistress, which were the source of ribald amusement for Ferrari and his tight circle of evening celebrants, not to mention the padre.

But for all the distractions with females, the task of winning major motor races occupied his every waking hour. Lampredi set to work on the "atmospheric" V12 in great secrecy, planning first a 3.3-liter version, with the intention of introducing the full 4.5-liter model at the 1950 Italian Grand Prix. In the meantime an increasingly sullen and discredited Colombo was laboring over the sports cars. They were commercially more important to the firm than the Grand Prix cars, yet in the mind of the boss they remained merely secondary means to an end.

The first major race for the team, excluding a few Formula Two events, was the spring running of the Mille Miglia. A mob of Ferraris appeared, led by Villoresi and Ascari in a pair of new 3.3-liter Lampredi V12s. These were clearly the fastest cars in the field, but both failed early when their rear axles broke (transmission and axle breakages again cursing the early Ferraris). The race was won by Milanese textile heir and lawyer Giannino Marzotto aboard a Colombo-engined Tipo 195S coupe—a victory that would be recalled as the one wherein the winner drove the entire distance elegantly attired in a double-breasted suit! He was the oldest of three Marzotto brothers who were not only top-ranking amateur drivers but first-class Ferrari customers.

Ferrari did not fare well at Le Mans. Nettled by Chinetti's victory the year before, Talbot converted a pair of its 4.5-liter Grand Prix cars into two-seaters and ran off with the race. Louis Rosier also bettered Chinetti's endurance record behind the wheel by driving all but two laps of the 24 Hour race. (He turned the car over to his son, Jean-Louis, for that brief interlude before returning to battle.) Sommer salvaged some honor by setting the lap record in a Tipo 195S coupe, but retired early in the going.

All of this sports-car racing was secondary to the impending confrontation with Alfa Romeo on the Grand Prix circuit. The new Tipo 275, with a shorter wheelbase and revised rear suspension, debuted at Spa after the considerably updated Alfa 158s had easily bested the old two-stage Colombo 1.5-liter machines at San Remo and Modena. With its stomach-turning 150-mph downhill bend through the village of Burnenville and its 190-mph Masta straight, Spa was considered to be the equal of the Nürburgring as a test of driving skill. Moreover, with its long uphill and downhill sections, raw horsepower would be a legitimate yardstick for how the new Lampredi effort would fare against the vaunted Alfas.

Not well, as it turned out. The car assigned to Ascari wheezed hard to keep up with the fleeing Alfas, and the driver had to marshal all his considerable talents to bring the car home in fifth place. But Lampredi was only beginning. At the end of July the team appeared at Geneva for the Grand Prix des Nations with a 4.1-liter Tipo 340 that boasted 310 hp and gobs of torque (this against Alfas that were said to be generating about 350 hp in 1950 trim). Villoresi nearly sat on a black kitten at a café prior to the race, which alarmed the extremely superstitious Ascari (with considerable justification, as it turned out). Ascari drove brilliantly to hold second behind Fangio until he fried a piston and retired. Gigi, aboard the older 3.3-liter machine, skidded on a patch of oil seven laps from the end and plunged into the crowd, killing three spectators and injuring twenty, including himself. The forty-one-year-old veteran suffered a broken femur and collarbone and serious head injuries that sidelined him for months.

The final clash between Alfa and the team from Maranello came at the Italian Grand Prix in September, where Ferrari produced a pair of full-size 4.5-liter machines for Ascari and longtime factory test driver Dorino Serafini, subbing for Villoresi. Alfa Romeo countered with a trio of flawlessly prepared 158s for Farina and Fangio—who were locked in a struggle for the first World Championship—plus

Luigi Fagioli and Piero Taruffi. This time Ascari was ready. He was a mere tenth of a second slower in practice than Fangio, thereby setting the stage for a rivalry that would establish the two men in a class by themselves in Grand Prix racing for the next five seasons.

As the cars were rolled out onto the boulevard-wide main straight at Monza, with the massive crowd chanting the names of their favorites, Lampredi was so overcome by the pressure of the moment that, in the presence of Ferrari, Colombo, Bazzi and team manager Nello Ugolini, he fainted! But he had every reason to feel the pressure. Here, in front of the madly enthusiastic *tifosi*, the wild-eyed Italian motoring press and eminences from the world of automobiles, his conviction that an unblown automobile could outrun a supercharged car of a type that had dominated Grand Prix racing since the early 1920s would finally be tested.

Ascari had a bad start but steadily moved up to take over second place from Farina, who was determined to win the World Championship over his rival and teammate Fangio. Then the rear axle broke again and Ascari was out. After abandoning his car on the track, Ascari trekked back to the pits and took over Serafini's sister car, which was ambling along in sixth. That triggered a furious drive back through the pack, with Ascari finishing second behind Farina.

Now the big Ferrari was gnawing at the heels of the old Alfa, but the proud Portello firm kept its winning streak intact. It had entered eleven races in 1950 and won eleven times. Alfa Romeos had also finished second in six Grand Prix and recorded four third places. Better yet, their longtime loyalist Nino Farina had clinched the World Championship, which he had sought with such fervor. To be sure, the supremacy of the Alfa Romeo 158/159 cars was eroding, but nearly another year would pass before those marvelous automobiles — first conceived in the Scuderia's modest workshops — would finally be driven from the Grand Prix landscape.

By the end of 1950 Colombo stalked back to his native Milan and into the arms of Alfa Romeo, where he set about seeking even more power from the machine he had helped create thirteen years earlier.

Lampredi, now alone as the chief designer in Maranello, concentrated on improving the Formula One cars, which now offered so much hope. He also began refining Colombo's "short" engine for sports-car use in unsupercharged form.

At this point, during the winter months of 1951, the Prancing Horse began to leave its hoofprints in every civilized corner of the globe. By then nearly seventy hand-built road cars, plus perhaps a dozen Formula One and Formula Two racing machines, had been turned out. Legions of the world's elite, including the Aga Khan, Prince Bernhard, King Leopold of Belgium and his wife, Princess Lilian de Réthy, Indochinese Emperor Bao Dai, the Shah of Iran, Juan Perón, Crown Prince Faisal of Saudi Arabia, the Dulles family and the Du Ponts, were either customers or about to become customers. Thanks to the brilliant coachwork being applied to the Ferrari chassis by the likes of Touring, Vignale and Ghia, and to the marvelous screeching of the machines themselves, the early production cars—the 195 and 212 Inters, 340 Mexicos and Americas and 375 Mille Miglias—had an electrifying quality about them that made even the rival Alfa Romeos, Jaguars, Aston Martins and Maseratis seem almost crude by comparison. Enzo Ferrari may have been distracted by winning Formula One races, and he may have considered the road cars simply sources of revenue, but the passenger cars, with their spiky, faintly feline nastiness, were the heart and soul of the early Ferrari mystique.

Chinetti actually held two of the fourteen dealer franchises around the world: at 252 East Sixty-first Street in New York (one of several rather temporary locations in Manhattan) and at 65 Avenue d'Iéna in Paris. In addition to Franco Cornacchia's prime Milan location on Via le Piave, dealers were located in London, Rome, Zurich, Algiers, Casablanca, Melbourne, Montevideo, São Paulo, Oporto, Florence (as part of a Studebaker dealership) and Brussels (where space was shared with the Belgian Nash distributor). For the most part these outlets carried little or no inventory and were dealers only in the sense that the Prancing Horse sign was carried in the windows.

Automobiles might be special-ordered if the stars were in the proper alignment and a delivery could be arranged through the rather ragged customer service office in Modena.

While the passenger-car side of the business was informal at best, Ferrari had by 1951 developed a strong, lucrative list of sponsors who supplied his racing team not only with money but with much-needed parts and technical support. Prime among them were such old associates as Pirelli, Champion, Shell Oil, Weber carburetors, Mondial pistons, Borrani wheels, Kiklos piston rings, Abarth exhausts, Marelli ignition, Livia valves, Houdaille shock absorbers and such newcomers as Vandervell bearings and Brevetti Fabbri, which supplied the factory with U-joints and drive shafts. In all, twenty-one manufacturers were associated with the Scuderia as suppliers and sponsors.

Ferrari had his eyes fixed westward, toward the Alfa Romeo racing shops at Portello. He knew that Colombo was laboring to find more power and handling with which to combat the new Ferrari threat. At the same time Lampredi developed a twin-spark-plug cylinder head that raised his V12s output to about 380 hp at 7,000 rpm. Colombo countered with the 159A, which cranked out 405 hp at just over 10,000 rpm. But fuel mileage dipped to below one and a half miles per gallon. Therefore tankage was increased to seventy-five gallons, adding unneeded weight. The cars, for all intents and purposes, were about equal, with the Alfa possessing more outright top speed but the Ferrari offering better throttle response and acceleration away from slow corners. The main difference lay in age. The Alfa was peaked out, the last erg of energy about to be sapped from its aging bones. By contrast, the Ferrari was essentially new and loaded with potential. Why it had taken four long years to reach a point where it was able to challenge the Alfa remains a mystery, considering the apparent vitality of the Ferrari operation.

The harsh fact is that over the years Ferraris often succeeded on the basis of quantity, not quality, and numerous race victories came in the face of limited competition. When one considers that wins at Le Mans were enjoyed only when major antagonists such as Jaguar, Mercedes-Benz, Porsche and Ford were not active, and that any number of Grand Prix teams produced faster automobiles, the championships won by the Maranello concern were as much a result of persistence and faithful attendance as they were of technical brilliance.

In 1951 the British firm of Jaguar fielded slim-bodied C-Type roadsters powered by essentially stock twin-cam 6-cylinder engines (still used forty years later in the XJ6 sedans!) and easily won the Le Mans 24 Hour race. The best Ferrari, driven by Chinetti and Lucas, straggled home in eighth. Some solace was gained from Villoresi's stirring win in the Mille Miglia, where he drove his 4.1-liter Vignale coupe to victory over a gaggle of smaller, hopelessly outdated and outclassed Lancias, Alias, OSCAs and small-displacement Ferraris. The Scuderia's record in this epic road race was superb, but it must be noted that during the 1950s it faced first-class competition only four times: against Mercedes-Benz in 1952 and 1955, Lancia in 1954 and Alfa Romeo in 1953. The Ferraris were soundly beaten in 1954 and 1955 and barely squeezed out victories in 1952 and 1953. In all four events the team enjoyed a massive numerical advantage, in terms of both cars and experienced drivers.

Sadly, Villoresi's victory celebration in Brescia was tempered by the news that Ascari had crashed shortly after the dawn start. He had spun into the crowd lining the highway between the northern villages of Lonato and Desenzano, killing a prominent local doctor and seriously injuring several others. Ascari complained that he had been blinded by the headlights of a car approaching from a side road. No matter, in keeping with the zany dictums of Italian jurisprudence, he was charged with manslaughter, while, ironically, the race organizers who ignored the most elementary crowd-control precautions went free. Ascari was finally cleared of the charges three years later.

During the opening races of the 1951 season, the two titans of Grand Prix competition, the Alfa Romeos and the Ferraris, struggled nose to nose. Alfa, thanks to the dazzling driving of Fangio, managed to keep its amazing winning streak unbroken until the second weekend of July. At that point the international racing circus converged on the flat, featureless converted aerodrome at Silverstone for the British Grand Prix. Added to the Ferrari team was a dour, portly Argentinian pal of Fangio's named Froilian Gonzales. The racing fraternity called him "Pepe." (While Fangio and Gonzales are generally ranked as two of the finest racing drivers of all time—and certainly the best pair ever to be produced by Argentina—one of their contemporaries may have been even better. Years later Nello Ugolini, whose knowledge of the sport during that period is unsurpassed, claimed that their country-man Oscar Galvez was faster than both Fangio and Gonzales, but lacked the blessing of the Perón government. He was therefore doomed to run in minor events in his homeland, rather than on the European Grand Prix circuit.)

On June 14, 1951, Pepe Gonzales rode his year-old, single-plug, 4.5-liter Tipo 375 Ferrari Grand Prix to the first World Champion-ship victory in the face of the full Alfa Romeo team. Ascari had bro-ken his gearbox and Farina's clutch had come apart, leaving the two Argentinians, Fangio and Gonzales, to battle it out for their respective teams. The superior torque of the Ferrari paid off on the track's flat right-angle corners, and Gonzales won by nearly a minute with Vil-loresi trailing in third.

This was a moment of grand-operatic significance for Ferrari. He had savored the prospect of outdueling his rivals since he had been cashiered by Alfa Romeo in 1939, and he celebrated it by writing some of his most mawkish prose: "I cried for joy. But my tears of en-thusiasm were mixed with those of sorrow because I thought, today I have killed my mother." He followed this up with a telegram to Satta at Alfa Romeo which said, in part, "I still feel for our Alfa the adoles-cent tenderness of first love."

Such florid twaddle is hard to swallow, even when dished up by an actor like Ferrari. Those bleatings, which might have come from a failed novel, are doubly absurd when one recalls that for five years the abiding ambition of Enzo Ferrari had been to flog his old company's cars on the racetrack. He may have "killed his mother," as he so melo-dramatically put it, but only after he had been mercilessly pummel-ing her about the head and shoulders for half a decade.

As it turned out, the old lady still had some life in her tired bones. Although Ascari won again for Ferrari at the German Grand Prix and led a one-two finish over Gonzales in the critical Italian Grand Prix, Fangio was victorious in the final race of the season at Barcelona, thereby securing the first of his five World Championships. The Ferrari team gambled on smaller 16-inch wheels to gain acceleration and thereby fried their tires trying to keep up with the Alias. Colombo, on the other hand, gained a measure of revenge by softening the 159A's suspension for the rough track and helped it finish the season in triumph. Gonzales struggled home second, well behind his countryman and completely frustrated over the tire problem. Angry words were exchanged between Ferrari and the Pirelli staff, each claiming that the other should bear the blame for the debacle. Ferrari maintained that the smaller 16-inch rubber should not have failed. Pirelli countered that their larger 17-inch tires—which both Ferrari and Alfa Romeo had employed for the entire season with good results—were perfectly suitable for Barcelona and should not have been replaced at the last hour. The dispute simmered on without resolution. Franco Rocchi, the expert designer who had just joined Ferrari (to begin a career that would span three decades), recalls that the argument between Ferrari and Pirelli following Barcelona was much more bitter than outsiders realized and blighted a relationship that had by then spanned over twenty years.

It had been widely predicted that the following season Alfa would resurrect the mid-engine, Ricart-designed 512, but the trend was clear. Lampredi's "atmospheric" 4.5-liter engine, with its massive torque and excellent fuel mileage, had, for the time being at least, doomed the supercharged power plant in Formula One. Following the Spanish Grand Prix, Alfa Romeo announced that it was retiring its overstressed and underfinanced 159s from international competition, thereby leaving the field clear for Scuderia Ferrari. A few feeble contenders hung about from Great Britain, including the stupefyingly complex V16 BRM, and Maserati seemed to be on the verge of organizing a return to motorsports, but their situation was uncertain. By all logical measurements, there was no one in sight to challenge the Tipo 375s and their contingent of expert drivers. Suddenly, after five years of indifferent results, Enzo Ferrari was king of all he surveyed.

12

It was now within his grasp. Since the first crude Tipo 125 had rolled onto the Abetone road, Enzo Ferrari had clung to the dream of becoming what he called a "grand constructor"—the classic manufacturer of complete racing cars and high-performance road automobiles for a select clientele of elite enthusiasts. Only a few men in history had achieved such an elevated status, most prominently Ettore Bugatti, England's W. O. Bentley and America's Duesenberg brothers. Ferdinand Porsche and his son Ferry were attempting to enter the elite circle with their tiny Stuttgart operation, but they would wait years before their tiny, Volkswagen-based sports cars would have an impact. The remaining Maserati brothers were struggling in Bologna with their nascent OSCA operation, but it had been long since proven that they lacked the business acumen necessary for success. The Orsis retained visions of grandeur for the husks of the Maserati operation, but they were spreading their efforts between industrial implements and limited offerings of thoroughbred automobiles.

The recent capitulation of Alfa Romeo had a devastating impact on the sport. It left the Formula One battleground with but one contender, the Scuderia Ferrari. There were no serious Italian rivals. The French Talbot-Lago operation was now suffering from financial problems and was cursed with prewar machinery that had not a prayer against the potent 375 Ferraris. The feisty Frenchman Amédée Gordini was ready to field a team of lightweight 1.5-liter cars, but a chronic lack of funds left his team with quick, but unreliable equipment. The British BRM program, while blooming with potential, was ravaged by bad organization and apparently was unprepared to do serious battle. A few tiny English and German makers harbored dreams of glory but nothing else. Alfa Romeo's withdrawal left Enzo Ferrari with an open racetrack for the entire 1952 season.

The FIA made a quick decision. Because of the lack of competition, the 1952 World Championship would be contested with Formula Two cars—the unblown 2-liter machines that served as a minor league of open-wheel racing in Europe. This news delighted Ferrari. His Formula Two 2-liters had been dominant for the past few years, and Lampredi already had a new engine on the dynamometer. It would be an extension of the designer's obsession with simplicity and reliability. The engine would be an in-line, twin-cam 4-cylinder similar in theme—if not direct design—to the powerful Miller-designed, Offenhauser-built 4s that had dominated Indianapolis competition since the early 1930s (with the notable exception of the Maserati triumphs of 1939–40, a memory which Ferrari intended to erase).

Franco Rocchi, who worked at Lampredi's side during those years, says flatly, "Lampredi was inspired by the Offenhauser. He adapted its simple 4-cylinder layout—with his own modifications—for Grand Prix competition." While technically different in many ways, the fact that Lampredi's new engine was theoretically based on an American power plant intended for dirt-track oval competition was long ignored by those seduced by the Ferrari mystique.

The new Lampredi 4-cylinder would be placed in a chassis that was essentially a lightened, downsized version of the big Tipo 375 Formula One car, with precious little thought given to exotic suspension components, special brakes or steering. Ferrari believed that the engine was 80 percent of a car's potential and that sophisticated chassis were frivolous. It would take years and many hard lessons to change his mind.

With the arrival of 1952, projects for the busy factory centered on the creation of the new Tipo 500 series racing cars and more road cars to be offered with both the big Lampredi V12 and the smaller Colombo engine. Moreover, a daring new venture would extend the Ferrari myth across the Atlantic Ocean in spectacular fashion. Luigi Chinetti was now firmly entrenched in larger quarters on Manhattan's West Fifty-Fourth Street and had persuaded Ferrari that an assault on Indianapolis was necessary to capture the American market in high-performance sports cars. The project made sense for a number of reasons. The "long" 4.5-liter V12 was useless for European competition and could easily be adapted for use at Indianapolis. Second, Ferrari savored the notion of winning that great race and thereby snatching the singular honor of an Italian triumph there from the Maseratis. The American organizers were known to be receptive to European entries, and Chinetti assured him that the skids would be properly greased. A specially designed car for Alberto Ascari was ordered up. Chinetti believed that he would be the exclusive entrant, only to discover that Ferrari had accepted orders for three more cars from Howard Keck, a wealthy California oilman; Colorado amateur driver Johnny Mauro; and the Grant Piston Ring Company. Chinetti was furious that his effort would be blurred and diffused by these three interlopers, but he was helpless in the face of what he considered to be an act of treachery. As it turned out, the trio of customer cars were slow and failed to qualify for the 500-mile race.

The Indianapolis adventure would end in failure, with Ascari barely making the starting field (after Lampredi made a last-hour flight to Indiana with a special manifold and carburetor setup). A wire wheel collapsed early in the race (the Americans were already using the stronger, lighter magnesium types, which Ferrari would resist for a decade), and Ascari spun out while running seventh and out of contention. The team returned to Italy pledging a return, but this never happened. Nino Farina would make several abortive attempts in cars with Ferrari engines, but the lack of a victory at Indianapolis would remain for the rest of his life an openly admitted disappointment for Enzo Ferrari.

While Alfa Romeo had disappeared as a major rival, the old nemesis from the north, Mercedes-Benz, had resumed the racing wars. The parent company, Daimler-Benz AG, had recovered sufficiently from the ravages of Allied bombing to begin building not only mundane passenger cars and trucks but a magnificent new sports car—a radical "gull wing" coupe (so named because the doors opened upward, like the wings of a bird). These 300SL coupes were to be entered in the major sports-car races of 1952, including the Mille Miglia. Power came from a 3-liter, single-overhead-camshaft, straight-6 engine that had been developed for the 300 series luxury sedans. (A year later Mercedes-Benz would equip the cars with direct fuel injection developed by their Stuttgart neighbor Bosch. This simple, ultra-efficient system had been in experimental use by the Mercedes-Benz racing department since 1934 and was about to come into general use on all types of racing cars. However, Ferrari would refuse to switch and remained loyal to Weber carburetors for another decade!)

The first race for the new 300SL Gull Wings was the 1952 Mille Miglia, where they directly challenged Ferrari on his home turf. The team was headed by prewar aces Rudi Caracciola and Hermann Lang, along with sports-car pro Karl Kling. Because Ascari was away at Indianapolis, Ferrari selected flinty, gray-haired Piero Taruffi to lead the Maranello defense with a new 4.1-liter Vignale-bodied 340 America roadster (featuring three "portholes" on the fenders of a type which had been introduced on the 1948 Cisitalia coupe styled by Pinin Farina and later adopted with typical Detroit overkill by Buick.) Backing him up were the Marzotto brothers, Gianinno, Paolo and Vittorio, and a privately entered 250S Vignale coupe for the spectacular but erratic Giovanni Bracco.

In an age in which racing drivers have become sullen, money-grubbing technoids, men like Bracco seem almost Neanderthal in their attitude toward the sport. But they brought to motor racing a brio and a kind of damn-the-torpedoes brashness that seem as out-of-date today as dueling over the honor of a lady. Bracco, whose family, like the Marzottos, had gained wealth from textiles, approached racing in a grand manner, driving with a maniacal style and conducting an off-track social life that would stagger a Cape buffalo. Like numerous wellborn Italian sportsmen, he considered the Mille Miglia his métier, a battleground where elegant amateurs like himself could compete with the Grand Prix elite. The year before, he had finished second in a smallish Lancia Aurelia B20 coupe with young Umberto Maglioli (who would later make his mark with the Ferrari team as a driver) as riding mechanic and navigator. The Aurelia, with its 2-liter engine, was half the size of Villoresi's winning Ferrari, but it handled flawlessly in the mountain passes and Bracco drove it like a madman. Over the route a shaken Maglioli recounted that he had lit no fewer than 140 cigarettes for Bracco during the thirteen-hour ordeal!

Bracco gave another display of bravado in 1952. The Mercedes-Benz of Kling worked flawlessly, and despite Bracco's best efforts, the German appeared headed for a decisive victory. Worse yet, at a final stop tires of the wrong size and compound were fitted to Bracco's Ferrari, which further slowed him. But the wild man from Biella would not be denied. With his navigator, Alfonso Rolfo, again feeding him cigarettes, and braced by a few snorts of brandy, Bracco rocketed over the rain-swept Futa and Raticosa passes and beat Kling by over four minutes. It was a brilliant drive, and the Italians celebrated the victory over the hated Germans for days. (Their jubilation would be brief. A month later the Germans' silver cars finished one-two at Le Mans, with the best Ferrari finishing fifth.) Following his Mille Miglia victory, Bracco told journalist Gino Rancati that he was inspired by visions of German troops executing Italian *Partigiani* near the end of the war. This forced him (surely with the help of the spirits and the nicotine) to drive all the faster!

The victory may have been more trouble than it was worth for the Scuderia. Bracco had made a deal to run the Mille Miglia on French-made Michelin tires, while the factory cars were fitted with those from their old ally, Pirelli. Assuming a win, the Pirelli advertising department had prepared a special campaign trumpeting their triumph. But Bracco's defection to Michelin caused an internal uproar and further soured the relationship.

Although Ferrari makes only passing reference to Bracco in his memoirs, this hard-living driver was not only a noted competitor in Italian events but a generous customer. (In 1962, ten years following his Mille Miglia win, he was invited to Modena to celebrate the anniversary. Enzo Ferrari gave him a wristwatch. Bracco mused, "A watch worth a few thousand lire. And in order to compete with his cars, it cost me six hundred million lire. But it doesn't matter.")

But Bracco had his moments of revenge. Tavoni recalls that he had arranged with Ferrari to drive a special Berlinetta coupe in the Tour of Sicily and the factory labored around the clock to complete the car and deliver it to Palermo. But Ferrari received news that Bracco did not start. The new coupe sat unattended and ignored as the flag had dropped. Furious that his quick service had gone for naught, Ferrari demanded an explanation. "You will understand, I'm sure," answered Bracco. "Before the race started I met a beautiful woman and fell in love. Knowing your affection for the opposite sex, I was confident that you would understand missing a simple race to spend a little time with such a lovely lady."

Happily for the Commendatore (a title he ceased to use, but one that was informally applied to him for most of his life), Bracco's zaniness was only a minor distraction in what was to be his most triumphant Grand Prix season. Alberto Ascari, now thirty-five and at the height of his considerable powers, led the Scuderia into what was expected to be a walkover against the feeble opposition. The team had been strengthened by the arrival of Farina from Alfa Romeo and that of old friend Nello Ugolini from Maserati, whose competition-management skills had earned him the nickname "Maestro." The June Monza Grand Prix saw their major threat removed from the field when Fangio crashed hard in his updated Maserati A6GCM and was seriously injured. He had driven for BRM the day before at Dundrod in Ulster. Flying to Paris, then grounded by fog, he borrowed a car and rushed across France and over the Alpine passes, reaching the track only hours before the start. Fatigued and irritable, the great Argentinian ace was placed at the back of the pack without the benefit of a practice lap. The flag dropped and Fangio began charging through the field, only to lose control on the second lap at the dreaded Lesmo right-handers and plunged into the trees bordering the circuit. His neck broken, he lay unconscious for several days, then awoke to face a five-month recuperation. This disaster removed from the scene the only driver consistently capable of challenging Ascari and the Ferraris.

With Fangio absent and with the Maseratis only beginning to be updated by Colombo, who had just arrived back in Modena after the Alfa racing operation had been closed, Ascari drove on to dominate the season in unprecedented fashion. He won eleven of the seventeen Grand Prix races the team entered. Villoresi and Taruffi won five more. Thus Lampredi's simple, neatly conceived little Tipo 500s won sixteen of the seventeen races they entered in 1952 (being beaten only at Rheims by an inspired Jean Behra aboard a French-built Gordini powered by what was rumored to be an illegal engine).

This was a period of enormous growth for Enzo Ferrari, in terms of both factory activity and prestige. He was honored with the title Cavaliere de l'Avoro Merito della Repubblica for his commercial successes. Ferrari would never use the title publicly, but privately was proud of the recognition and considered himself to have been elevated to an exclusive group by the appellation. Surely the second Cavaliere award was more meaningful and significant than the one awarded at Pescara nearly three decades earlier. Ascari's triumphs raised Ferrari and his team to the level of national superstars. The dominance of the automobiles carrying the *Cavallino Rampante* made them hot properties on three continents. The wily Ferrari, with three decades of negotiations with race organizers behind him, drove hard bargains for starting money from promoters in Europe as well as North and South America. Thanks to Chinetti and to a group of rich, enthusiastic sportsmen led by Jim Kimberly, Temple Buell and John Edgar and an intense, talented young Southern California driver named Phil Hill, the Ferrari mystique was spreading in the United States. Hill was a superb competitor, possessed of a singular wit and intelligence. His achievements with the Maranello cars would lead to dozens of sales and the beginning of a Yankee-based Ferrari fever that is unabated to this day. "The enthusiasm of the Americans for Ferrari cars was unequaled anywhere else and is *ninety percent* responsible for the success and the mystique surrounding the cars, a fact, ironically, that Ferrari himself never understood or cared about," says Roger Bailey, a renowned team manager, race engineer and former mechanic who worked in the Ferrari competition department during the late 1960s.

Still, the North American market remained a relatively low priority for the provincial Ferrari. Despite his race victories and the Indianapolis campaign, Chinetti remained something of an outsider, relegated to selling automobiles in a faraway land and essentially disenfranchised from the daily politics of the factory. Each negotiation was a tedious struggle, a battle of numbers, broken promises and misplaced trust. A perfect example: In 1952 he had gone to Maranello to take delivery of a pair of new V12 340 Mexico coupes. They were to be entered in the Mexican Road Race, the vaunted Carrera Panamericana de México—one for himself and co-driver Jean Lucas, a second for Bill Spear, a rotund, bespectacled, intensely competitive American sportsman. A year earlier Chinetti had teamed with veteran Italian road racer Piero Taruffi to win the twisting, flat-out, 1,936-mile open-road race that ran from the Guatemalan to the Texas border. Their victory had been the first of international magnitude for Ferrari in North America and sales had spurted. Chinetti had returned to Italy to arrange for delivery of the two new Mexico models in hopes of repeating the victory and thereby further enhancing the U.S. market. Ferrari was enthusiastic. To a point.

There was the question of an outstanding balance of $2,200. That much was owed on the two cars and he would not release them until it was paid. But surely, Chinetti had argued, a tiny balance like that for the company's own exclusive American importer . . . Surely it could be overlooked, at least until the balance could be cabled in a few days. Surely an understanding could be reached between old friends . . . The argument gained in decibels as the two men drifted into the cobbled courtyard inside the factory. Chinetti was returning to his hotel in Modena. He wanted to arrange transport to Mexico on the next ship. Time was critical. He needed the automobiles. Funds would be forthcoming, he shouted. "Just trust me and give me the damned cars!"

The argument was over. Ferrari pulled himself to his considerable height and his soaring Roman nose jutted out. His arm swept forward as if he was directing a cavalry charge. Chinetti instinctively stepped back from this imposing, red-faced proconsul and found himself standing on the edge of the Abetone road. Ferrari's deep, dramatic voice thundered across the small brick courtyard like a cannon shot. "Close the gate!" he screamed. A guard in a rumpled uniform leapt forward and slammed the iron barrier in Chinetti's stunned face. Enzo Ferrari had spoken. And the two cars had not been released before the additional balance was received and deposited.

Modena was becoming a kind of Casbah of high-performance automobiles. In addition to the powerhouses of Ferrari and Maserati, all manner of specialty manufacturers were located in the city, energized by, as Enzo Ferrari described it, "a psychosis for racing." Now the hulking, burnt-umber building on the Viale Trento e Trieste was serving as a customer service and delivery center for both passenger and racing cars, as well as the family residence. Around the corner and across Garibaldi Square was the Albergo Real. That hostelry, which was run by an ex-madam, lay a few blocks to the west along the Via Emilia from the other decent hotel in town, the Grand. Between them, they served as headquarters for the endless trail of aspiring drivers, designers, visionaries, pretenders, rich sportsmen eager to exchange their money for a brief dash through the spotlight, car traders, dealers, hustlers, promoters, charlatans, phonies, journalists and romantics of all kinds who had been drawn to Modena by the siren song of the Ferrari V12s and the guttural bark of the Maseratis 6s.

One Modena regular was Alejandro De Tomaso, an Argentinian enthusiast and sometime racing driver who had fled to Italy after, according to a colorful rumor, patriotically attempting to bomb Perón's palace. In Modena, De Tomaso would meet an American heiress named Isabelle Haskell, who briefly fancied herself as a racing driver and was a steady Ferrari customer. After they were married, De Tomaso used Haskell's abundant funds to help start his own car company in Modena in 1965 and then to seize control of a debilitated Maserati firm a decade later. De Tomaso was but one of the legions of colorful characters who could often be found headquartered at the Real and the Grand during the 1950s.

The guests were for the most part decked out in the de rigueur uniform of the sporting fifties: Lacoste pullovers, lightweight slacks and Gucci loafers, accented with Rolex watches. Their flashy women were laden with the wares of Pucci and Gucci, two exemplars of the spectacular Italian postwar fashion revolution. These aspiring pace-setters of international style had come as supplicants to visit the lordly Ferrari, who, by contrast, remained attached to Old World frumpi-ness. His wardrobe never varied: a white shirt and muted tie, suspend-ers and a gray or brown business suit with a tiny Prancing Horse pin in the lapel. The expensive Swiss chronograph he carried on his wrist also displayed a tiny Prancing Horse on its face. Chilly days brought out a tired, deep red woolen overcoat and a rumpled gray fedora. The iron-gray hair, always perfectly trimmed, was closely clipped on the sides and swept back in the traditional mode. Ferrari's only departure from the fashion of an old Modenese *padrone* was the occasional donning of a yellow pullover sweater and a blue jacket (the colors of the city and its football squad) for public appearances to introduce a new car model. Otherwise, this man, now so rooted in time and place that he never slept out of his own bed for the final forty years of his life, was indistinguishable from a thousand other senior citizens that plied Garibaldi Square on their daily rounds.

It was Enzo Ferrari, handcuffed by the past, yet goaded by the future, who single-handedly had given Modena its newfound reputa-tion as a center of world-class glamour cars. To be sure, the Orsis were there, but they were businessmen, not ego-driven visionaries. They would soon falter in the race and give way to mundane financial de-mands while Ferrari's commitment was total, unswerving, and emo-tionally complete. He would whine about retiring on occasion, over dinner in the Fontana restaurant or among fellow members of the Biella Club—an informal social group composed of motorsports enthusiasts—but the threat was momentary and meaningless. Periodic public threats to leave the arena were simply a way of collecting more sponsorship support. Those who were close to him—Tavoni, Ugolini, Amarotti, Giberti—knew full well that his automobile business was not a means to an end, it *was* the end, the very essence of his being. In the final analysis, the fast cars, the Grand Prix victories, the factory itself were all secondary to what they stood for, the towering ego of the man whose name they carried, Enzo Anselmo Ferrari.

Overall, his personal style was an extreme contrast to the automobiles he created. They were flashy, high-styled, daring, brash, openly outrageous; he was drab, mundane, semi-monastic. In many ways Ferrari was an automotive couturier, but unlike his predecessor Ettore Bugatti, who fairly blossomed with elegance and artistic panache, or compared with his counterparts in fashion like Coco Chanel or Christian Dior, Ferrari remained a simple man with simple tastes. In his automobiles there seemed to be a transference of his powerful ego and the need to express himself artistically, but in personal terms he was muted and introverted to the point of obsession.

As his world focused more each day within the narrow route between the apartment and the factory in Maranello, Ferrari's hatred of the Orsis and their Maseratis intensified. To be sure, massive threats to his empire were forming in faraway Stuttgart and even in England, where former friend Vandervell was creating a formidable team, but it was the impudent growl of the Maserati 6s that echoed through his shuttered apartment windows each day.

Now their primal shrieks were even filtering through the sylvan, tree-shrouded environs of the western neighborhoods. The Modena autodrome had been opened on the edge of the city. A short, dangerous, hedge-lined, roughly rectangular circuit had been laid out around the perimeter of the acreage, which also served as a landing strip for light aircraft and contained several soccer fields. Both Ferrari and Maserati were using the autodrome for extensive private testing, although stretches of public highways like the Abetone road and the Autostrada del Sol were still employed for flat-out high-speed runs.

It was during this heady period, when powerful, outrageously noisy automobiles were ripping along the roads of Italy, that the modern mystique of fast driving was created. With romantic, woolly characters like Bracco and Villoresi and Farina and Taruffi and Ascari forever skidding around corners and blasting through villages behind the wood-rimmed steering wheels of their bombastic red machines, the entire Italian population was infused with this "psychosis" for speed.

Still, there was a severe, humorless edge to Modena that countered the rage for fast cars. The Communists were entrenched in the political power structure, and continue to be relevant to the city's politics to this day. Following the end of the war, northern Italy had made a shift to the left. This had been in a sense nullified nationally by the De Gaspari centrist government in Rome. But the labor unions, the petit bourgeoisie and the intellectuals, in reaction to the reality that the upper classes had for the most part supported the Fascists, moved into the ranks of the socialists and the Communists, who retained the strength they had gained as the leaders of the partisan guerrilla units. Any number of northern industrial centers, including Milan, Bologna and Modena, harbored large, committed Communist parties (although the Russians soon discovered that they were an extremely independent, highly nationalist lot who ignored the dictums from Moscow).

The nature of the Italian body politic led Ferrari into a display of theatrics after engaging in a dispute with the Modenese city government over some long-forgotten slight. Knowing that the mayor—now conveniently converted to Communism—had been an enthusiastic Fascist in the 1930s, Ferrari entered a meeting wearing a black shirt— the hated symbol of Mussolini's thugs. Dramatically tearing away the garment and tossing it at the mayor, Ferrari shouted, "You wore this long before I did!" From then on this citizen, now a powerful factor in the local economy, was seldom jostled by the local government. On the other hand, as Ferrari became more prosperous, his disputes with the labor unions would grow. This would lead to strikes, but they often took the form of shouting matches and symbolic walkouts lasting no more than a few hours. Considering the volatility of his personality and the pugnacious attitude of northern Italian labor leaders (many of whom recalled Ferrari's profiteering from the war effort), it is a testimony to the man's leadership qualities that work stoppages weren't more common or more violent.

But for all the notoriety, racing victories and increasing customer sales, the company (formally called Auto Costruzione Ferrari) was being run in an organizational sense like one of the dirt-floored custom shops scattered around the city. Money remained tight, and Ferrari was constantly meeting with bankers, independent financiers and wealthy patrons to stay in operation. As his business increased, so had his overhead, and while the 1950s brought him world fame in the car business, his bank accounts remained slim. This prompted some chaotic—and often hilarious—business practices.

It was during this time that a prosperous Frenchman, Jacques Goddet, the publisher of the giant sporting weekly newspaper *L'Equipe*, ordered a little 212 Inter coupe for his wife. It was to be her birthday present and the entire car was to be in her favorite color, blue. The Pinin Farina coachwork was to be blue, as was the leather upholstery and the Florentine custom-made travel luggage for the trunk. The money was handed over and the task of assembling the exotic machine was begun. After months of work, the car was returned from Pinin Farina's Turin body plant and final detailing was undertaken at the old Scuderia building in Modena. It was at this point that a visiting Texas oilman strolled into the shop and spotted the car. He made what Mario Puzo might have described as "an offer that could not be refused," and Ferrari sold the Frenchman's car to the American. Now, with the delivery date nearing, a mad scramble ensued to find a replacement. A similar machine was located, but there was a problem: It was red.

When Goddet arrived, the Scuderia was vacant. Ferrari was nowhere to be found, nor were any of the senior staff. Only a lowly floor sweeper was in the place and he of course pleaded total ignorance to the apoplectic new owner. There being no rational alternative, having traveled all the way from Paris to get the automobile, the seething Frenchman drove away with his red substitute. Months later Ferrari received a letter. It said, in essence, "Dear Monsieur Ferrari, you make a wonderful automobile and my wife is delighted. There is but one thing you must understand: you are hopelessly color-blind!"

Despite these lapses, Ferrari's little production line turned out forty-four customer cars in 1952. But what truly mattered to him, of course, was maintaining the vise grip on Formula One that he, Lampredi and Ascari had established. Few changes would be made in either the race cars or the composition of the team, although a cocky, flaxen-haired Englishman named Mike Hawthorn would be added to the driving lineup—the first non-Italian to be accorded such an honor since the late, lamented Raymond Sommer (who had died in September 1950 during a minor event at Gardours).

Hawthorn's selection to join the Scuderia was greeted with much jubilation by the British motoring community. He was the first Englishman to be invited to join a first-rank continental operation since the late Dick Seaman. At that point the British were suffering from a massive inferiority complex in Grand Prix racing. Their best effort with BRM had been a failure and their other specialty builders were simply too small and underfinanced to field competitive cars. Ferrari derided them as *garagistas* and assumed that they were incapable of challenging the might of the Italian "grand constructors." While his fellow British enthusiasts fantasized about his anointing, Hawthorn quickly learned that while his move to Maranello might bring glory, it would assuredly not bring riches. Ferrari's deal was standardized, even for stars like Ascari and Villoresi. He kept all starting money, all sponsor and contingency funds and half the prize money. The driver got the other half of the purse (generally modest), although it was expected that 10 percent of that share would be given to his mechanics. Therefore, aside from the prestige accrued from driving for the vaunted Prancing Horse and the opportunity to line one's shelves with silver bowls and trophies, remuneration from the Ferrari team was strictly pocket change.

Facing them was a reenergized Maserati team, headed by a totally fit Fangio and Gonzales, who had defected to the crosstown rivals on nearby Via Ciro Menotti during the routine game of postseason musical chairs that to this day causes race drivers to endlessly migrate from team to team. With Colombo having great incentive to tweak the aging A6GCMs to a level where they could match the 500s of the company that had dumped him, Ferrari had legitimate reason to believe that the 1953 Grand Prix season would be a brutal intramural struggle between the two Modenese teams.

He was wrong. Maserati, which had ridden a financial roller coaster since its formation, was on a swooping downslide. The Korean War had ended with a truce in the summer of 1953, and the company, which had made a major commitment to the machine tool business, found orders at a standstill. Cutbacks were made, and the Orsis began spending more and more time in Argentina, where the Perón government appeared to offer new opportunities for commerce. Even with the genius of Colombo on their side, Maserati was too crippled to offer serious competition in the short term. Ascari opened the season with wins in the Grand Prix of Argentina, Holland and Belgium, which, coupled with his late-1952 string of six victories, gave him nine consecutive triumphs in World Championship events—an achievement never equaled before or since. Hawthorn was to win the French Grand Prix at Rheims after a monumental struggle with Fangio, and Ferrari went on to dominate the rest of the calendar, losing only at Monza, where Ascari tangled with Onofré Marimon on the final lap and let Fangio through for Maserati's only win of the season. Colombo had succeeded in giving the Maseratis horsepower (about 195) roughly equal to the Ferraris', but the cars were deficient in road holding and handling. Therefore, as the year wound down and the 2-liter World Championship ended, Alberto Ascari easily repeated as king of the sport, with Ferrari having won twenty-eight of the thirty-one races contested under the formula. Until the domination by the McLaren-Hondas during the 1988–90 seasons, no single team ever enjoyed such superiority over the competition.

Still, it was not enough for Enzo Ferrari. His streak in the Mille Miglia remained unbroken with Giannino Marzotto's lusty Tipo 340MM nipping Fangio's new Alfa Romeo 6C-3000 CM coupe by twelve minutes to win overall. But Le Mans was a disaster. Jaguar returned with updated C-Types fitted with disc brakes (which Ferrari refused to consider using) and smashed all records for the race. The sleek British racing-green cars finished first, second and fourth, while the best Ferrari straggled home fifth, almost a hundred miles behind in total distance. The front-line car was a 340MM coupe for Ascari and Villoresi, fitted with a 4.5-liter Indianapolis engine. It set the lap record, but the clutch gave out in the face of the extra horsepower and the car retired short of the halfway mark. The 24 Hours of Le Mans was a critical selling tool for the production *gran turismos*, and with ever-increasing market activity by Jaguar, Aston Martin, Maserati, Alfa Romeo, Lancia, Mercedes-Benz, Porsche and the American-built Cunningham, failure in that highly visible race could mean disaster in what was becoming a crowded bazaar of exotic automobiles.

More troubles lay ahead. The Fédération Internationale de l'Automobile had long since announced a major Formula One change for 1954–60. The power plants would be 2.5 liters, unsupercharged—a mere 500 cubic centimeters larger than the existing formula. But with Mercedes-Benz already announced as a contender and a somewhat revitalized Maserati completing a Colombo-designed single-seater to be called the 250F, word was received in Maranello that Lancia, after a three-decade absence from competition, was returning to Formula One. The automobiles were being designed by Vittorio Jano, and intelligence produced via Ferrari's squads of informers indicated that the cars—to be called D50s—were radical in the extreme. Ferrari's response to this new firepower was simple. Prior to the final race of 1953, held on the hometown streets and avenues of Modena, he announced to the world that his money woes had become intolerable and that he was withdrawing from competition.

Cynics have maintained that these periodic retirements were unvarnished theatrics and that Ferrari never entertained thoughts of quitting. But there is no question that he was a moody man, and it is possible that he occasionally faltered. After all, his financial problems were legitimate, and at fifty-six he was in his fifth decade of one kind of a war or another—either the real thing or one fought with angry, lethal automobiles.

A new foundry had been added to the factory and the coffers were no doubt depleted. However, the addition of the capacity to cast his own engine blocks and transmission and differential housings was a major breakthrough. For years Ferrari's pattern makers had been forced to depend on outside foundries, adding both to the cost and to the time required to create fresh designs. Now he could call upon the cadre of artisans in Modena already skilled in the art of sand-casting, which had remained unchanged since the glory days of Bernini. This facility and its elite retinue of craftsmen would provide an enormous advantage to Ferrari over the years to come. Literally hundreds of exquisitely crafted aluminum-alloy engines—12s, 8s, 6s, 4s, even a twin—would be poured out of the molds of the foundry. To this day the men who work in the essentially medieval conditions of the Maranello foundry form a link to the origins of the Ferrari car building philosophy and to the finest traditions of Italian bronze sculpture.

More bad news arrived. The greatest of the racing warriors he had ever known was dead. On a steamy August day Tazio Nuvolari had breathed his last following an agonizing year of struggling to stay alive. To his old friend and riding mechanic Decimo Compagnoni, Nuvolari had lamented, "I, who could dominate any car, am incapable of controlling my own body." He died in the arms of his wife, Carolina, on August 11, 1953, after making a final request that he be buried in his driving uniform. Hearing of the little man's death, Ferrari drove north to Mantua to comfort Nuvolari's long-suffering wife. He lost his way in the rabbit warren of streets in the ancient center of the city and finally, in desperation, stopped to inquire at a plumber's shop. An old man came out. Typically wary of outsiders, the Mantuan circled Ferrari's car, seeking the number plate. Sighting the letters "MO," which signified Modena, he shook Ferrari's hand and before directing him to Nuvolari's house said, "Thank you for coming. A man like that won't be born again." Until his own death, Ferrari dreamed of finding another driver who, in his mind at least, possessed the fire, the passion and the driving skill of the beloved Nivola. (A grief-stricken city transferred the Nuvolari home to Carolina and offered her a hefty lifetime living allowance, based on the mistaken belief that she was suffering from a fatal tumor. As it turned out she lived for twenty years on the largesse of the city.)

If indeed Enzo Ferrari was even momentarily serious about not competing in 1954, it was soon forgotten in the flurry of plans to ready new cars for the 2.5-liter formula. Lampredi was at the height of his creative powers and appeared intent on retaining the simplicity of his winning Tipo 500 4-cylinders. He even tried reducing the already modest 4-cylinder to a mere twin. This he accomplished thanks to the speed with which the Ferrari pattern makers and foundry men could create new engine blocks.

The 2.5-liter 2-cylinder was tried, and after several prototypes were rumored to have vibrated off the dynamometer and exploded, the idea was scrapped. But Lampredi insisted on another 4-cylinder for Formula One, with a straight-6 variation for sports-car competition. Meanwhile it was known that Mercedes-Benz was coming with a fuel-injected straight-8, and Lancia had already won the Mexican Road Race (Fangio up) with a 3.3-liter V6 similar to the V8 they would use during 1954. The opposition would be infinitely more potent than that which Ascari had steamrolled during the past two seasons.

Dino, now twenty-two years old, was beginning to exert a small, but significant influence within the factory. The Lampredi 4-cylinder was adopted for a number of sports cars, including the so-called 750 Monza. The machine was an evil-handling beast, but was shrouded in a lovely Scaglietti body which Dino is credited with designing. Ferrari himself was proud, writing to Laura, who was visiting family in Turin, that the car was being applauded for its good looks. "Even Pepino Verdelli is pleased," he noted. Despite his widely known frailty, Dino was spending more and more time at Maranello and had grown particularly close to a twenty-nine-year-old factory test driver named Sergio Sighinolfi. Dino often accompanied Sighinolfi on wild test rides in the nearby hills. These were bright moments in the life of a young man whose health seemed to be deteriorating by the day and whose drab and unhappy home in Modena offered him little joy.

As the year ended, Ferrari's relationship with his stars, Ascari and Villoresi, began to unravel. The two men were being wooed by Lancia. Fangio had signed on as Number One with Mercedes-Benz. Stirling Moss, now evidencing the brilliance that would propel him to all-time greatness, was aligned as an independent with Maserati, which left the Ferrari ranks temporarily unmanned. The flashy, prematurely balding Londoner might have been available to the Scuderia, had he not been the victim of a major faux pas in September 1951. Already a prodigy in British club racing, the twenty-two-year-old Moss had been invited by Ferrari to race a works Formula Two car in a minor event at Bari, on the Adriatic Coast. In company with his father, Alfred, a prominent dentist and former driver who had finished fourteenth in the 1924 Indianapolis 500, the eager youth made his way across Europe and down the boot of Italy, only to be tartly informed that the Commendatore had changed his mind and his car was to be driven by Piero Taruffi. No other explanation was offered and Moss returned to England swearing never to drive for the Scuderia. This pledge would almost be broken, but tragedy would intervene.

Two issues complicated Ferrari's negotiations with Ascari and Villoresi. During a series of meetings in Modena, Ferrari waffled on exactly what kind of car he would field against the new opposition, but seemed to be leaning toward a simple update of the Tipo 500s. This seemed inadequate to Ascari, who knew more about Jano's D50 project than Ferrari realized. Moreover, both Ascari and Villoresi were dissatisfied with Ferrari's meager financial deals, about which Villoresi later remarked, "We were blockheads to accept in the first place."

While Ascari and Villoresi negotiated with the Commendatore independently, it was a foregone conclusion that they would operate as a pair for whatever team signed them. With two championships to his credit, "Ciccio" was no longer the eager kid begging for a ride.

Villoresi, who had never been an admirer of Ferrari since his brother's death, felt limited loyalty to the Scuderia, although he was inclined to follow in the footsteps of his younger—and faster—friend and protégé. Both men dealt with Ferrari as an equal, offering none of the bowing and scraping that he had come to accept from the aspiring drivers who were appearing at his door. As a group, Ferrari disdained the rich pretenders who wanted to race for him, and he tended to let them freeze or fry (depending on the season) in the Maranello gate shack longer than normal. Calling him simply Ferrari, which he accepted from men he considered his social equals, Ascari refused to sign a contract in late December. (Villoresi claims he already had signed a deal with Lancia.) Immediately after the first of the year Ferrari produced a carefully worded press release indicating that his two stars would be racing elsewhere in 1954. Shortly thereafter both were welcomed into the arms of Lancia, where they would drive sports cars until Jano completed the vaunted D50 Grand Prix cars later in the season.

After two years in which the Scuderia could not seem to lose a race, Ferrari and his band of worthies plunged into a twenty-four-month slump during which they could barely buy a victory. Lampredi's new cars were unmitigated disasters. This happened despite the return of Alberto Massimino and Vittorio Bellentani from Maserati after aiding Colombo in the creation of the lovely, but only partially successful 250F—a single-seater of classic beauty that Maserati intended to sell on the open market.

With the departure of Ascari and Villoresi, Ferrari signed Gonzales, Hawthorn, the Frenchman Maurice Trintignant and the still feisty Farina, who was edging up on his fiftieth birthday. This foursome was engaged at the standard Ferrari pay scale—an insultingly low rate he kept in effect for many years because the supply of drivers yearning to serve the Scuderia always exceeded the demand.

Despite fevered revisions that spanned the next two seasons and created the ultimate in evil-handling Ferrari Grand Prix cars, the Tipo 553 "Squalo" and the Tipo 555 "Supersqualo," the team was cursed for two dreary seasons while the Mercedes-Benz W196s steamrolled everybody and carried Juan Manuel Fangio to his second and third World Championships. Worse yet, Ferrari was being thrashed in the major sports-car races by the radically improved Lancias and Mercedes-Benz 300SLR roadsters. Ascari drew first blood in the 1954 Mille Miglia by decisively beating Vittorio Marzotto's Ferrari 500 Mondial in his 3.3-liter Lancia D24 sports-racing car. Some honor was salvaged at Le Mans, where Gonzales and Trintignant narrowly won with a big 375MM. It was this race in which the American Briggs Cunningham ran a privately entered 375MM with special water-cooled hydraulic brakes. The car retired with rear axle problems, ending a contentious relationship between the American and Ferrari. Cunningham, an easygoing, elegant amateur sportsman, had been influential in promoting Ferrari fortunes in the United States. Not only had he purchased three of the Scuderia's automobiles, but his prominence led to the sale of similar cars to a number of wealthy associates. Yet when Cunningham informed Ferrari that his Le Mans car had broken an engine rocker arm and requested replacement, Ferrari imperiously replied that his rocker arms *never* failed and therefore no spare would be forthcoming. Considering this a betrayal of trust, Cunningham sold the car and ultimately formed a team composed of D-Type Jaguars and Maseratis that consistently crushed the Ferrari opposition in the small, but extremely influential world of American amateur sports-car competition.

Ferrari's snubbing of Briggs Cunningham was hardly unusual. While his relations with paper-manufacturing heir Jim Kimberly were cordial, most American customers found him to be arrogant and openly disdainful of them and the market they represented. In many cases this merely enhanced his image and increased their desire to obtain one of his automobiles at any cost. Ferrari was quick to understand this quirk of inferiority in the Yankees and exploited it to the fullest. The more he abused them, the more they rhapsodized about his quirky, regal demeanor (this from a man who was the embodiment of Italian lower-class behavior). The Americans were hopeless dupes for the Ferrari mystique. They were prepared—even grateful—to be gouged in the showroom when buying the automobiles and then to be driven to the brink of bankruptcy trying to maintain them. Luigi Chinetti was equally perceptive in spotting this weakness, and hundreds of otherwise intelligent and reasonable men and women staggered out of his New York facility after having been ignored, abused and fleeced for thousands after their Ferrari received a simple tune-up. Europeans, being more confident of their station, were less inclined to tolerate such nonsense, although today virtually every important European merchant king—from Chanel to Gucci to Hermes to Cross to Vuitton to Turnbull & Asser—has embraced the sales philosophy employed by Enzo Ferrari from the beginning: treat an American like a hick and you'll own him for life.

Ferrari's road cars were little more than detuned race machines. They were abominations to drive anywhere but on open roads at full throttle. They were noisy, they fouled spark plugs, they were impossible to start on cold mornings and they had notoriously weak clutches. The bodies, while beautiful, tended to rust and leak. Worse yet, the cars boiled over in city driving. At one point an American who was selling the cars in status-starved California went to Modena to complain that Ferraris simply could not be operated in Los Angeles traffic without overheating. Ferrari reacted in mock amazement and immediately summoned a new coupe to demonstrate the contrary—that Ferraris could be puttered through the hottest, narrowest, most stifling Modenese alleys with ease. A trip was undertaken with Ferrari at the wheel. The Californian watched the temperature gauge as they meandered through the city. So did Ferrari. Each time the needle began to edge toward the boiling point, Ferrari would glide to the curb and point out a famed local landmark or get out and sip an aperitif at a sidewalk café. He would tarry long enough for the engine to cool down and then proceed. When the temperature began to escalate again, the process would be repeated. While the Californian understood the charade, he returned home happy, content that he had been honored by the Commendatore with his charming little ruse!

By the middle of 1955 the racing program was in a shambles. Stirling Moss, who had transferred from Maserati to Mercedes-Benz, won the Mille Miglia in stunning fashion aboard a rocket-fast 300SLR roadster. Fangio, driving alone in a sister car, was second. Umberto Maglioli, a twenty-seven-year-old doctor's son from Biella who was to become a top sports-car driver for the Scuderia, finished third, nearly an hour behind the winner. A month later at Le Mans a similar 300SLR driven by the Frenchman Pierre Levegh crashed into the crowd lining the main straight, killing Levegh and somewhere between eighty-eight and ninety-six spectators (accounts vary). With Franco-German relations still strained from the war, Mercedes-Benz withdrew from the race and at the end of the 1955 season were to park their big-time racing cars forever. But even with the Stuttgart menace gone, Ferrari's sports cars—for the most part ill-handling 4s and 6s from the slumping hand of Lampredi—were, even on their best days, barely equal to the task of running with the D-Type Jaguars and the revitalized 300S Maseratis.

Ferrari's relationship with the Italian press had turned into open warfare. Fickle in the best of times, many of the major sporting papers were openly critical of the Ferrari racing efforts and passionately denounced the repeated losses to the hated Germans. Stridently jingoistic, the writers and editors expected Ferrari to continue his domination of 1952–53 in the name of the entire nation, and seemed to imply that his failures bordered on the treasonous. Always sensitive to his treatment in the press, Ferrari battled back with artfully worded publicity releases and annual press conferences—which often deteriorated into shouting matches—as well as a column he periodically wrote for the motoring press. But this was a losing battle and the sniping was to continue until the Scuderia could field another winner on the Grand Prix circuit.

As the external criticism increased, internal pressure on Lampredi and the rest of the engineering staff reached intolerable levels. Daily meetings were held by Ferrari to track any progress—either actual or presumed—that might lead to improvements in the performance of the dreaded Squalos and Supersqualos. The sessions were often held in a meeting room known as the "chamber of horrors," where the walls were lined with shelves containing the broken bits of old racing cars. When a part fractured it was not tossed away but rather placed on display, where men like Lampredi, Massimino, Bellentani and Amarotti were to be perpetually reminded of the failure.

Gonzales, the steady *cabezón*, was lost to the team for 1955, as was the gritty, hard-driving Hawthorn. The Argentine ace had been disheartened by the death of his friend Onofré Marimon in the German Grand Prix the year before and, following a crash in Ireland during a sports-car race, had decided to retire from the European campaigns. Hawthorn had lost his father in a highway crash and decided to accept an offer from Tony Vandervell to drive one of his rapidly improving Vanwalls for the 1955 season. While he tried to explain that this would afford him more time to operate the family's automobile business in Farnham, Ferrari considered his departure a violation of faith, and the parting was—as so often was the case—noisy and unpleasant. (The separation would be brief. After winning Le Mans for Jaguar and driving a few races for Vanwall, Hawthorn had a falling-out with the testy Vandervell and would return to Ferrari by mid-season.) But with the departure of those two worthies, the Scuderia's underpowered, cranky machines were left in the hands of Maurice Trintignant, a steady, but unspectacular thirty-eight-year-old vintner from the south of France; the feisty but rapidly slowing Farina; and Franco-American playboy Harry Schell, whose driving was generally more enthusiastic than effective. As the 1955 season had begun, Ferrari faced the might of the much-improved Mercedes-Benz operation, plus the vast potential of the new Lancias. He had applied the lash to his engineers, but, with finances running thin, there was little choice but to soldier on with the overstressed Lampredi 4-cylinders and hope for the best. What transpired was a series of bizarre twists that would lead Ferrari out of his doldrums.

The months of May and June 1955 were among the blackest in the history of motorsports. The major disaster, of course, was the Levegh crash at Le Mans, which caused enormous aftershocks in the world of motorsports. Switzerland banned automobile racing, thereby erasing contests at Berne and Geneva from the international calendar. The German and Spanish Grand Prix were canceled. Two weeks before Le Mans the two-time winner of the Indianapolis 500, the brilliant, gritty Californian Bill Vukovich, had been killed in a multi-car crash while leading the 500-miler. This tragedy, plus the Le Mans disaster, prompted the American Automobile Association to cease sanctioning races in the United States, thereby ending an association with the sport that had begun in the early years of the century.

But the carnage at Le Mans and the loss of the American champion paled in comparison to the tragedy that struck Italy on the sunny afternoon of May 26. Four days earlier Alberto Ascari had been about to take the lead in the Monaco Grand Prix when he lost control of his superb new D50 Lancia at the harborside chicane and tumbled, at 100 mph, into the Bay of Hercules. He had been quickly rescued by scuba divers and escaped with bruises. He was kept overnight in the local hospital as a precaution. Upon returning to his Milan home, Ascari had decided on the spur of the moment to drive the few kilometers out to Monza, where early practice for the Supercortemaggiore 1,000-kilometer race was underway. After a relaxed lunch with friends at the grandstand restaurant, Ascari had gone to the pits to visit his young friend and Lancia teammate Eugenio Castellotti, who was preparing to take a trial run in his new, still-unpainted 750 Monza Ferrari. Impulsively, Ascari had asked Castellotti if he could drive a few slow laps. "Just to see if my back is not too stiff," Ascari said. The young driver consented, and Ascari climbed aboard after doffing his suit jacket and borrowing Castellotti's helmet. Ascari was an extremely superstitious man and never drove without his own blue helmet, which perplexed Villoresi as he watched his old friend ease the Ferrari down the pit road. On the third lap Ascari increased speed, then inexplicably lost control exiting a gentle left-hander called Vialone on the back part of the giant circuit. The Ferrari flipped violently, slamming the great driver to the tarmac. Mortally wounded, Alberto Ascari died in Villoresi's arms on the way to the hospital.

Italy was plunged into mourning. Numerologists and students of coincidence went into an orgy of calculation following Ascari's death. Someone recalled that Alberto and his father, Antonio, had both died on the twenty-sixth day of the month. Both men, as well as Antonio's patron saint, St. Anthony of Padua, had been thirty-six years old at the time of their deaths. The saint and Antonio Ascari had been born on the thirteenth of June; Alberto had lived 13,463 days, or precisely three days longer than his father. The day of their demise, the twenty-sixth, was twice thirteen, et cetera, et cetera.

There is no evidence that Enzo Ferrari was consumed by superstition (although he was known to avoid the number 17), but the loss of Alberto Ascari was surely an ominous blow. Like the rest of the nation, he was thoroughly shaken by the hero's death, which had come at the prime of his career. But as was the case with so many moments of bleakness in his life, Ferrari saw new opportunity arise, phoenix-like, out of the disasters. Mercedes-Benz, already shocked by the enormity of the Le Mans crash, and having proven its superiority in Grand Prix competition, announced that at the end of the 1955 season it would follow Alfa Romeo into retirement. That removed one major antagonist from the field, but another was soon to follow. The Lancia management and team were shattered by Ascari's death. Villoresi was distraught and, at age forty-six, openly spoke of quitting forever. (He raced on, however, before crashing badly at Rome in 1956 and finally retiring for good in 1958.) Worse yet was the financial condition of the company. The development of the racing cars had been unexpectedly costly.

That, plus modest sales of its passenger cars, was moving the firm's leader, Gianni Lancia, toward a decision to cancel the promising D50 Formula One program. Castellotti, inflamed with a desire to honor his fallen teammate, persuaded the company to enter one of the D50s in the Belgian Grand Prix at Spa. He managed to put the car in the pole position in the face of the full Mercedes-Benz might, but the D50 expired after fifteen laps. At that point the promising, but unfulfilled dreams of a Lancia Grand Prix car expired. (The company would soon fall out of the Lancia family's hands and into the holdings of a cement company magnate. It is now part of the Fiat empire.)

In the midst of the Lancia misfortune Ferrari's situation worsened as well. Pirelli announced that it was reducing its racing program—

perhaps prompted by rumors that tire failures had caused Ascari's death (never proven) and the Le Mans tragedy. Beginning the following season, the usual discounts of 12 to 25 percent offered to racing teams—and in the case of major accounts like Ferrari, the outright gift of tires—would be canceled. Pirelli tires would be purchased at full list price or not at all. Ferrari was deeply indebted to Pirelli for the supply of racing tires and considered the firm's withdrawal an act of unforgivable treachery. He angrily announced that never again would a Ferrari be fitted with Pirelli tires (soon forgotten, of course) and that he would have to seek help from a foreign manufacturer. But from whom? Mercedes-Benz was the prime customer for German Continentals. Dunlop was in the throes of irrational chauvinism and refused to supply tires to anyone but a British racing team. Michelin was not interested in cooperating with an Italian rival, and Firestone, the only American company making racing tires, did not have sufficient time to develop a road-racing tire. Desperate, Ferrari sought out his old pals at Englebert. Working through the Princess Lilian de Réthy, King Leopold's wife and a close friend, Ferrari was able to obtain a long-term contract with the Belgian firm. Had the princess not intervened, it is possible that Ferrari would actually have had to make good on one of his numerous threats to quit competition.

Suddenly the international Formula One scene was in chaos. Italy's most popular driver was dead. The brilliant D50s were sidelined and apparently doomed to the same dusty corner of a museum as the 158/159 Alfa Romeos. Mercedes-Benz was also about to depart, which left the two best racing drivers of the day (and perhaps of *any* day, in the minds of many)—Juan Manuel Fangio and Stirling Moss—at liberty. It was a moment of opportunity for Enzo Ferrari for which he was uniquely qualified. He put his considerable powers of persuasion to maximum use. Once again he began to bleat the old song of poverty. Open threats to stop racing were resumed. Knowing that this would leave only Maserati as an Italian representative in international motorsports, Ferrari was counting on support from the press and from the automotive industry to continue his efforts. His current cars were failures, and he understood full well that unless financial and technical assistance was forthcoming, his operation was doomed. It was a moment of both uncertainty and opportunity. During the month of July 1955 he entered into intense three-way negotiations with the Agnelli family, which controlled the mega-conglomerate Fiat, and Gianni Lancia. After hours of heated argument Ferrari negotiated an extraordinary deal. Six of the superb Lancia D50s, several of them incomplete, were to be given to Ferrari along with tons of spare parts, plans and tooling patterns. Jano would be included in the arrangement as a design consultant. It was a case of déjà vu for Enzo Ferrari. As had happened in 1934, when Alfa Romeo presented his sagging Scuderia with its arsenal of P3 *monoposti*, he was now the recipient of a half dozen of the most advanced racing cars on the planet, plus the genius who created them. Moreover, he had extracted an arrangement from Fiat whereby he would receive a stipend of an estimated $100,000 annually for five years to support the new program. It was truly a providential gift. There was widespread celebration in Modena and Maranello when the fleet of bellowing, streamlined Lancia transporters chugged into the Ferrari works on July 26, 1955. On hand were mobs of press and representatives of the Italian Automobile Federation, which had been involved in the negotiations.

The Ferrari mechanics sweated as they rolled the six exquisite machines into the humid afternoon air. Smiles brightened the faces of those assembled. Yet it was something of a Pyrrhic victory for Ferrari, an admission that his own machines were lacking and he needed an infusion of outside talent. For Lancia it was a capitulation that had come on the brink of success. Had Ascari lived and sufficient funding been available, there is little question that the D50s would have dominated the upcoming 1956 season.

Moreover, the transfer of the cars would trigger the loss of Aurelio Lampredi. Bruised emotionally by the defeats of the past two years and the increasingly brutal verbal assaults of Ferrari, Lampredi had no choice but to view (quite correctly) the arrival of the Lancias as a repudiation of his most recent efforts with the 4-cylinder Grand Prix cars. Sullen and isolated, Lampredi—the man who brought Enzo

Ferrari his first truly successful Grand Prix car and two World Championships—left Maranello that September to take a senior design position with Fiat. Lampredi's departure was part of a massive bloodletting at the factory. Nello Ugolini, apparently exhausted by the incessant ranting of the boss and the nightmarish company intrigues, quit and joined Maserati.

Ferrari unloaded the creaking Farina as well as Trintignant and Schell, neither of whom had performed with much distinction, although Trintignant had won at Monaco with a 625 after Moss and Fangio had retired and Ascari had sailed into the harbor. The talent drain continued as Franco Cornacchia, the Milanese sportsman who had held the Ferrari concession for northern Italy since 1951 and who, with Chinetti, had been instrumental in promoting the marque, announced he was dumping the franchise and taking up with Maserati.

Still, the net gain for Ferrari was phenomenal, despite the loss of Lampredi et al. and the rift with Pirelli. In the Maranello courtyard was the rolling stock to create a new renaissance, a return to the domination of 1952–53, not only for himself and his factory but for the rather blemished escutcheon of Italian motorsports.

Yet Enzo Ferrari was a man who was incapable of saying "thank you." He either had reached a point in his life where he appeared to believe that all he received was his just due or simply lacked the good manners to offer simple expressions of gratitude. Whatever the reason, ordinary statements of goodwill seldom if ever issued from the mouth of Enzo Ferrari.

In fact, as he stood with Romolo Tavoni and team manager Mino Amarotti in front of the immobile D50s, a wry smile crossed his face. He turned to his associates and whispered, "They ought to thank me for accepting this junk!"

13

The jumble of change that had swept through Enzo Ferrari's professional life at the end of 1955 appeared on balance to present new opportunities. But they paled in the face of the disaster that was developing in the tiny apartment above the old Scuderia garage on the Viale Trento e Trieste. Dino was dying. An enigmatic disease was ravaging his young body. He was spending more and more time in bed and in the local hospital, constantly tended by his mother, Laura, and visited, after the press of business, by his father.

Alfredino Ferrari was by all accounts a fine young man, open and affable with friends, but shy and withdrawn around strangers. American racing driver Carroll Shelby spent many hours with the youth during the summer and autumn of 1955 and recalls him losing weight while his legs became so stiff that he had difficulty walking even within the constricted environs of the Scuderia. Despite later claims of attentiveness by Ferrari during this period, Shelby insists that the boy was left to fend for himself, idling away the days with the mechanics and customers who jammed the bustling old Scuderia building. By now all manufacturing had been transferred to Maranello, and the old works was employed strictly as a new-car delivery center. Dino, unlike his father, could speak English, which brought him closer to Shelby and the rich Americans and Britons who were appearing in Modena.

Dino had tried hard to exploit his strong interest in automobiles to the extent his failing health would permit. He had enrolled in the local Corni Technical Institute, which had become something of a farm system for aspiring Ferrari technicians, then taken a year with the economics faculty at Bologna University before the rigors of a full-time academic life forced him to withdraw. It is part of the Ferrari mythology that Dino then took a correspondence course from the University of Fribourg in Switzerland and completed a thesis on the design of a 1.5-liter 4-cylinder engine with an interesting valve layout.

He was allegedly helped first by Lampredi and then by young Andrea Fraschetti, a talented engineer and draftsman. But there is a problem: no one named Ferrari is on record at Fribourg during the 1950s, nor is there one on the rolls at the University of Freiburg in Germany or at Switzerland's premier technical school at Lausanne. Moreover, Fribourg is noted as a school for the humanities, not engineering, which places Dino's entire background as a fledgling engineer in limbo.

While Dino never raced because of his health, his father provided him with a number of cars which he drove with enthusiasm—first a tiny Fiat 1100 and finally a 2-liter Ferrari in which he and his close friend Sergio Sighinolfi sported about in the nearby hills. But as the gloom of the Modena winter of 1955–56 gathered around him, his tall, rangy frame was too debilitated for driving and he was forced to spend more and more of his hours confined to his room. It was during this period that Enzo Ferrari later claimed that the young man created the V6 engine which was to bear his name and which was considered one of the firm's most successful designs. According to Ferrari, Dino's final bedridden months were spent doodling over the layout for a V6 engine, and each evening, after work, he, Vittorio Jano and Dino would spend hours discussing the technical nuances of the new power plant. Dino's actual contribution has been debated at length. Most biographers have taken Ferrari's version at face value. Others, less gullible, have noted that Jano—an eminence in design for thirty years—designed the Lancia Aurelia sports coupe in 1950 and the Lancia D24 sports-racing car that won the Mille Miglia and the Mexican Road Race. Both machines carried superb V6 engines. Lampredi was also working on a V6 before he departed, making the Ferrari contention that the V6 rose full-blown out of his ailing son's brain highly suspect. In retrospect, there seems to be little doubt that Lampredi and Vittorio Jano, not Dino Ferrari, were the prime movers of the project, regardless of the understandable revisionism of a distraught father.

With Mercedes-Benz out of racing and six fresh D50 Lancias residing in the race shops, the 1956 Grand Prix season could be approached with considerable optimism *if* a decent team of drivers could be collected. Stirling Moss was soaring closer to all-time greatness. He seemed to embody some of the flair and dash of Nuvolari and was the first choice. But Moss was still piqued about the Bari incident and chose to drive for Maserati, at least until the revolutionary new British Vanwalls could be completed.

Fangio, the three-time World Champion, was at liberty. He was, on balance, the finest, fastest driver in the world. Although forty-six years old, Fangio retained uncanny powers over a speeding automobile. If any single man could propel Ferrari back into the winner's circle, it was this taciturn, cold-eyed, bowlegged Argentinian. Ferrari began a protracted and difficult negotiation to obtain his services. There was trouble from the start. Ferrari was accustomed to dealing with drivers directly, without intermediaries. This was to his advantage. Most were naïve in the ways of business ("blockheads," by Villoresi's account) and could easily be bamboozled into driving for pittances. But Fangio was different. He had come from Mercedes-Benz, where drivers were paid handsomely, and was not about to accept Ferrari's menial sums—gift-wrapped in the potential "glory" to be gained by a drive for the Scuderia.

Fangio brought to the meetings a wily agent named Marcello Giambertone, an Italian race promoter and general gadfly in the sport. Ferrari was furious at Fangio's effrontery in coming with an intermediary. Worse yet, Giambertone hammered out a lucrative deal for his client. It is said that Fangio received about 12 million lire a year for driving, plus a list of small perks that battered Ferrari's wallet. But whatever the deal, the relationship between the Commendatore and his star driver began on a sour note and never improved.

To counterbalance the tough-minded Fangio, Ferrari was able to attract a marvelous collection of wild-eyed youngsters brimming with passion to race—and perhaps die—for the glory of the team. Leading them was the darkly handsome Eugenio Castellotti, scion of a family of landed aristocrats from Lodi, who was so handsome that Ferrari came to call him *il bello* (the handsome one). Fiercely competitive and vain to the point of wearing elevator shoes, Castellotti had purchased at least a half dozen Ferraris and raced them as an enthusiastic amateur before gaining a modicum of success with the ill-fated Lancia team the year before. He was Italy's grand hope to succeed the fallen Ascari.

Another rich Italian immediately became Castellotti's most strident rival. The Roman Luigi Musso was the son of a prominent diplomat. Six years older than Castellotti, Musso was cursed from the start by his southern origins. Italy was fragmented by insane local chauvinism. City hated city, province hated province and region hated region. The Romans were at best marginally acceptable to the northerners, but natives of fetid places like Naples and Salerno (not to mention Sicily and Sardinia!) might as well have come from darkest Africa. This provincialism only heightened the rivalry between Musso and Castellotti, who were already engaged in a struggle to become Italy's premier racing hero. This rivalry would be fanned to a superheated intensity by the old "agitator of men," Enzo Ferrari.

Nationalism was rampant in all of European racing. The cars still carried the colors of their native states: Italians red, British green, French blue, Germans silver, Belgians yellow, and so on. Numerous companies, including Dunlop, the English tire company, Michelin from France and Continental from Germany, were reluctant to do business with any team that was not composed of fellow countrymen. The Italians, led by a jingoistic press, were screaming for revenge for the German onslaught of 1954–55. Their counterparts in England were wallowing in the shame of the BRM debacle and pressuring men like Moss to race only in cars painted British racing green. The French, who had not won at their own Le Mans race since 1950, had jiggled the rules with the introduction of a complicated "index of performance" that would permit slow, underpowered home-built machines like Dyna-Panhards and Gordinis to gain a modicum of honor. The day when racing cars would become little more than rolling billboards for commercial sponsors was a decade away. Racing for the glory and honor of one's homeland was primary in the minds of all but a few hardened hired guns like Fangio.

Joining the two Italians would be slight, blond Peter Collins, a crisp Englishman who had displayed some notably aggressive driving at the wheel of various English single-seaters before being discovered by Ferrari. At twenty-five, Collins was the embodiment of solid British schoolboy sportsmanship: a plucky but ethical player on the field, a dashing and quite charming gentleman off it. Like the rest of the team, Collins fully expected to run a full schedule for the Scuderia, competing not only in the Formula One wars but in the numerous major sports-car contests around the globe, which included the 1,000-kilometer races at Buenos Aires and the Nürburgring, the 12 Hour of Sebring, Florida, the open-road Targa Florio, the Mille Miglia and of course Le Mans. Today drivers specialize for the most part in one type of racing, but in the 1950s top drivers performed with equal skill over twisting public roads aboard heavy, full-bodied sports cars or on closed circuits with lithe single-seaters. They were, in the classic English sporting sense, true "all-rounders."

Also added to the lineup for major sports-car races was the Belgian Olivier Gendebien, a former Resistance hero and ace horseman who had dazzled Ferrari with his brilliant drive to seventh place aboard his personal 300SL in the 1955 Mille Miglia. This patrician man was less naïve than his eager associates and would never be consumed by the intrigues of the Scuderia. Perhaps he took to heart an admission by Ferrari over lunch. Reflecting on the vagaries of the Italian personality, Ferrari had mused to Gendebien, "We Italians are all tricky comedians."

Another member of the sports-car team who failed to see the humor of it all was twenty-nine-year-old Californian Phil Hill. He had come to Maranello immersed in the romance of European motorsports. He was endowed with great talent and an appetite for culture that included enthusiasm for the opera and antiquities, but his eagerness to please became his Achilles' heel. Ferrari was to spot this weakness and play a shameful mind game of carrot-and-stick with the vulnerable and temperamental driver—even after he had rewarded the factory with a World Championship. Hill had run well in the ill-fated Le Mans race of 1955 and, sanctioned by the ever-loyal Chinetti, was given an introduction into the hallowed precincts of Maranello. But regardless of his obvious skills, he would molder as a backup driver for nearly three years.

Quite the contrary took place with a colorful Spaniard known—albeit briefly—to the world as Count Alfonso de Portago. This twenty-eight-year-old nobleman was the original punk aristocrat. The son of a former Irish nurse, "Fon" Portago had been born, it was said, five centuries too late. He was the quintessential adventurer, lover and general rakehell of a type Ferrari loved. Portago was a brutal, uncaring driver who got around the track on nerve alone. He worked hard to dress down, favoring leather jackets, long hair and a stubble of a beard and generally smelling of garlic and onions. Although he was married to a wealthy American woman, Portago carried on a number of flashy romances in Europe with the likes of international model Dorian Leigh and actress Linda Christian. Accomplished at polo, bobsledding and chasing—both steeples and females—Portago was attracted to automobile racing for its pure, brazen danger. He managed to break a leg in a 1954 Silverstone crash while developing a bravado driving style that intrigued Ferrari. Fluent in four languages and sufficiently infused with his royal heritage to believe he had a shot at the Spanish throne should Francisco Franco step down or be deposed, he was a favorite of Ferrari's from the start. It was Portago's *garabaldino* style that fascinated Ferrari, as well as his outrageous, audacious approach to life.

Enzo Ferrari himself, of course, was a notably crude man, which may explain his fondness for flashy "gentlemen." Sessions at the dinner table were punctuated by his endless sexual references, snide asides about his rivals, loud belches, crotch scratching and noisy snorts into a large linen handkerchief that was always part of his wardrobe. Those close to him claim that his more refined public persona was formed—or, more correctly, re-formed—during this period when a journalist from Cremona named Dario Zanesi arrived in Modena to write a profile for *Resto del Carlino*. He was appalled by what he found. Rather than a patrician lord regally overseeing a duchy of artisans, he discovered a "Modenese *paesano*" running a small Italian factory that happened to produce fast cars. In a desire to confirm the fantasies of his readers, Zanesi painted a new portrait of the Commendatore—"a new suit," as he called it. The old paisan became a stately, dignified, aloof, cultured doge who presided over his empire with a firm but fair hand. The social improprieties, the screaming fits of temper, the cursing, the hectoring of his staff, and the *man* himself were washed away in a sea of flattering hyperbole. Those near him recall that the Zanesi story had a profound influence. Always the actor, Ferrari seemed to assume the persona of the man described in the story, at least in public. Over time he was to develop an aura of dignified grandeur to the outside world, although to the tight circle of engineers, flunkies and occasional friends, he remained the same blustery, manipulative, ill-mannered *scemo* as before.

Ferrari learned early in the going to tell people what they wanted to hear. His public utterances, often couched in terms of self-pity and the specter of financial ruin, were crafted to elicit sympathy and respect for this allegedly lonely, martyred, eternally loyal Italian who was carrying the banner of national honor with indomitable pride. It is often said that his greatest skill was his ability to recognize talent. This was probably true, although within that talent pool he demanded unflinching allegiance and a willingness to subordinate self to the Ferrari name. While he was a master at verbal deception, so too were his staff. They had no other choice, because criticism was never tolerated. On one occasion Romolo Tavoni returned from Holland and casually noted that an important figure in international motorsports had remarked that it was time Ferrari ceased acting like a small artisan and became a serious businessman. Furious at the slight, Ferrari demanded the name of the offender. Tavoni had been told in confidence and refused. "Leave the office for five minutes and come back with the name," Ferrari ordered. Tavoni duly returned, but refused to talk. "Go home for four days!" bellowed Ferrari. Tavoni again returned from his exile but refused to break the confidence. "One month!" Ferrari screamed. Tavoni, who was by then a critical element in the organization, was finally summoned back. More interrogations led nowhere and finally the incident was forgotten.

One man who understood how to deal with Ferrari was an immigrant American named Tony Parravano. A native of Puglia, a village in the hill country east of Naples, Parravano had come to the United States as a young man and had made his way in the construction business, first in Detroit and finally, after 1943, in the South Bay district of Los Angeles. By the early 1950s Parravano had become a powerful, if somewhat shady figure in Southern California housing development circles. Rumors spread regularly that he was backed by the Mafia, but this did nothing to impede his success. Parravano was fascinated by fast cars, especially the twin Modenese breeds of Ferrari and Maserati. He became something of a patron to the bevy of gifted road-racing drivers that were rising out of the same amateur Southern California sports-car circles that had spawned Phil Hill. In 1955 Parravano had sent Carroll Shelby, a leathery Texan who was winning regularly in California, to Italy. His assignment was to purchase cars for Parravano's collection and to develop a relationship with the primary players at both the Ferrari and Maserati works. That led to a number of trips by Parravano himself, wherein he made the acquaintance of Ferrari. Money was extremely tight and a number of loans were arranged between Ferrari and Parravano, totaling as much as $300,000. The offer to Parravano would be in automobiles and spare parts, all of which would be sold to him at half price—even though Luigi Chinetti was ostensibly operating in New York as the exclusive importer for the cars. This arrangement with Parravano clearly circumvented Chinetti's franchise and no doubt cost him considerably, in terms of both money and prestige. It would not be the last time that, in the words of Shelby, "the Old Man cheated Luigi and generally treated him like shit."

This conduit of sports-racing cars to the West Coast of the United States was obviously beneficial to Ferrari, both in terms of his image and monetarily in a period of extreme financial peril. (Parravano told a friend that the situation was sufficiently critical that Ferrari spoke of seeking an alliance with Maserati, although the seriousness of such a remark is open to debate.) The Ferraris that Parravano brought to California—especially the monster 4.9s—did much to spread the Ferrari mystique among the status-oriented Hollywood set. It also produced a backlash of innuendo when Parravano disappeared a year later, the victim of a presumed gangland hit. While rumors periodically spread that Parravano had fled to South America or Africa, the conventional theory is that he was probably murdered because of an indiscretion against the Mafia. This in turn led to occasional implications that Ferrari was somehow being financed by the mob, or perhaps, in some bizarre hookup of logic, was somehow distantly connected to the incident. There is no evidence whatsoever that Enzo Ferrari ever had serious dealings with the Italian Mafia, either in Italy or in the United States. While such a connection is possible, no historian or former associate, no matter how critical he might be, has ever suggested that Enzo Ferrari had any linkage with the underworld. If the rumors persist, it is based on the poisonous stereotype that all successful Italians are somehow in bed with the Mafia and in part because of the mysterious disappearance of Tony Parravano.

With Jano, Fangio and a retinue of eager young drivers on board, Ferrari faced the 1956 season in excellent shape. The Lancias had been modified, with their fuel storage removed from the bizarre pannier side pods and mounted in a conventional tail tank. The precise reasons for this change are unclear, other than perhaps to make the handling a bit more predictable, but most engineering critics maintain that the alteration was regressive. It is possible the modification was made for reasons of pure ego—a cheap and relatively simple way for Ferrari to put his mark on a machine that was clearly foreign to him in every sense.

The rift with Pirelli had forced the Scuderia into a full-time association with the Belgian firm of Englebert, who in turn supplied tires for the racing team in a rather unorthodox manner. Each spring a truck would arrive from Belgium loaded with the team's annual supply of rubber. The tires would be off-loaded into a Ferrari warehouse, where the new inventory was expected to last the full season. This was a laughably crude system that within a decade would be rendered obsolete by Goodyear, Dunlop, Michelin and Pirelli, who would custom-build special tread and compound configurations for individual racetracks and weather conditions.

Central to Ferrari's thoughts was success in Formula One. For this he spent an unprecedented amount to retain Fangio and it paid off; the Argentine maestro won his fourth World Championship with relative ease, although he needed help from Peter Collins to achieve the goal. As the season wound down, both Collins and Fangio were in contention for the title. At the Italian Grand Prix, Fangio's car broke and his chances to repeat as World Champion appeared doomed. Then Collins stopped at the pits and, in a supreme act of sportsmanship, turned his car over to the old master, who went on to finish second behind Stirling Moss's fleeting 250F Maserati. This selfless act gave Fangio a clear path to the title and endeared the Englishman to Ferrari, who for once saw a driver subordinate his personal desires to that of the team. Collins explained his extraordinary graciousness thus: "I'm only twenty-five years old and have plenty of time to win the championship on my own."

Castellotti and Fangio had begun the season by winning the 12 Hour race at Sebring, Florida, for the Scuderia, and aside from yet another loss to Jaguar at Le Mans and defeats at the deft hand of Moss and his Maserati at the Monaco and Monza Grand Prix, most of the important events of the season fell to Ferrari. Castellotti won the rain-soaked Mille Miglia with a courageous drive, leading home a Ferrari sweep of the first five places. In fifth was Gendebien aboard a slower

250 GT coupe that he recalls with considerable disdain. The windshield had leaked terribly during the earlier Tour of Sicily and the Belgian had demanded that it be repaired prior to the Mille Miglia. This was ignored and he labored through 1,000 miles of rain; cold and soaked, with minimal visibility. He finally spun in the treacherous Futa pass, sufficiently bending the bodywork so that his navigator and cousin, Jacques Washer, had to kick open the door to extract himself at the finish. Yet Ferrari, who was at the Bologna control about a hundred miles south of Brescia, was unconcerned about Gendebien's troubles. "He said, 'You *must* win the grand touring class.' His voice and his manner was almost hypnotic. It was almost impossible to resist," recalls Gendebien.

There is little question that Fangio was in the midst of a bad patch during his stint with Ferrari. Perhaps spoiled by the workmanlike atmosphere at Mercedes-Benz and his rather relaxed times at Maserati, Fangio was repelled by the intrigues that swirled around the team. He was too old and too wise to fall victim to Ferrari's exercises in psychological warfare. Moreover, he became convinced, probably without justification, that he was being provided with second-rate machinery and that the young lions like Collins and Castellotti were favored. When he was relegated to fourth place in the French Grand Prix after a fuel line ruptured, Fangio and his supporters darkly hinted at sabotage—despite the fact that mechanical failures were common during the entire season and seemed to indiscriminately afflict all of the drivers at one time or another. Ferrari was later to recall that Fangio's version of the 1956 season was "sort of a thriller, a concoction of betrayals, sabotage, deceit and machinations of every kind—all perpetrated to lay him in the dust." Ferrari quite lucidly reasoned that it would have been insanity for him to have hired the best racing driver in the world and then actively worked to keep him from winning. Ferrari was too much the pragmatist for that. He wanted his investment in Fangio to pay off with a World Championship. The clash between the two men was more a matter of personal style. On the one hand, Fangio expected the same brand of intense, unemotional professionalism that had been the hallmark of the Mercedes-Benz effort.

Ferrari, on the other hand, sought a kind of hysteria and mindless loyalty in his drivers that Fangio was simply unable to provide. Suffice it to say that the Argentinian left the Scuderia immediately after the last 1956 race, to transfer his services across town to Maserati. The two men remained at arm's length for the rest of Ferrari's life, despite later theatrical displays of public affection and forgiveness.

The tension that marred the Fangio relationship was trivial compared with the drama that was unfolding at the little apartment above the Scuderia. As the Le Mans race dribbled to an end (with the third-place Gendebien-Trintignant Tipo 625 LM being the only Ferrari of the six entered to finish), Dino was on his deathbed. During the final week of June mechanics in the Scuderia workshop witnessed a new Ferrari, face contorted with grief, sobbing openly as he descended the steps from the apartment. The end came on June 30, 1956, when the pleasant young man's kidneys failed and he slipped away. His father and mother were sick with grief. Tavoni was with Ferrari the following afternoon when Eraldo Sculati, the team manager, called from Rheims to announce that Collins had won the French Grand Prix. Tavoni attempted to hand the telephone to Ferrari, who tearfully waved it off, claiming that his interest in racing cars was forever dead.

A funeral cortege, numbering perhaps a thousand mourners, formed up outside the old Scuderia building on the Viale Trento e Trieste and snaked its way to the cemetery at San Cataldo. Ferrari, standing tall and proud, was in the vanguard, as was a thoroughly shaken, black-shrouded Laura. Chugging along in the midst of the marchers was a tiny Fiat sedan carrying mother Adalgisa, now in her mid-eighties and unable to make the long trek in the steaming humidity.

What followed was a dirge of grief and posthumous glorification of Dino that bordered on the bestowal of secular sainthood. A massive Romanesque marble tomb was constructed in the cemetery at San Cataldo, financed in part with city of Modena funds, to serve as the family crypt. The remains of Dino and his grandfather were laid there, and places were reserved for Enzo, his mother and Laura as well.

The illness that killed Dino was to have a supreme influence on Ferrari although the exact nature of Dino Ferrari's sickness remains a mystery. Various biographers have claimed the illness was leukemia, multiple sclerosis, nephritis or muscular dystrophy. The latter is most commonly mentioned, and Ferrari himself spoke of the "dystrophy" that afflicted both his son and his wife. He also said that a "nephritis virus" killed the boy, although no such ailment exists in medical annals.

People who knew Dino in the final years of his life recall him as a cheerful young man who was perfectly normal except for being extremely thin and generally fatigued and having severe stiffness in his limbs. "Dino was a good guy, always smiling," recalls Shelby. "His old man never spent any time with him and the kid didn't have much to do."

If Dino was afflicted with muscular dystrophy, of which there are several types, it was probably the most common variety, Duchenne, which exclusively afflicts males aged three to ten and generally causes death in the late teens or early twenties. However, the final years of life usually bring complete debilitation, and it seems unlikely that Dino could have remained ambulatory—albeit with difficulty—up to the final months of his life if he had this disease.

It is known that Enzo Ferrari contributed generously to various muscular dystrophy causes following Dino's death, and one prominent American television producer and avid Ferrari collector recalls that he gained Ferrari's friendship *only* after he revealed that he had made a sympathetic documentary on the disease.

Dino's death—whatever the cause—marked a pivotal change in Ferrari's life. From that point on he became more reclusive, more embittered, more cynical, more stoic about the other broken bodies that suddenly seemed part of his daily existence.

For decades Ferrari would visit his son's grave as part of his morning routine, which, like clockwork, included a stop at the barbershop and a short session with his aging mother. All V6 engines and the cars that carried them would be labeled "Dinos." Ferrari's stark workplace at Maranello became more shrine than an office, with a brooding picture of Dino facing his desk and acting as the cloying centerpiece of the room. For years he was to wear a black tie as a memorial to his only legitimate son and was to confer a number of awards and scholarships in his name and to donate substantial sums to various muscular dystrophy causes.

Later that year the little trade school that had been started in the Abetone road stable was expanded into a formal technical school associated with the Corni Technical Institute under the auspices of the Italian Ministry of Education. Ferrari wanted the school named for his dead son, and due to his efforts it would be formally transformed into the Istituto Professionale Statale Alfredo Ferrari.

When Dino Ferrari died, his father was fifty-eight years old. Thanks to the aid of Fiat and Lancia and the largesse of numerous rich customers around the world, he was beginning to enjoy legitimate prosperity. He and Laura, with whom a kind of "entente cordiale" had been established, were accumulating real estate in Modena and environs and plans were afoot to expand the factory, which had produced eighty-one cars that year, and to convert the small stable across the Abetone road into a restaurant and small inn. It was to be called the Cavallino—a hostelry that was to become almost as famous as the factory itself. But the loss of Dino overrode all else for a time. Ferrari repeatedly threatened to quit racing during the summer of 1956 and even those who had witnessed such theatrics before were inclined to believe him.

As is the case with many Italians, Ferrari wallowed in mourning with an intensity far greater than the attention he had given the boy in life. Death often brings the illusion of perfection, and so it was with Dino. His flaws were erased and the young man enjoyed the luxury of never having to displease or disappoint his father. Ferrari, quite simply, could create a new and perfect Dino. His grief, prodigious as it was, evoked a measure of cynicism among those who knew him well. It was so protracted, so maudlin, so operatic, that associates began to question its motives.

With Dino gone but hardly forgotten, Ferrari still had another son to contend with. Piero had just turned eleven and was still living with his mother in Castelvetro. It is not known whether the two boys ever met or whether Dino even knew of his half-brother's existence, but there is no question that at some point in the late 1950s Laura Ferrari discovered her husband's second life. Her reaction could hardly be described as tranquil. She was clearly the loser. With her only son dead, Ferrari himself had seemingly hedged his bets by siring a second offspring—a substitute heir, as it were—who remained concealed to all but a few of Ferrari's closest associates.

Then death struck the family again, albeit obliquely, a little more than a month after Dino's death. Sergio Sighinolfi, who had been one of the pallbearers at the funeral, had been grief-stricken by the loss of his friend. The two young men had become extremely close as Dino's health had declined, and the hours spent with Sighinolfi had become increasingly pleasant diversions for Dino. Then came the news on August 9 that Sergio had lost control of the Ferrari he was testing in the Apennine hills and had been killed. It had happened on a curve he had screamed through a hundred times on other test runs. This time he was a trifle off line and plunged over a steep bank to his death.

Enzo Ferrari was deeply affected by the loss. The young man's closeness to his son had transcended his simple employee's status. While Ferrari seldom blinked over the death of one of his drivers, he went to special lengths to comfort Sighinolfi's family. Money, the exact amount of which is unknown, was donated by Ferrari to help alleviate the suffering. This was a rare, if not unprecedented act of generosity on his part.

This was a period of enormous tension for Ferrari, which only intensified his lament for the fallen Dino. He wrote to his friend Rancati two weeks after the funeral, "In thinking deeply about the competition season I had decided to leave to others the task of defending the prestige of Italian automotive creation in other countries. In our lives we must learn to renounce some things which we value very highly and I believed that, after having lost my son, I did not have anything of greater value to give up."

The stories surrounding Ferrari's career have it that he ceased attending races following Dino's death. Much of the mystique that Ferrari created around himself in later life centered on this refusal to appear at motorsports events, other than a traditional afternoon during qualifying for the Italian Grand Prix at Monza. That he started this practice after Dino died is simply not true. Nor is it correct that he stopped after the Dino V6 Formula Two car ran its first race at Naples in the spring of 1957, as some claim. Enzo Ferrari was surely at the May 1957 Mille Miglia, directing his team from the Bologna control, and it is probable that the last formal race honored by his presence was the non-championship Modena Grand Prix later that season. That was fifteen months after his son's death. His self-enforced exile has never been sufficiently explained. Some suggest that he became uncomfortable in the increasingly unruly crowds that swarmed around him—although he tolerated the mobs of screaming *tifosi* who engulfed him on the Monza practice day he chose to attend. After the Mille Miglia was canceled, only the Targa Florio and the Grand Prix of Italy remained as world-class events in the nation. Never what could be described as a peripatetic traveler, Ferrari was able to remain at home and still maintain contact with his racing managers via long-distance telephone. It became normal for him to spend two to three hours a day in communication with the track during practice and qualification periods, discussing suspension settings, gear ratios, tire compounds, race strategies, et cetera. Following the race itself, lengthy postmortems with the engineers and team managers were conducted. As international communications became more sophisticated, capped by television, telex, computer terminals and fax machines prior to his death, it became possible for him to know as much about the event in the comfort of his own headquarters as if he was leaning on the pit counter. Besides, to a consummate public relations man, incessant speculation about his absence was a much stronger hype to the mystique than if he had become a familiar presence.

Interestingly, drivers were seldom if ever included in the long-distance interrogations and conferences. With few exceptions, drivers were considered to be a low-priority commodity. From the time he entered the sport, Ferrari understood that there existed a ready supply of young men eager to risk their lives, their reputations and their family fortunes behind the wheel of a first-class racing car. As long as his machinery was ready to race, the driver source would take care of itself.

Even as 1956 wound down and he knew that Fangio would never return to the stable, a surplus of drivers floated around him, all yearning for glory. In addition to the two great Italian hopes, Castellotti and Musso, he had the determined young Englishman, Collins, and the wild man, Portago, plus Gendebien and Hill, both of whom ached for Formula One rides. And waiting in the wings was a classy German noble, Count Wolfgang "Taffy" von Trips, who was sure to give a good account of himself if offered the opportunity. Moreover, Mike Hawthorn was ready to return. He had once again fought with Vanwall impresario Tony Vandervell (after the boss insisted on driving his race car, through traffic, from the hotel to the track prior to the French Grand Prix and fried the clutch) and had notified Maranello that he was ready to deal once again.

The Italian press would come to call this group *il squadra primavera* (the spring team).

If there was a favorite driver among this youthful retinue, it surely was Peter Collins. This garrulous, totally loyal Englishman had endeared himself to Ferrari by turning over his car to Fangio at Monza, and for a while it seemed as if he was being groomed as a replacement—at least in an emotional sense—for the much-lamented Dino. Collins was given the use of a villa Ferrari had purchased on the Abetone road a few yards north of the factory, and as the months passed, it became an extension of the Ferrari household. Enzo stopped there periodically on the way to work, and Laura, chauffeured by the loyal Pepino Verdelli, would travel regularly to do Collins's laundry and tidy up his quarters. At the time the likable bachelor was between girlfriends, which suited Ferrari perfectly. Women (other than his own) were not welcome members of the retinue. Ferrari considered them a distraction, not only to the drivers but also to the mechanics and the entire operation.

This, of course, had little bearing on the amorous pursuits of the team's stars. They were dashing international celebrities, and glamorous women were traditional fixtures on the international Grand Prix scene. Castellotti was involved in a much-publicized romance with a stunning actress named Delia Scala. She was exerting increasing pressure on him to retire from racing and get married, which infuriated Ferrari. Musso, who was married with children, was carrying on an affair with a beautiful woman named Fiamma Breschi. Hill, who was trying to totally concentrate on his race driving, was being dogged by an American female journalist, whose presence, by universal agreement, was an unsettling distraction for him. As for Portago, Ferrari accepted his endless liaisons as part of the bargain and made no effort to tame the wild Spaniard. But on the whole these women were viewed as flashy camp followers who diverted attention from the single purpose at hand: crushing the opposition on the racetrack and glorifying the name of Enzo Ferrari.

Death visited the operation once again in the autumn of 1957. Andrea Fraschetti, eager to prove his skills at the wheel as well as at the drawing board, took one of the new prototype V6 Dino F2s to the Modena autodrome for a day of testing. By then he thought he knew every ripple on the ragged rectangle. The pits were little more than open sheds designed to protect the crews from the simmering summertime sun and the dank winds of winter. On any given day factory Ferraris and Maseratis might be present for tests, as well as private teams and a few amateurs hoping to showcase their talents. Ferrari was a regular at the autodrome, although it is not known if he was present when Fraschetti climbed aboard the curvaceous, red-skinned single-seater. The lap record at Modena, unofficial as it was, remained the Holy Grail for any number of hot shoes. Jean Behra, the gnarled, thick-necked Frenchman, was very quick at Modena with his Maserati 250F, as were Castellotti and Musso aboard their Ferraris. Owning the fastest time around the circuit, which measured no more than a mile, was a badge of honor, coveted not only by the two rival factories but by every serious driver who ever drove there. There were two nasty spots: a tight little chicane at the end of a long straight and a sweeping left-hander that could be taken flat out. Fraschetti lost it at full speed. The car spun across a stretch of parched brown grass and tumbled upside down. The brilliant engineer was mortally wounded and died the following day.

But as was the case with fallen or departed drivers, the replacement of engineers was never a problem for Ferrari. Legions of talented designers stood in the wings, eager to demonstrate their creative powers to the Commendatore. A few days after Fraschetti's death, Ferrari offered the job of technical director to a portly, talented, opinionated thirty-two-year-old Tuscan named Carlo Chiti. He had been brought to Ferrari from Alfa Romeo by an old friend, Giotto Bizzarrini, who had worked at both factories as an engineer and test driver. Chiti was one of the new breed of engineers who were convinced that high performance was linked to the new technologies of suspension geometry, weight distribution, braking and aerodynamics. The "old school," as espoused by Alfa Romeo and its chief disciple, Ferrari, believed that brute power was the all-important element. Horsepower and torque, the keys to top speed and acceleration, were the only components of race-car design that counted. This ossified conviction, maintained as canon law by Ferrari, would nearly drive him out of business before Chiti and other progressives changed his mind.

Carlo Chiti's ascendancy to the chief engineer's job severed one more link to the past. A few old-timers remained—Bazzi, Rocchi, Bellentani, Massimino, Jano—but the key roles were now held by younger men who owed nothing to the aged legends of the prewar Scuderia and its musty, encumbering traditions. Ferrari himself remained shackled to the past, with his antiquated convictions about race-car design and his endless, maundering search for another Nuvolari. But the new breed of drivers and designers who now held sway were postwar men, ready to engage in battle against the modernists from Germany, Great Britain and the United States without being enslaved by the glorious ghosts of the 1930s. Breaking this bond between Enzo Ferrari and his past would be difficult. Stubborn spirits haunted the halls and work galleries of the Scuderia. The young lions did not have carte blanche. They would prove their worth only by driving the machines of the Prancing Horse faster and farther than those who had gone before. In their fevered attempts much would be sacrificed. And more blood would be shed than anyone dared to imagine.

14

Robert Daley, who covered European sports for *The New York Times* in the 1950s before returning home to embark on a successful career as a novelist, was one of the few journalists to be favored with several exclusive interviews with Enzo Ferrari. While he never even made a pretense of wanting to visit America, Ferrari had slowly been made acutely aware of the blossoming market across the Atlantic and therefore often accommodated writers like Daley more than he might their counterparts from Rome or Milan.

Daley observed that Ferrari appeared to be happier when he was losing, which jibes with mechanics' observations that the race shop on Monday was more serene following a defeat than a victory. But why? Was not winning the central object of the exercise? Logically that had to be the case, because only victories would beget prize money, starting funds, sponsorship backing and car sales. But Ferrari explained it differently to Daley, employing a logic whose purpose may have been merely to generate intriguing newspaper copy. "There is always something to learn. One never stops learning. Particularly when one is losing. When one loses one knows what has to be done. When one wins one is never sure." Perhaps he was sincere in a broad context, but the entire thrust of his life centered on crushing defeats of his adversaries, thanks to the superiority of his legendary automobiles and, in a secondary sense, the courage and devotion of his drivers.

At one point Ferrari began his customary bleat to Daley about how he did not attend races because he could not bear witnessing the suffering of his beloved automobiles. Daley asked, "You mean you suffer too much for the car, not the driver?" Ferrari paused for a moment, eyeing Phil Hill, who was also present, and responded reflexively, "The driver too, of course."

Following the interview Hill became sullen and confused. At that point he still had the idyllic notion that the Scuderia was a cadre of warrior knights bonded by honor and led by the Arthurian father figure of Ferrari. "I guess we like to think he loves us because we are all so brave and drive so fast. But deep down I suppose all of us know he cares more about his cars than he does about us."

It was not long after the beginning of 1957 that Peter Collins fell from grace. He had left Modena for a swing to America following the Argentine races that traditionally kicked off the season. He was to be the guest of bespectacled, gravel-throated Masten Gregory, the son of a Kansas City banker, who, when he was not crashing, could drive a racing car at staggering speeds. On the way to Kansas, Collins and Gregory stopped in Miami, where they reestablished an acquaintance with Louise King, a lovely actress who was starring in a road version of *The Seven Year Itch*. King had met Collins a year earlier at Monte Carlo, but had been unimpressed. This time it was different, and the Miami rendezvous led to an intense, giddy courtship and a mid-February marriage. When the couple returned to Modena the relationship with Ferrari became steely and distant. Clearly, Ferrari viewed the marriage as an act of treachery. Implying that the new wife had altered Collins's approach to the sport, he wrote, "Collins still preserved his old enthusiasm and skill, which was still outstanding, but a change nevertheless became evident in his happy character. He became irritable."

Quite the contrary. Friends recall that Peter Collins was deliriously happy with his new wife. If there was any discontent, it was with the Scuderia's race cars—generally inept rehashes of the old Jano Lancias—and with Ferrari himself, who Collins apparently felt was overly—and perhaps ingenuously—distracted by the loss of Dino. Then an argument arose over Collins's purchase of a small Lancia Flaminia. Ferrari was irritated that Collins's choice had not been one of his own machines, despite the Englishman's explanation that he could not afford such an extravagance. This was corrected when Ferrari took the Lancia in partial trade for a shimmering blue 410 Superamerica coupe with bodywork by Pininfarina—a gesture surely intended to regain the devotion of the distracted driver. The offering of the car (which would play an ironic role a year later), plus the provision of a small villa outside Maranello for the couple, patched things up until a year later when the Collinses purchased a yacht and went to live in the Monte Carlo harbor.

Castellotti's duties were also being diluted by affairs of the heart. He was spending more and more time with Delia Scala, much to the delight of the adoring Italian public, and to the distress of rival Musso, who was being elbowed out of the spotlight by his more visible and successful teammate. In mid-March the two lovers were holidaying in Florence when Castellotti received a call from Ferrari. Jean Behra was lapping the Modena autodrome at alarming speeds with the new 250F Maserati. The track record, held with pride by Ferrari, was in peril. Castellotti must come immediately to defend the honor of the team. Angry at having his holiday interrupted but intent on remaining the fastest man ever to drive the autodrome, Castellotti departed from Florence at dawn on March 14 and sped north over the snowy spine of the Apennines. Waiting for him at the autodrome was Ferrari, a bevy of staff members, mechanics and a new Tipo 801 2.5-liter Formula One car. Hitching up his helmet, Castellotti climbed into the cockpit and fired the engine. He jammed the gearbox into first, let out the clutch and skittered onto the track. He rumbled around the gloomy acreage for a few laps to warm up the engine and brakes, then stood on the power. He swooped by the pits at over 100 mph, leaving Ferrari and his crew to crane their necks in his wake as he charged into the nasty little chicane. They stiffened at the sound of screeching, anguished rubber, followed by the hollow thump of the steel-and-aluminum body slamming against an unyielding wall. Then silence. The footsteps of the younger mechanics sprinting down the tarmac toward the chicane were pitiful punctuations to the awful crash that had left the mortally wounded Eugenio Castellotti bashed and battered against the concrete base of a small grandstand.

The press went insane with speculation about what caused the accident. Was it a dog that meandered into his path as he braked for the chicane? Did the throttle stick? What was left unsaid was the probable cause: An exhausted driver, bloated with pride and emotionally enslaved to Ferrari, had simply charged into the corner too fast, locked up the brakes and lost control.

Of all the indictments that can be laid at the feet of Enzo Ferrari, negligence in building cars is not one of them. Unlike some other builders, he never sacrificed strength in the name of speed. His cars were always strong and reliable—in many cases several hundred pounds heavier than the opposition. Drivers complained about the mind games he played to psych them into taking chances to win, but never that he knowingly stinted on mechanical components that might have failed in the heat of competition. Still, the call to Castellotti to defend what was a meaningless lap record on an obscure racetrack in a backwater Italian city was pure ego gratification for Ferrari. He was yanking the string of the driver, calling him back to duty from the frivolity of what he considered to be an irrelevant romance. Luigi Villoresi remarked later, "Castellotti's death ended what was left of my friendship with Ferrari. For the sake of his pride, he asked Italy's best driver to risk his life."

Ferrari dismissed the accident as "stupid" and explained it away on the basis of Castellotti "going through a conflicting and confusing time emotionally and it is probable that his end was brought about by a momentary slowness in his reactions." Ferrari noted that on the morning of the crash, "He was bitter and absent in his manner," although of course neglecting to assume any responsibility for his driver's bad attitude—considering that it was he who had summoned him there in the first place.

But the worst was yet to come. The national mourning over Castellotti had barely subsided when the annual craze surrounding the Mille Miglia began to intensify. Ferrari had every intention of repeating the victory of Castellotti the year before and set about forming up a formidable four-car team. Hawthorn, who had happily rejoined the team and linked up once again with his longtime pal Collins, chose not to run in the dangerous, uniquely Italian event, while Hill was not selected, apparently on the basis of his presumed novice status. Piero Taruffi, now just beyond his fiftieth birthday and still obsessed with the idea of winning the legendary race that he had first run twenty-seven years earlier, had gained a ride in one of the new 3.8-liter V12 Tipo 315 sports-racing cars, replacing Castellotti. Joining him would be Collins and likable, enthusiastic Wolfgang von Trips, who had shown considerable potential in both Formula Ones and sports cars. Fon Portago was called in as a last-minute substitute when Luigi Musso took ill and was assigned the Roman's larger 4.1-liter Tipo 335—a very fast machine that was surely intended to carry Musso, the new Italian standard bearer, to victory. Olivier Gendebien was once again given a slower 250GT coupe, which galled him no end.

Gendebien recalls an incident prior to the race that is revealing in terms of how Ferrari tried to manipulate his drivers and to play one off against the other. Gendebien wanted the Musso car. He felt he deserved one of the sports cars, having proved his worth in the Grand Touring class for two consecutive years. He had a meeting with Ferrari to plead his case. Ferrari would not budge; he would drive the 250GT and that was the end of it. Fuming, Gendebien stalked out into the factory courtyard as Portago entered the office. A few minutes later Portago came out. He walked over to Gendebien and lit a cigarette. "Ferrari says you want my sports car, but the son of a bitch says it doesn't matter. He said you'll beat me anyway, no matter what I drive."

Open-road races had long been considered too dangerous by most civilized nations and had long since been banned. Even the Mexican government had canceled the notorious Carrera Panamericana de México after eight people—four drivers, two riding mechanics and a pair of spectators—died in the 1954 event. But the Mille Miglia only gained in popularity among the Italian people. Estimates of the crowd that turned out to line the route soared to as high as 10 million. Thousands of police and Army regulars labored in a futile attempt to keep the mobs off the course, but drivers still had to be steeled to drive flat out into packs of fans—moving, wobbling walls of flesh—that would part like the Red Sea as they sped through. In the rural sections, children skittered across the highway and often rode their bicycles on the shoulder. Strutting young men tried to show their mettle by attempting to touch the fenders of the speeding cars—an Italian counterpart to running the bulls at Pamplona. Some farmers refused to adjust to the annual madness. Competitors would sometimes crest a hill to discover a tiny Fiat sedan, a tractor or a farm wagon wobbling along in the opposite direction.

Yet the notion of fast cars racing on real roads, around twisting mountain hairpins and through narrow city streets, was the embodiment of every man's fantasy to charge down an open road flat out, unencumbered by laws or moral and social impediments of any kind. For all its insanity, the Mille Miglia was the ultimate automobile race, encompassing, as Belgian journalist Jacques Ickx put it, "an entire lifetime condensed into a few hours." Enzo Ferrari said of the epic, "No driver could ever say he achieved his victor's laurels if he had not won at Brescia." To have done so meant that a man had faced down the specter of violent death for half a day over some of the most difficult and unforgiving roadways in the entire world.

Even an audacious man like Portago was daunted by the contest. "I don't like the Mille Miglia," he told American writer Ken Purdy earlier in the year. "No matter how much you practice, you can't possibly come to know the thousand miles of roads as well as the Italians, and, as Fangio says, if you have a conscience, you can't really drive that fast anyway. There are hundreds of corners in the Mille Miglia where one slip by a driver can kill fifty people. You can't keep the spectators from crowding into the road; you couldn't do it with an army. It's a race I hope I never run."

Not only would he drive it, he would do so with perhaps the most powerful car in the race, and with no practice. He and his co-driver, Edmund Nelson, attempted a reconnaissance run in a private car, but thumped a bridge parapet a few miles outside Brescia and never completed even the first stage of the hideously complicated route. Unlike Denis Jenkinson, who had guided Stirling Moss to his record-shattering win in 1955 by unwinding along the way an eighteen-foot roll of detailed route instructions, Nelson was as rank a novice as Portago. He had met Nelson, a forty-two-year-old Air Force veteran who had never returned to his South Dakota home after the war, while he was operating an elevator at New York's Plaza Hotel. It was Nelson who had introduced his pal to bobsledding and who had served as his sidekick, bodyguard and father confessor as the young nobleman caroused his way through the high life of Rome and Monte Carlo. Nelson knew nothing of motor racing and was simply along for the ride to point out the hundreds of distinctive red arrows that had been planted on the roadside to mark the route.

Some say Portago had a premonition of disaster. He did write a note to Dorian Leigh prior to the race, confessing, "As you know, in the first place I did not want to go in the Mille Miglia . . . Then Ferrari told me I must do it . . . That means that my early death may well come next Sunday."

Aligned against the five Ferraris were no fewer than 293 other cars, ranging in size from ridiculous 750-cc Fiat-Abarth sedans with a top speed of no more than 100 mph to the monster 4.5-liter Maserati V12 of Moss and Jenkinson. Each car was assigned a number that coincided with its starting time. Portage's carried 531, telling the crowds it had been launched at 5:31 A.M. The starters had been leaping off the Brescia starting ramp since eleven o'clock the previous evening, so that for at least the first 500 miles the route would be mobbed with slower automobiles, many of them careening around the myriad corners and loping down the middle of the narrow, tree-lined straights.

Moss, who started last, was eliminated only seven miles into the race when his Maserati's brake pedal snapped off when he slowed down for a corner, and he and Jenkinson ended up in a farmer's field, mercifully unhurt. Collins was the early leader in his 3.8-liter Tipo 315, averaging 118.4 mph on the level run between Brescia and Verona. But as the cars swept south along the craggy Adriatic coast, Taruffi, known to the *tifosi* as the "Silver Fox," began to move up to challenge, as did Trips. Portago was driving with apparent discipline, keeping the brute beneath him in check and staying only a few minutes ahead of Gendebien's 3-liter *gran turismo* coupe.

At the Rome checkpoint a scene unfolded worthy of a Hollywood director's most lurid dreams. As Portago's car was checked over and he was about to pull away, a beautiful, tawny-haired woman burst out of the crowd and ran to him. It was Linda Christian. Portago rose up in the Ferrari's seat, pulled her into his arms and kissed her passionately. He released her, then kissed her again and slid back behind the wheel. The big V12 shrieked, the tires gnawed at the cobbles and he sped away, flinging one final wave at the sobbing woman.

He and Nelson probed through some light snow in the higher elevations of the Apennines south of Bologna, managing to navigate the dangerous Futa and Raticosa passes without difficulty. A light drizzle was enveloping Bologna as the Ferrari bounced off a few curbstones and cat's-eyes that marked some of the more treacherous two-lanes, without apparent damage, but by the time Portago skidded to a stop in a light drizzle at the final Bologna fuel stop, he had fallen back into fifth behind the fleeing Gendebien. Ferrari was there, and told Portago that Collins's transaxle was breaking and he would not finish the race. Taruffi's rear end was grinding ominously as well and he was running at a reduced speed. But, he said, Gendebien was beating him, as he had predicted. (A few minutes earlier, during Gendebien's stop, he had told the Belgian that Portago was catching him.)

The inference was clear: Portago must drive faster or risk the shame of being beaten by a much slower automobile. As he was about to depart, a mechanic spotted trouble on the left front; a low control arm was bent and the tire was rubbing on the bodywork. A fresh set of Engleberts were rolled out but Portago waved them off and sped away. Ahead lay the flatlands of the Po Valley, where the enormous power of the 4.1-liter V12 would gobble up not only Gendebien but the crippled machines of Collins and Taruffi.

Moreover, since 1954 a special prize had been offered for the final, ultra-fast stages of the race. Called the Gran Premio Tazio Nuvolari, it was awarded to the driver recording the fastest time over the final eighty-two miles of the race between Cremona and Brescia and passing through the famed driver's hometown of Mantua. The record was held by Stirling Moss and the Mercedes-Benz 300SLR at just over 123 mph. Even if outright victory was out of reach, surely the Nuvolari Prize could be Portago's. After all, he had the most powerful car, capable of 180 mph on the straight, flat, freshly paved stretches beyond Mantua.

They had passed Collins's broken car on the roadside beyond Parma. Surely Gendebien's little coupe could not be far ahead. And Taruffi, the old man, was easing along with his crippled machine. Portago unleashed the full power of the big V12. The flashy red machine began to stretch its legs beyond Mantua. Except when negotiating a few gentle bends leading into the dingy villages dotting the route, he could run with the throttle wide open and the tachometer needle edging toward the red line. Ahead, through the stands of copper beeches and poplars that lined the road, he and Nelson spotted the tall Romanesque campanile at Cerlongo, the single landmark visible from any appreciable distance across the unremitting flatlands. Scattered among the trees were clumps of people, local farmers and their children who had turned out to view the passing parade of berserk machinery.

The road snaked out of Cerlongo and began a five-kilometer string-straight run to the village of Guidizzolo. The few people scattered along the roadside, little more than blurred human pickets to the drivers, had seen first Taruffi limp through, then a relaxed Trips, who had graciously backed down to give his senior teammate the victory (an act of largesse that Taruffi would deny to his dying day, but one that experts tended to agree actually happened).

A scream in the distance. A shimmering pinpoint of red blossomed into a race car rocketing toward a knot of locals who had drifted out from a side road a kilometer south of Guidizzolo to watch the leaders. The braver ones edged onto the macadam while children poked their heads through the legs of their parents, covering their ears against the banshee wail of the approaching car. Portago, hunched out of the slipstream for extra speed, came at them as straight and true as a large-caliber bullet until, a few hundred feet away, the Ferrari twitched to the left, catching a stone kilometer marker.

Suddenly out of control, the car came at them like a pinwheel of death, spinning and gyrating, first into a ditch, then sailing clear over the first row of onlookers to disintegrate against a pole. Shreds of aluminum and steel flailed into the crowd as the car, amidst evil thuds and booms of impact, flung its two passengers into the trees and pounded itself into a rumpled, steaming, inert, upside-down lump in a deep roadside drainage ditch. All was silent, save for the moans and screams of the dying and wounded. Twelve people lay dead, including Portago, who had been sawed in half by the scything hood of the Ferrari, Nelson and ten locals, including five children. Twenty more lay seriously injured.

It was the end. An outraged cry of anger, frustration and anguish came from every corner of the nation. Newspapers which hours before had given hour-by-hour updates of the race were now in full cry.

A typical headline roared across the front page of the influential *Corriere d'Informazione*: "La Mille Miglia, Cimitero di Bimbi e di Uomini, BASTA!" ("The Mille Miglia, Cemetery of Babies and Men, Enough!")

The Vatican was in high dudgeon as well, demanding an end to the event. Within hours sufficient votes had been accumulated in the Chamber of Deputies and the Senate to ban the race forever. The national mood was to lash out, to pinpoint a perpetrator and to lynch him for the carnage. The scapegoat would be Enzo Ferrari.

Logic demanded that the organizers of the race, the Automobile Club of Brescia, and the patriarch of the event, Count Aymo Maggi, be held responsible, if anyone. After all, death had been part of the Mille Miglia for years. In 1938 a Lancia Aprilia had crashed in Bologna after crossing a tramline and killed ten, including seven children.

The race had been banned for a year, but had resumed full tilt in 1947. Worse yet, the rules had been altered in 1954, waiving the provision that the cars be production-based touring machines with two people on board. From then on the race was dominated by all-out sports-racing cars, often driven solo by a Grand Prix ace like Ascari or Castellotti, ripping across the countryside at speeds nipping at 200 mph. Portago's prediction of inevitable doom was perfectly accurate, although, with typical Italian passion, blame was laid at the feet of the wrong man.

Not long after the last patches of blood had been washed away from the scene of the disaster and a series of fervid post-race dramas had been played out, the final blow in the name of perverted justice was struck: Enzo Ferrari was charged with manslaughter. The indictment read, in part: "Enzo Ferrari, born in Modena on February 20 [*sic*], 1898, and resident therein, charged with manslaughter and causing grievous bodily harm by negligence in that, as principal of the firm of Ferrari, of Modena, specialists in the construction of cars for road and track racing, in the 24th Mille Miglia, he did adopt for the cars of his Scuderia, and specifically the one bearing license plate BO81825 and race number 531, driven by Alfonso Cabeza de Vaca, Marquis de Portago, tires made by Englebert of Liège, Belgium, of which the constructional characteristics and the manner of employment (tread of approximately 8mm thickness and inflated to 2.5 kg per square centimeter) were unsuited to the performances of said cars, which at full power developed maximum speeds of no less than 280 kph, whereas the aforesaid tires allowed, at the most, a speed of 220 kph; and the overheating consequence of this overinflation having caused the central portion of the tread to strip, with the bursting of the whole tire, the car ran off the road, causing the death of nine spectators and the two drivers."

The charge was as silly and irrelevant as the one that had been brought against Ascari in 1951. No matter, Italian justice operated in strange and convoluted ways and it would not be the last time a racing driver or car manufacturer would be similarly charged for an accident that killed spectators while those who failed to protect them properly would slip the noose. Four and a half years of legal wrangling, public posturing, cruel innuendos and unfounded accusations would pass before Ferrari was finally cleared of the alleged crime. Surely, the court had to recognize that the cars of Trips, Taruffi and Gendebien all finished the race on Engleberts similar in design and pattern to those on Portago's automobile (with Gendebien winning the Nuvolari Prize as well!). While tire failure was partly the cause of the accident, it was Portago's refusal to change rubber at the Bologna fuel stop, or to correct the suspension and body damage, that was at the root of the problem. If Ferrari bore any blame, it was permitting Portago to proceed with the damaged machine after goading him, subtly at least, to catch Gendebien. Within the team that sentiment was whispered and from then on Gendebien's wife, a soft-spoken, gentle woman, referred to Ferrari only as "the assassin."

Legal troubles were only part of Ferrari's misfortunes during the 1957 season. A month later the Scuderia was blitzed at Le Mans for the third year in a row by the Jaguars. This time the green cars from Coventry captured the first four places, with the first Ferrari home in fifth, 227 miles in recorded distance behind the winner! The Collins–Phil Hill Tipo 335 lasted three laps before burning a piston, and the Hawthorn-Musso sister car went on for five hours before being forced to retire. The Grand Prix season was an even greater disaster. As the much-modified Jano Lancia-Ferraris grew older, the even more aged Maserati 250FS found new life in the hands of Fangio and the determined Behra.

But perhaps the worst complication was to be found inside the Maranello factory itself. In true Machiavellian form, Ferrari's favorite modus operandi was to create two-on-one situations, rivalries and alliances that were imbalanced, thereby setting up a dynamic instability that would inevitably shatter itself with its own emotional force. The long friendship between Hawthorn and Collins, which dated back to their British club-racing days, was a natural to set against Musso, now expected to carry the mantle as Italy's reigning racing champion after Castellotti's demise. Musso was never a particularly stable person, and his life was already complicated by a disposition to gamble heavily and the existence of the lovely Fiamma Breschi. Moreover, Collins and Hawthorn were clever, witty, raucous men who never tired of taunting Musso about the various complexities of his life and their open threat to deny him the position of Number One on the team. It would have been one thing to have been beaten out by an Italian like Castellotti; it would be quite another to be bested by a pair of Englishmen who were often disdainful of Italian food, Italian women and the general lifestyle south of the Alps.

Musso responded to this enormous pressure by driving like a madman. Still, he was, in terms of experience, the novice on the team and over the short Grand Prix season simply could not match the speed of his two English colleagues. They remained constant companions and referred to each other as "mon ami mate." Musso must have gained some quiet satisfaction at the German Grand Prix, where Fangio gave them both a driving lesson in a race that equaled Nuvolari's great 1935 upset of the Germans on the same imposing Nürburgring.

Fangio, aboard his 250F Maserati, started the 22-lap, 312-mile German Grand Prix with his fuel tank half full. This permitted his lighter car to scoot away from the heavier Ferraris of Hawthorn and Collins. By the halfway mark, at twelve laps around the Ring, he had a thirty-second lead. What was planned to be a quick pit stop for fuel turned into a farce. Italian pit stops were often comic exercises, and Fangio's crew put on a classic demonstration. By the time the shouting, gesturing and mad sprinting around the idled machine was complete and the tank loaded up, a total of fifty-two seconds had been consumed. Hawthorn and Collins had whizzed by and, assuming that Fangio was lost in the ruck, began a frolic around the vast track. They played a game of 100 mph tag, exchanging the lead at will and running the straights nose to nose. They would race on the final lap, but until then it was to be an easy ride through the Adenau woodlands. That is, until Romolo Tavoni's watches signaled trouble. Fangio had summoned up every erg of his towering talent and was beginning a demonic charge toward the lead. On the sixteenth lap the Ferrari advantage tumbled to thirty-three seconds. Eight more seconds fell on the seventeenth. The Brits received a signal to stop playing and bear down. But it was no use. The "Trucker" was rolling. On each succeeding lap he broke and re-broke the record. On the twentieth he slashed six seconds off the mark he'd set on the previous round!

Up ahead, the "mon ami mates" were driving as if the devil himself was aboard the Maserati. They sailed through the twisting downhill of the *Hatzenbach* with their tires smoking, then hammered, eyeballs bulging and palms dripping sweat, through the terrible, blind humps of the *Flugplatz*. But still Fangio reeled them in. Two Rip Van Winkles hounded by the Headless Horseman, they skidded and bounced around the ghostly place, the shriek of their Jano V8s slowly being battered away by the baleful yowl of Colombo's venerable straight-6. On the twenty-first lap Fangio stuffed the Maserati inside the wavering snout of Collins's 801 and took second. The pass was so brutal, so decisive that the flying stones from the Maestro's wheels shattered one of Collins's goggle lenses and he was forced to fall back. Hawthorn was next. On the long, flat downhill toward the humps of the Adenau bridge, Fangio thrust alongside the Englishman and took the lead. He broke away from the Ferraris and powered onward, possessed of skill and daring he would never find again, to win by four seconds. The victory, which would be his last in international competition, clinched his fifth World Championship. It qualified him, in the minds of many, as the greatest racing driver who ever lived.

The 1957 season was to end without a Formula One victory for the Scuderia—an intolerable humiliation coming on top of the Mille Miglia disaster and the death of Castellotti. Throughout the season Chiti and the engineers labored on with the V6 Dino, which appeared at the non-championship Modena Grand Prix in September with displacement increased to 1.8 liters and plans on the drawing board to run it as a full 2.5-liter Formula One car in 1958. The cars, driven by Musso and Collins, were beaten once again by the Maseratis, but it did not matter. Win or lose, Enzo Ferrari was on hand to say goodbye once and for all to the nettlesome rivals from across town.

The Orsis, battered by their collapsed business in Argentina, a failed effort to sell cars in America and increasing costs of competition, were again to give up motorsports. From that point on, the Scuderia Ferrari was to become the preeminent, exclusive and hopefully dominant standard bearer of Italian honor in international motor racing. The question was, would it be sufficient to stave off the rising tide of fast cars swelling up across the Channel in Great Britain?

Ferrari was fully aware of the potential threat posed by old rival Vandervell and his radically aerodynamic Vanwalls. For 1958 the Englishman had announced a formidable team composed of the icy, analytical, abundantly talented Tony Brooks, a young prodigy named Stuart Lewis-Evans and Moss—who would contribute his talents to other teams until the Vanwalls were ready later in the season. What Ferrari did not comprehend was the tectonic shift in the sport about to be triggered by a small-time Surbiton garage owner named John Cooper, who with his father had cobbled together bits salvaged from a scrapyard to begin building tiny 500-cc racing cars after the war. By 1956 the Coopers were producing excellent Formula Two cars with their Coventry-Climax 4-cylinder engines mounted midships, behind the driver. Such a configuration reduced weight (by eliminating the drive shaft), lowered the car (by permitting the driver to recline with his feet stretched out in front of him) and aided traction and cornering by placing the mass of the engine more in the center of the chassis. Both Cooper and his friendly rival, Colin Chapman, were enthusiastic converts to the notion of the "rear engine" (actually "mid-engine," although such a design is commonly referred to as a "rear engine" simply because the power plant is behind the driver). In reality a mid-engine car has its engine positioned inside the wheelbase, whereas a rear-engine machine, à la the Volkswagen, has it mounted behind the rear wheels.

A new breed of driver, including Roy Salvadori, the Australian midget-car ace Jack Brabham and the New Zealander Bruce McLaren, to name a few, were quick to adopt to the radical new machines. Coopers started to dominate in the Formula Two ranks and began making forays into Grand Prix competition. In 1957 Salvadori had finished a surprising fifth in the British Grand Prix, the first driver ever to score points in modern Formula One competition with a rear-engine machine.

Despite this performance, Enzo Ferrari remained disdainful not only of the English *garagistas* and their home-built cars but of the new rear-engine technology as well. When Chiti noted their nimble handling and high cornering speeds, Ferrari dredged up his old saw about "oxen pulling the cart" and the subject was summarily dismissed. In the meantime the Coopers and Chapmans of the world were forging ahead with shifts in design that were to leave "grand constructor" Ferrari looking like a backyard tinkerer.

Still, the introduction of the new 246 Dino produced much optimism for the 1958 season. Featuring a new, lighter, more rigid small-tube frame designed by Massimino's staff and a 270-hp 2.4-liter version of the Jano V6, the cars, in the hands of Collins, Hawthorn and Musso, were expected to dominate the proceedings once again. The aged 801s had been scrapped and Ferrari's only concern lay in a change in the formula that had been made, making the use of aviation gas, as opposed to methanol, mandatory. The Dinos were operating perfectly on the new fuel, while reports filtering in through Ferrari's vast and effective network of spies indicated that Vandervell's staff was having difficulty converting their 4-cylinder powerplants to run efficiently on gasoline. Therefore, with Maserati's role reduced to a few independent teams and Vanwall apparently not yet ready to do battle, Ferrari's fortunes were certain to rise.

The joy was short-lived, to say the least. In fact, the vaunted Formula One Dinos were rendered obsolete before they completed their first race. The season opened at Buenos Aires on January 15, 1958, and it was to generate shock waves that reverberated all the way back to Maranello. Vanwall and BRM chose not to enter, having not yet solved the conversion to aviation gasoline. Therefore the only Ferrari opposition was composed of a gaggle of 250F Maseratis and a single 2-liter Cooper entered for Stirling Moss by Scotch distillery heir Rob Walker. The little machine, dubbed the "spider" by Fangio, was something of a joke in the presence of the muscular red Ferraris. Its

4-cylinder Coventry-Climax did not even carry a proper racing pedigree, having been converted from a design intended to power fire-engine pumps. In its most perfect state of tune the little engine generated no more than 190 hp, or 80 shy of the Dinos. Worse yet, the car carried bolt-on magnesium wheels, so that a tire change would consume many minutes. Convinced that the Cooper could not go the distance on a single set of Dunlops (recalling perhaps the heavy tire wear of the prewar rear-engine Auto Unions, which were, in this case, technically irrelevant), the Ferrari strategists discounted the Cooper's chances for success despite Moss's impressive lap times in practice. The debacle started early when Collins's car broke its drive shaft on the starting line. Then Hawthorn made a lengthy stop at the pits to diagnose the cause of dropping oil pressure. This left Musso as the only man left to challenge Moss, who was driving masterfully, using the patches of oil and rubber in the corners to act as cushions to reduce tire wear. The little car buzzed onward, with Musso seemingly mired in second place, waiting for the Cooper to glide into the pits for new rubber. As the race droned on, it became apparent to Tavoni that Moss was planning to run nonstop, and Collins began frantically waving a pit board at the Roman, ordering him to speed up. After several laps Musso responded and tried to run down the Englishman. But it was too late. The Cooper slipped home the winner by just over two seconds, generating a shouting match in the Ferrari pits in which Musso claimed he hadn't been properly informed of the situation and had never seen Collins's signals. Hawthorn, who struggled home third, sided with his friend, further opening the rift between the Italian and the two Englishmen.

After a number of successful minor outings in Italy and Great Britain, the Ferraris were hauled to Monaco for the next full-blown round of Formula One competition. This time they faced the full might of the potent Vanwalls, which were returning after winning the final two races of 1957 at Pescara and Monza. Also on hand were three more of the nuisance Coopers, in the hands of Brabham, Salvadori and Maurice Trintignant, who had taken over Walker's dark blue machine after Moss had transferred to Vanwall. Once again, one of the insane mites triumphed. Hawthorn and Moss dueled hard during the early stages of the race, which wound its way around the harbor of the magnificent Mediterranean principality. The cars skidded past the shimmering casino and rocketed through the long ocean-side tunnel as they had done since 1929. But both the Ferrari and the Vanwall broke down, and Trintignant, the infuriating tortoise, eased through to win. Back in Maranello, Enzo Ferrari was apoplectic. Not only had the devilish Coopers won two World Championships in a row, but British cars, the heretofore useless *macchina inglese*, had won the last four, if the late-1957 Vanwall wins in Italy were included.

At least Ferrari was able to regain a modicum of honor at Le Mans, where his combination of Olivier Gendebien and Phil Hill rode their 250TR open roadster to a hard-fought win in abominable weather. The Jaguars, which had been so nettlesome in years past, were absent, and the only serious threat was posed by the Aston Martins, one of which finished in second place. Back in third, showing amazing durability and speed, was a 1,600-cc Porsche—a harbinger of things to come in endurance competition. Ten Ferraris had entered, but only three finished, putting the lie to the Commendatore's statement to a journalist: "My cars must be beautiful. But more than that, they must not stop on the circuit. For then people will say, 'What a pity, it was so pretty.'"

The 24 Hours was notable for the arrival of several new faces from North America. Dan Gurney, a brash, broad-shouldered Californian with but a handful of races under his belt, had evidenced enormous talent and had been brought to Le Mans by Chinetti and his North American Racing Team—which was serving as an on-again, off-again arm of the factory effort. Gurney was in the pits when his co-driver, Bruce Kessler, crashed into an independently entered D-Type Jaguar during the seventh hour, killing the other driver and seriously injuring himself. The other Chinetti car was entered for a pair of Mexican youths, eighteen-year old Ricardo Rodriguez and sixteen-year-old Pedro (who was not allowed to drive because of his age), thanks to the sponsorship of their father, the hugely wealthy Don Pedro Rodriguez, whose fortunes were based on Acapulco property and a chain of high-class brothels run for the Mexican aristocracy. Young Pedro's place was taken by Jean Behra's brother, Jose, but the car retired with overheating just past halfway.

With the official Jaguar team out of racing and the Aston Martins running only a limited schedule of major races, Ferrari had little to fear in the major sports-car leagues. A few well-driven Maseratis could occasionally offer trouble and the gnatlike Porsches were showing amazing grit despite their tiny air-cooled engines, but overall the major sports-car races—those long-distance struggles at places like Sebring, Buenos Aires, the Targa Florio and the Nürburgring that counted toward the World Manufacturers Championship—were owned by Ferrari until the blitzkrieg mounted by Ford in the mid-1960s.

Unhappily for the proud Italian motorsports supporters, the same domination would not be enjoyed in Formula One. Following Trintignant's shocking win at Monte Carlo, the humbling process continued amidst the sand dunes of Zandvoort, where the Dutch Grand Prix was run. The hated lime-green Vanwalls won again as Stirling Moss led from start to finish while Hawthorn battled his evil-handling Dino into a distant, infuriating fifth. The car was so erratic on Zandvoort's faster bends that even newcomer Cliff Allison's tiny 2-liter Lotus-Climax finished ahead of him. Furious over the situation, Hawthorn wrote a sharply critical note to Ferrari himself. Had Ferrari been in a stronger position there is little doubt that he would have canned Hawthorn for such impudence. But he was wise enough to understand that the Englishman was among the best in the business and if he could not make the Dinos behave, no one could. He answered with an uncharacteristically humble and soothing note saying that hard work was underway to correct the car's ailments and he hoped it would bear fruit at the Belgian Grand Prix.

It did not. Tony Brooks, whose silky, courageous style was perfectly suited to fast racetracks like Spa, won easily, while Collins's car succumbed to terminal overheating and Musso made a hairy, 160-mph spin into a field at the end of the Masta straight. Hawthorn did manage to set the fastest lap, which salvaged a smidgen of honor for the team, but finished second even after his engine blew up as he exited the final corner.

The team returned to Maranello in tatters. Chiti was quick to recognize that the British, with their rear engines, space-age body-work, coil-spring suspensions, independent suspensions, magnesium wheels and disc brakes, were stealing a march on the firm and its ty-rannical leader. Only one significant alteration in engineering phi-losophy had come from Ferrari in all the years of business: the Lampredi-induced switch to a large-displacement, naturally aspirated engine almost ten years earlier. But locked in Enzo Ferrari's brain like the catechism in the Pope's was the notion that the engine was everything—that raw power overcame all deficiencies in the rest of the machine. To be sure, his racing cars gave nothing away in terms of the power plant, nor would they for another twenty years. But the hated "spider" of the Coopers, with its ridiculous junkyard engine and its bizarre, arachnid-like shape, was driving the sport in an en-tirely new direction, and the man who represented himself as the doyen of fast automobiles was being left in the dust.

The two English drivers moved closer together as the factory staff chose sides, each contingent trying to shift the blame elsewhere. In this maneuvering Musso was left alone, suffering the added burden of being Italy's great hope to reclaim lost glory and being forever re-minded of his deficiencies as the defender of the faith. His driving became even more ragged, more possessed by the daunting demon of the Commendatore, who was always expecting more speed, more victories, more glory for his red machines. When the American In-dianapolis contingent made its second annual visit to the Monza speed bowl for the event called the Race of Two Worlds (which the European teams had boycotted a year earlier), Musso appeared with a war-weary 375GP car fitted with the 4.1-liter V12 that had been scavenged from the Portago wreck. With it he set the fastest time in qualifying at a stunning 174 mph, skating between the guardrails on the banking like a marble in a drainpipe. He was overcome by fumes in the actual race, but he removed all doubt that he was prepared to

An exhausted Enzo Ferrari in an Alfa Romeo ES following his second-place finish in the Circuiti del Mugello, Italy, July 24, 1921. His equally fatigued riding mechanic, Michele Conti, is at right. (AP Photo)

The Ferrari birthplace at 85 Via Camurri (now Via Paolo Ferrari—no relation) in Modena. Ferrari's father carried on the family business on the lower floor; the family lived upstairs. (Yates Collection)

The massive Ferrari home on Garibaldi Square in Modena. (Yates Collection)

The Ferrari family crypt in San Cataldo cemetery, Modena. (Yates Collection)

The 1949 Ferrari 166MM—Touring Barchetta Lusso—Tipo MM 49-166 Mille Miglia (MM). Manufactured from 1949 to 1952, these cars scored many of Ferrari's early international victories, making Ferrari a serious competitor in the racing industry. (Steve Rossini)

Enzo Ferrari with the brilliant designer and engineer Aurelio Lampredi. (The Revs Institute for Automotive Research)

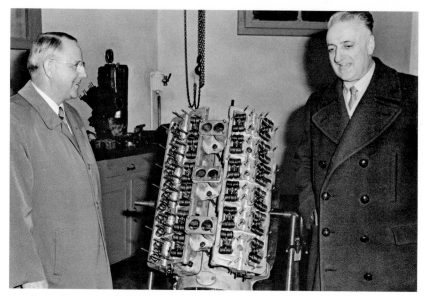

Gerry Grant, president of the Grant Piston Ring Company, and Enzo Ferrari examine a Ferrari engine block for the 1952 Indianapolis 500. (The Revs Institute for Automotive Research)

A nattily dressed Enzo Ferrari greets Robert Braunschweig, editor of *Automobil Revue*, at the factory in 1953. (The Revs Institute for Automotive Research)

Enzo Ferrari with Luigi Villoresi and Alberto Ascari at Monza, June 1953.
(The Revs Institute for Automotive Research)

A jaunty Enzo Ferrari (with the reading glasses he seldom wore in public)
catches up on the racing news at the Italian Grand Prix, September 1953.
(The Revs Institute for Automotive Research)

Peter Collins in his Ferrari 335 Sport at the Ferrari factory's main gate in 1957, taking it for a short test run up the Abetone road before the Mille Miglia. (Klemantaski Collection)

Enzo Ferrari consults with Martino Severi (at right) and Peter Collins (in car) before the Mille Miglia in 1957. (Klemantaski Collection)

The 1960 Ferrari 400 Superamerica SWB Cabriolet by Pininfarina was the last word in sporting elegance. It offered the very best in terms of luxury and performance. (Steve Rossini)

Laura Ferrari at the Grand Prix of Portugal 1960, in Porto. (The Cahier Archive)

Laura Ferrari at the British Grand Prix in Aintree, Merseyside, England, in 1961. She is with race winner Wolfgang von Trips (left) and Phil Hill, who finished second. (The Cahier Archive)

Enzo Ferrari and Battista Pininfarina, his longtime friend and coach builder. (Klemantaski Collection)

British ace driver John Surtees and Enzo Ferrari in the pits of the racing track at Monza on September 4, 1964. Surtees had bettered the track record in an eight-cylinder Ferrari car. (AP Photo)

The 1965 Ferrari 250LM (Le Mans) won the 1965 24 Hours of Le Mans driven by Jochen Rindt and Masten Gregory. (Steve Rossini)

The Ferrari 250 GTO. In 2014, a 250 GTO became the world's most expensive car in history, selling at auction for $38 million. (Steve Rossini)

Enzo Ferrari consults with (from left) Luigi Musso, Eugenio Castellotti, and Peter Collins at Maranello in 1956. (Klemantaski Collection)

The 1969 Ferrari Dino 246 GT, a V6 mid-engine sports car. It was the first Ferrari model produced in high numbers. It is lauded by many for its intrinsic driving qualities and groundbreaking design. (Steve Rossini)

Brock Yates and Dan Gurney with the Cannonball Ferrari 365 GTB/4 Daytona driven to victory in the 1971 Cannonball Baker Sea-to-Shining-Sea Memorial Trophy Dash. The duo traveled from New York to Los Angeles (2,863 miles) in thirty-five hours and fifty-four minutes. (Yates Collection)

Enzo Ferrari during a news
conference in Maranello,
Italy, on September 15, 1980.
(AP Photo/Ferdinando Meazza)

Enzo Ferrari and longtime team engineer Vittorio Bellentani at the Fiorano
test circuit with a 1976 312T2 F1 Grand Prix car. (Klemantaski Collection)

Enzo Ferrari (seated) at Maranello in 1987. Also pictured (at left) is Ferrari's longtime aide-de-camp, Franco Gozzi. The dark-suited man in the center is Modena hotel magnate Giorgio Fini. (Yates Collection)

A smiling Enzo Ferrari at Fiorano for a Goodyear television commercial in 1987. (Car and Driver Magazine)

Luca Cordero di Montezemolo, president of Ferrari, during the fiftieth anniversary of Ferrari celebration in the United States in 2012. (Robin Platzer / Getty Images)

Piero Ferrari, son of Enzo Ferrari, attends the Museo Casa Enzo Ferrari opening press preview on March 9, 2012, in Modena. (Daniele Venturelli/ Getty Images)

Sergio Marchionne, chief executive officer of Fiat SpA and Chrysler Group LLC, November 15, 2012. (Bloomberg / Getty Images)

The 2015 Ferrari LaFerrari (F150), a limited production hybrid, 6.3L V12, producing 950 bhp, allowing the car to exceed 217 mph. (Steve Rossini)

Jody Scheckter driving the 1980 Ferrari 312 T5, a Ferrari Flat-12, at the 1980 Watkins Glen Grand Prix, where he finished eleventh. (Steve Rossini)

Enzo Ferrari during a briefing in his private office in 1983. The walls are stark and bare, save for a picture of Francesco Baracca. (Car and Driver Magazine—Larry Griffin)

sacrifice everything—including his life—for the honor of the Scuderia and the nation he represented.

He gave the rest of himself at Rheims two weeks later. The weekend in the lovely champagne region of northeastern France involved a normal gathering of the racing clan. Away from the racetrack it was an exclusive, fun-loving group composed of bright, rather brittle young adventurers and their bevy of wives, mistresses and girlfriends. In the retinue were a small group of journalists, photographers and hangers-on who moved back and forth across the four continents which hosted the big races like an elegant, high-living circus troupe. The accepted quarters for the racing crowd while at the French Grand Prix was the Hotel Lion d'Or at Rheims, and as practice started on the sunny weekend, the elegant hostelry was packed with various team members and their women. Musso was there with Fiamma Breschi, and Collins had of course brought along his wife, Louise, who was now the lightning rod for a dispute that was growing between Ferrari and his once-favored English driver.

Ferrari had decided that Collins was losing his edge. He speculated, without basis, that the young man's marriage was affecting his urge to win. The situation was further complicated by the Collinses' decision to move out of their Maranello villa and take up residence on a yacht berthed at Monte Carlo. This was interpreted by Ferrari as an act of treason, and he responded by ordering Tavoni to take Collins off the Formula One team and assign him to a Tipo 156 car for the supporting Formula Two race. Collins was furious at the demotion and was joined by Hawthorn, his pal and the only man on the team who was capable of running with the Vanwalls of Stirling Moss and Tony Brooks. Again working in tandem, the two Englishmen faced down Ferrari, and Collins was returned to the first team.

Further tension was added to the operation when Musso received a telegram from an unknown source urging him to drive at the maximum in the race. The inference was clear. His mounting gambling debts were causing trouble, and the giant prize for winning the French Grand Prix—about $50,000—might erase his indebtedness.

Luigi Musso was at Rheims not only to carry the honor of Italy but to recover some much-needed solvency in his personal life.

On the eleventh lap, Hawthorn, who had shattered the lap record in qualifying with his Tipo 246, took an immediate and commanding lead on the sprawling, roughly triangular, six-mile collection of three French Routes Nationales. But Musso was challenging. As the two cars streaked past the odd stack of pits and grandstands planted incongruously in the middle of the impeccably tended vineyards, Musso had moved into Hawthorn's slipstream. Ahead lay a flat, open right-hander that a few men, including Fangio and Hawthorn, had learned to take with the throttle buried on the floor. Musso tried it as well, although lapping a slower Maserati may have forced him a few centimeters off line as he entered the sweeping curve. Suddenly the car began a long, arcing spin to the left before its outside wheels clipped the brow of a ditch. This sent the Ferrari into a series of sickening 150-mph tumbles that immediately flung the hapless Musso to his death. Hawthorn watched the accident in his rearview mirror but grimly marched on to victory in a race that was memorable not only for Musso's tragic death but for the magnificent Juan Manuel Fangio's retirement.

Once again Italy was shorn of a hero. Ferrari moaned, "I have lost the only Italian driver who mattered," although his relationship with a key survivor was just beginning. According to French journalist Bernard Cahier, who with his American wife, Joan, was with Fiamma Breschi during the few tearful days following the race, Musso's family and widow arrived en masse at Rheims to revile the poor woman and even to seize the jewelry that Musso had given her. When and how Enzo Ferrari got involved is unknown, but at some point in the summer of 1958 he took Breschi under his wing and, in the broadest sense, added her to his retinue, which already included his wife, Laura, and his mistress, Lina. The precise relationship is unclear, except that it is known that Ferrari set Breschi up in a small boutique, first in Bologna, then in Florence.

For the next few years he visited her regularly on Thursday afternoons. A number of his associates recall that he "always returned with a smile on his face." Considering that he had already celebrated his sixtieth birthday Enzo Ferrari's dance card was full. Not only did he have his daily confrontation with Laura, but Lina and Piero needed constant tending, as did Fiamma—who had been strategically located well apart from the other protagonists in the growing drama. Clearly, something had to be done to simplify the situation.

In the meantime there was a championship to be pursued. Musso's death left Hawthorn and Collins in complete command of the team, and they responded by running one-two in the British Grand Prix at Silverstone. The Vanwalls refused to handle properly on the table-level circuit, and Collins drove like a master, beating his pal Hawthorn by half a minute. This shellacking hardly bothered Hawthorn, who slowed down on his cool-off lap to snatch a pint of ale from a photographer friend standing at trackside at Beckett's Corner and returned to the Ferrari pits waving the mug.

The joy of the "mon ami mates" would be short-lived. At the next race, the German Grand Prix, Peter Collins crashed off the narrow, undulating circuit beyond the woolly dip called *Pflanzgarten* while chasing the leader, Tony Brooks and his Vanwall. Collins and Hawthorn had just been passed by the uncannily smooth Brooks. Collins may have recalled how he and Hawthorn had been so unceremoniously reeled in by Fangio a year earlier. As the three cars—Brooks's Vanwall and the Ferraris of Collins and Hawthorn—sailed into a climbing, off-camber right-hander, the Tipo 246 of Collins skated wide. The car's left rear wheel clipped the dirt banking on the outside of the corner, and suddenly Hawthorn's windshield was filled with the vermilion gyrations of a berserk Ferrari. Somehow he sped through the dust and flying debris, only to have his clutch fail on the distant reaches of the fourteen-mile track. Heartsick, he stood by his broken car waiting for word on Collins. It was nearly an hour before he was hauled back to the pits and learned that his mate had been taken to a Bonn hospital, where he died of massive head injuries. Ironically, both he and Luigi Musso had lost their lives on the eleventh lap of their last races, driving cars bearing the number 2.

Ferrari was stunned by the news. In a few short weeks he had lost two more of the drivers on his vaunted "spring team." Again the press was at full cry. The Vatican rose up again. Its official newspaper, *Osservatore Romano*, labeled him an "industrial Saturn . . . who continues to devour his own sons." Another major Catholic newspaper would, a few months later, call for racing of all kinds to be banned. The Jesuit priest Leonardo Azzollini, writing in *Civilta Cattolica* demanded that "all speed contests, whether on the track or on the road, and however organized, ought to be stopped." These attacks on Ferrari and the entire sport of automobile racing would continue to be issued by the Vatican until the mid-1960s.

In one shattering moment on a shadowed stretch of German road Louise Collins's idyllic life had ended. Distraught, she went back to Modena, seeking emotional shelter from Collins's friends on the team. She went to see Ferrari, who joined her in shared grief. He recommended that she return to a Grand Prix race as a way of cleansing the frustration and bitterness she felt toward the sport. If she would come to the Italian Grand Prix at Monza, Ferrari told her, he would attend as well, accompanying her on a round of visitations with old friends at the famed race. She agreed to meet him there on the weekend of September 7.

The intervening race at Oporto in Portugal brought another victory to Moss and his Vanwall, and gave him a slight lead over Hawthorn for the World Championship. The race further proved that the archaic drum brakes on the 246s were a hopeless disadvantage against the more advanced disc types being employed on the Vanwalls and the revitalized BRMs. Some members of the team, including Phil Hill, were convinced that Collins's death was caused in part by the overheated and seriously faded drum brakes. A fix was suggested that Ferrari himself grudgingly accepted. Ironically Peter Collins had fitted a set of British-made Dunlop disc brakes to his Ferrari road car, which had been returned to the factory after his death. At the urging of Hawthorn, and with the enthusiastic backing of Chiti and the more progressive members of the engineering staff, the disc brakes were taken from Collins's road car and fitted to his Dino. Two Dunlop brake experts were summoned from England to supervise the conversion, which involved extensive machine work on the hubs and modifying the Borrani wire wheels.

The Italian Grand Prix was by any measurable standard Ferrari's most important race. Here he showcased his best against the opposition and the team was annually whipped into an emotional frenzy. Success would be celebrated like the Second Coming by the national press; failure would produce suggestions of exile and worse. To fill his decimated ranks, Hill, who had been languishing in sports cars and Formula Twos, was finally given a chance to drive on the Formula One team. He was to join Hawthorn, the jocular, well-liked, urbane Trips and Gendebien, who, like Hill, had been labeled as a sports-car expert and was to yet see limited duty on the Formula One varsity.

As was to become his well-publicized tradition in years to come, Ferrari made his appearance on the final day of practice, then returned to Modena. Louise Collins dutifully appeared on race day and was met by a number of friends, including the now retired Piero Taruffi and his wife. But Enzo Ferrari was nowhere to be found. Nor would she ever receive an explanation as to why he chose to break the engagement.

Tony Brooks won the race for Vanwall, no doubt producing enormous satisfaction for Tony Vandervell, who watched with delight as his sensuously shaped green machines bested the Ferraris on their home turf. Hawthorn straggled home second with a failing clutch, having signaled a charging Hill (who had driven brilliantly in his debut and had led the opening five laps) to hang behind him in fourth place. This act of teamwork gave Hawthorn a crucial point lead toward the championship. Although he had won but a single Grand Prix, Hawthorn's consistent high finishes had placed Moss in the position of having to win the final race in Morocco five weeks hence. Moreover, the flashy Moss would have to set the fastest lap—which was also worth a point in the totals—to squeeze by Hawthorn for the title.

Moss in fact won the race over the screaming, sand-swept Ain-Diab circuit at Casablanca, but once again Hill played the perfect teammate and let Hawthorn through to second place and the World

Championship for Ferrari despite the fact that Moss had won four races to Hawthorn's one and the Vanwalls had clearly been the fastest cars for the entire season. Moss's and Vandervell's frustration over the title eluding them was transformed to outright sorrow a few days later when their friend and fellow Vanwall driver Stuart Lewis-Evans died from the serious burns he had sustained in a crash late in the race. Lewis-Evans's death was simply the capstone to what had been a tragic and frustrating year for Hawthorn. The World Championship, eked out over a more deserving Moss by a fluke of scorekeeping rather than outright speed, hardly compensated for the loss of Collins. Mike Hawthorn was a superb race driver and eminently qualified to be the titleholder, but the victory was disputed by the mobs of Moss supporters. Exhausted from five years of intense world-class competition and disheartened by the carnage that had surrounded his climb to the top, Hawthorn made good on his repeated threats to retire following the Casablanca race.

Enzo Ferrari liked Hawthorn more than most of his drivers and worked hard to dissuade him from quitting at the cherubic age of twenty-nine. He would later describe him as "disconcerting on account of his ability and his unpredictableness. Capable of facing up to any situation and getting out of a tight corner with a cold and calculated courage and an exceptional speed of reflexes, he was nevertheless suddenly liable to go to pieces." Ferrari also noted, enigmatically, that Hawthorn had a "rather absent manner," although friends recall him as a jovial, hard-living, hard-drinking fellow who, until the final months of his career, approached motorsports with an exceptional exuberance and boyish enthusiasm. Sadly, that joie de vivre would cost him his life. He returned to England and his varied business interests, some of which necessitated repeated trips from his home in Farnham to Greater London. One late January morning in early 1959, Hawthorn climbed into his new and powerful 3.8-liter Jaguar sedan and headed off toward the city. On the way he encountered Rob Walker, the wealthy sportsman and racing-stable owner, who was traveling the same way aboard his Mercedes-Benz 300SL Gull Wing. Some say the two stopped at a pub for a few pints before continuing their journey—which steadily increased in speed. As they raced over a hump called the Hog's Back and onto a long downhill four-lane bypass around the city of Guildford, Hawthorn hammered the Jaguar well beyond 120 mph, leaving Walker in his wake. What happened next remains a mystery. Some believe Hawthorn clipped a slow-moving truck; others say he lost control on the slick pavement. All agree that the popular champion, who had learned to drive on the same Surrey highways only eleven years earlier, died instantly as the Jaguar was bent nearly in half against an unyielding English oak. The last of Ferrari's "spring team" was gone, as were any hopes of competing with the English upstarts and their rakish new automobiles. Clearly, massive changes lay ahead for the men of the Prancing Horse.

15

Death had wrapped itself around Enzo Ferrari like a Modena winter fog. It had arrived in the spring of 1956 with the passing of Dino and had refused to lift itself from a man whose life—considering the danger of his chosen enterprise—had been amazingly unbloodied. Until Castellotti had died, no man had experienced a fatal crash at the wheel of a car carrying the Prancing Horse on its cowling. But now that was all changed. In three years no fewer than sixteen people—including the spectators along the road to Guidizzolo—had died as victims of a Ferrari. As if that wasn't enough, his vaunted spring team had disappeared, meaning that an entire new squad of drivers would have to be assembled for the 1959 season.

Criticism from the Vatican and the national press was unrelenting, although their motives were different. The Church had taken the position that motorsports were essentially immoral and should be banned. The newspapers, on the other hand, were harassing Ferrari for his failure to win. With the disappearance of Maserati, the Scuderia Ferrari had become a national institution, a quasi-official representative of Italy on the international racing circuit. The Maranello operation had long since been elevated from a simple business enterprise into a machine of raw jingoism for the masses. To the public the success or failure of a Ferrari car was wrapped in national honor, especially if an Italian driver was at the wheel. But following the yelps of grief and frustration that surrounded the deaths of Castellotti and Musso, Enzo Ferrari made a calculated decision: while paying lip service to the desire for an all-Italian driving team, he was thereafter to seek the best drivers, no matter what their nationality. And besides, if an Englishman or a German or an American was to be killed, he would be spared the endless hysterical bleatings of his countrymen. That had to be worth peace of mind, if nothing else.

Chiti and the engineering staff had made little headway in convincing him that the British rear-engine cars were the wave of the future. The 1959 Grand Prix automobiles were to be moderately updated Dinos with—mercifully—disc brakes from England (the first major non-Italian mechanical components besides tires to be employed by Ferrari). Coil springs and more supple suspensions would also be adopted, making the late-1959 and 1960 Dinos among the most advanced but utterly obsolete automobiles ever to be fielded in Grand Prix competition.

Two men, diametrically opposed in personality, would be chosen to lead the team. Jean Behra, late of Maserati, was tough and headstrong, a blunt little man with one ear shorn off from an earlier crash. With him, in the broadest sense only, was Englishman Tony Brooks, a spare, courtly former dental student from a good family who embodied all the chin-out pluckiness and pleasant nature of the classic RAF Spitfire pilot. Behra was a bull rider behind the wheel, Brooks a deft toreador who many believe was one of the fastest drivers in the history of the sport. The chemistry between the pair was terrible from the start. Behra spoke only French. Brooks spoke English and a smattering of Italian. He was clearly the quicker driver, but Behra, perhaps because of his age and experience, considered himself Number One on the team, although Ferrari refused to designate such a position.

Joining them would be Phil Hill, whose talents continued to be denigrated by Ferrari, along with the sports-car ace Gendebien and Cliff Allison, a young English garage owner and gentleman farmer with a bright future. Also in the wings was another Luigi Chinetti find—tall, smiling Dan Gurney, who appeared to possess towering talent, although he had driven in fewer than two dozen races. Gurney had arrived in Modena with his wife and two young boys to be given a test at the autodrome. Like so many other prospects before him, he had hung around the Real Hotel for a few days awaiting the call, then on a chilly morning had driven his little Volkswagen the few kilometers west on the Via Emilia to the vast, fog-shrouded track. Awaiting him were Ferrari, Tavoni, a cluster of mechanics and a few omnipresent local journalists who seemed to camp out at the track. Lined up along the pit counter were a pair of sports-racing two-seaters and a Formula One car. Ferrari was dressed in his heavy winter topcoat and fedora and said little as Gurney angled his rangy body into the cramped cockpit of one of the sports cars. He was quick. Brilliantly quick. Clearly a prodigy—eager, brave and abundantly talented. Ferrari was good at recognizing such skills and immediately signed him to a modest contract—$163 a week and 50 percent of the prize money and sponsorship funds. (No one ever explained either the source or the amount of the latter and to his knowledge Gurney never received a lira from such sources. During the season he drove for Ferrari, he made about $7,000.)

Ferrari came to call Dan Gurney *il grande Marine*, and had he not abandoned the Scuderia for BRM the following season, it is likely that he would have become a major star for the Maranello operation. When he packed up for England, Gurney was a mere novice, but his ability was clear to everyone. He had left believing that the obsolete Ferrari technology would make the team increasingly noncompetitive against the new wave of cars from Great Britain. Had he stayed, he would have benefited from the burst of success Ferrari enjoyed in 1961 and he surely would have seriously contested both Hill and Trips for the World Championship. Gurney would, of course, go on to gain credentials as one of the greatest drivers of all time.

Yet another American arrived to take his place the following year. Richie Ginther was but one of a mob of superb drivers to come out of Southern California road racing in the 1950s. He had acted as Phil

Hill's riding mechanic during the 1953 and 1954 Mexican Road Races before gaining a reputation of his own. The slight, brush-cut, taciturn Ginther would find a home in Modena, and although he was never to gain immortality on the track, his reputation as a test driver is unsurpassed. It was Ginther, in fact, along with Chiti, who would be credited with the discovery of the rear-deck spoiler—an aerodynamic aid that was an invaluable adjunct to stability at high speed. However, the first version was probably fitted *not* to increase road holding but to divert exhaust fumes from the cockpit. The discovery may have been purely accidental.

In the midst of this mini-invasion from America, Phil Hill was the odd man out. He had been in and around the team for nearly four years and despite his obvious ability could not seem to gain Ferrari's full approval. He was a nervous man by nature, perhaps too sensitive to be taken seriously by the cynical, hard-bitten Maranello racing crowd. He enjoyed opera and classical music and was well informed on a broad range of subjects outside racing. This may have unfairly given him the image of a dilettante. He had learned to speak Italian fluently and was able to view the political maneuverings within the team with a jaundiced eye. "They were an absurd mob," he recalls. "Ferrari was surrounded by flunkies, all seeking some sort of favor or approval." Unlike Brooks, who never took up residence in Modena and refused to become immersed in the Machiavellian schemes of the Scuderia, or Behra, who remained a pugnacious outsider, Hill vainly attempted to work within the system even as he was cynically aware of the nonsensical theater that much of it really was.

A man who learned to deal with the intrigues was another junior member of the team, Count Wolfgang von Trips. A charming German nobleman who quickly understood Ferrari's vulnerability to aristocracy, Trips was a close friend of renowned fashion designer and amateur driver John Weitz. The German quickly recognized that there was a great diversity in the quality of the cars being fielded for various races. Some would be shabbily repaired wrecks or mutants scavenged from several older machines. No two were alike, and Trips—or "Taffy" as he was called by his friends—told Weitz he would pay off certain members of the team management to make certain that he got the best car. Internal bribery was simply a way of life within the Byzantine confines of the team, and even outside contractors were not immune. By then Luigi Chinetti had established a formidable arm of Ferrari in America with his North American Racing Team (NART) and was fielding excellent efforts in long-distance events like Sebring and Le Mans. His cars were generally first-line machines or late-model hand-me-downs. But much diligence was required to be sure they weren't recycled wrecks or carried feeble engines and welded-up gearboxes. Payoffs were necessary to ensure a constant flow of decent parts. "We had no choice," muses Chinetti's son, Luigi Jr., known widely as "Coco." "We paid so they wouldn't screw us."

Thanks in no small part to Chinetti's efforts to woo a steady stream of gullible Americans, the production-car business was beginning to thrive. About 248 passenger cars were built in 1959, with a majority of the bodywork supplied by Pininfarina (formerly Pinin Farina) and Scaglietti. About 40 percent of production was heading to North America, which often prompted Giberti to shout to his staff, "Please finish the cars for Chinetti. We need the Americans' money!" By now Pininfarina was the principal designer, with Scaglietti sharing the duties as the primary passenger-car coach builders. Other *carrozzerias* were sometimes employed for special one-off examples, but most of the vivid and beautiful Ferrari shapes (and there were many from this era) came from those two shops. The cars remained detuned racing machines with sexy skin and elicited little interest on the part of Ferrari. He continued to travel via 1100 Fiat with the loyal Pepino Verdelli at the wheel. He was beginning to obtain significant real estate holdings in Modena and as far away as Bologna. He bought an immense four-story Romanesque villa around the corner from the old Scuderia. It faced the Largo Garibaldi on the Via Emilia and was one of the largest homes in Modena, certainly rivaling in size and splendor that of the Orsis down the street to the east. Ferrari, Laura and his mother moved into the rambling edifice, taking up but a tiny segment of the massive sprawl of rooms. Mama Adalgisa was strategically located in her own apartment on the third floor, well away from Laura.

Another, less visible but equally desirable acquisition was made at Viserba, on the Adriatic coast a few miles north of Rimini. This would prove to be a refuge from the oppressive heat, humidity and ravenous mosquitoes that engulfed Modena each summer. Laura in particular would come to spend most of the midyear months at Viserba, with her husband a more frequent visitor than is generally known.

The move to the big house at 11 Largo Garibaldi forced an alteration in Ferrari's social habits. Heretofore he had been a regular at the bar of the Real (soon to be purchased by the Fini family, who owned one of the finest restaurants in Modena), but now he moved his celebrating and woman hunting to the Grand, a block away and beyond Laura's curious eyes. The Grand was directly across the Via Emilia from the Orsi manse, yet but a block away from Ferrari's new house. He was becoming increasingly territorial, like an aging lion. His daily drives to Maranello, occasional trips to Castelvetro and Lina and Piero, some test sessions at the nearby autodrome and his Thursday-afternoon visits to Fiamma in Bologna were his only serious movements. In the morning the routine was the same: a visit to his mother, a shave at the barbershop, a stop at Dino's tomb and a morning of office work at the Scuderia. Before noon he and Pepino would motor off toward Maranello, where lunch would be taken at the old school/mule stable across the road, which was slowly being converted into a small restaurant and inn called the Cavallino.

"There was a Ferrari by day and a Ferrari by night," recalls an old lady friend. "By day he was the *capo*, all business. But at night he was different. Then it was the women. Ferrari loved the fuck!" Ferrari's appetite for the female sex was prodigious, even after he reached his sixtieth birthday. The Grand Hotel was a center of action in Modena. The bar was often filled with beautiful women, both professional and amateur, and Ferrari reveled in the scene. When it came down to a contest for a selected female between Ferrari and one of his minions— including drivers—the Commendatore always pulled rank. Moreover, he was not without a sense of humor in such situations. One evening a woolly Argentinian who would become a car builder in his own right snared a prostitute and hauled her off to an upstairs room. Ferrari gathered up a small group of conspirators and packed the victims' doorsill with newspapers and set them afire. They stood in the hall reeling with laughter as the lovers burst through the flames, wrapped only in sheets.

Much of this tomfoolery took place under the ragged auspices of the Biella Club, the old social group which had for a long time operated within the motorsports community surrounding Ferrari. It was still populated by drivers, wealthy customers, hangers-on, a few members of the press and assorted outriders who gathered periodically to carouse at places like the Grand and the Real-Fini. That Enzo Ferrari was a regular attendee at such functions goes without saying.

On one occasion Ferrari gathered up a gaggle of bicycles for his fellow Biella members and organized an impromptu race through the streets of Modena. The American journalist Peter Coltrin was less adept on two wheels than his Italian rivals and clipped a curb. He took a mighty spill, much to the amusement of his pals. It was a small group of Americans, including Coltrin, who did much to create the Ferrari mystique during this period. A Californian who settled in Modena and married a charming local woman, Peter Coltrin was one of dozens of expatriate automobile enthusiasts who had been overpowered by the romance of the Modena car scene. Repelled by the chromed arks that infested the American roads in the Eisenhower era, they had arrived in Italy to find the place brimming with fast cars, lovely scenery, elegant women and tables groaning with delicacies, all for a pittance in an era of powerful postwar dollars.

Coltrin was joined by a pair of female writers: tall, elegant Logan Bentley and her slight, dark compatriot Diana Bartley, who walked with the aid of a crutch. They, plus Denise McCluggage, a fine writer and excellent driver in her own right, who acted as a personal champion of Phil Hill's, wrote reams of copy for the American automobile magazines of the day—*Road & Track, Sports Cars Illustrated* (soon to be *Car and Driver*)—plus *Town & Country*, the New York *Herald Tribune* and *Sports Illustrated*, rhapsodizing about this noisy Nirvana for automobiles. These writers, plus Griffith Borgeson and Henry Manney III, an irascible, bearded, tweedy Californian whose ironic humor graced the pages of *Road & Track*, created, often unwittingly, the Enzo Ferrari persona which has fascinated, charmed and deluded Americans to this day.

In the midst of all this enthusiasm for the good life Modenese style, many of their dispatches managed to ignore the fact that the 1959 Grand Prix season was an unmitigated disaster. The team won but twice. Brooks was victorious on the ultra-high-speed circuits at Rheims and AVUS, where big power was the arbiter rather than nimble handling. It was at Rheims where Behra left the team after punching Romolo Tavoni. The feisty Frenchman, brimming with frustration over the team politics and the endless row over who was the Number One driver (clearly Brooks, although never officially designated), came unglued after his car burned a piston and stopped on the twenty-ninth lap. Behra had driven superbly and had broken the lap record twice before retiring. The day was unbearably hot, and he vented his anger and exhaustion by slamming the team manager, Tavoni, with a hard right. Such an act of insubordination might have been forgiven had it come from a superstar like Fangio or Ascari, but Behra was more than expendable and he was heaved off the team. He was killed at AVUS a few weeks later when his Porsche skittered over the lip of the 43-degree north curve and his body was flung against a flagpole.

These were violent times for Tavoni. He was surrounded by incessant arguments, involving not only truculent drivers like Behra, but moody engineers, cranky, strike-prone mechanics and Ferrari himself, whose demands for success were as loud and regular as the chiming of the Duomo bells. These disputes were seldom settled in orderly meetings behind closed doors. One day passersby on the Abetone road were shocked to see two men in front of the factory gates screaming at one another and heaving stones between gaps in the traffic. It was Ferrari and Tavoni!

Despite the endless bickering and the regular thrashings being administered by the despised English *garagistas*, Brooks somehow managed to maintain sufficient concentration to remain in contention for the World Championship until late in the year, when the first postwar American Grand Prix was run on the Sebring airport circuit. However, he was bumped at the start by teammate Trips and called at the pits to check for damage. This was considered an unforgivable sin by the Ferrari loyalists, who thought Brooks should have soldiered on. But Brooks, who had experienced several serious crashes due to mechanical failure, would not relent. Always analytical about his driving and not about to be swallowed up in the emotional vortex of the Scuderia, Brooks checked the car, then charged from fifteenth to finish third. That finish doomed him to second place in the World Championship, with Australian Jack Brabham winning the title for Cooper. The Italian press flailed Brooks for not soldiering on with the damaged car.

The only bright spot in the otherwise bleak competition landscape was Chiti's successful 1.5-liter V6 Formula Two car. It bode well for the future. In 1958 the FIA had announced, without warning, that 1961 Formula One engines would be reduced from 2.5 liters to 1.5 liters. This meant that the Chiti development would be well suited, provided the boss could be persuaded to shuck his oft-repeated "ox pulls the cart" nonsense and get on with the design of a proper rear-engine race car.

The opening year of the mad 1960s was infuriating for Enzo Ferrari. His cars were flogged mercilessly by the Coopers and Lotuses. Brooks and Gurney had left the team, which was now manned by Hill, Allison, Trips and a mildly deranged Belgian, Willy Mairesse, a dour, sleepy-eyed sliver of a man with a withered arm who seemed to wear his death wish on his sleeve. The antithesis of his gentle countryman Olivier Gendebien, "Wild Willy" Mairesse was notable only for the number of crashes he survived until he finally finished himself off at Ostend in 1969 with an overdose of sleeping pills.

Once again, the four-cam Dino engine produced gobs of power (around 280 hp at 8,500 rpm), but the nose-heavy, bulky machines were simply no match for the lighter, more sophisticated British cars. The first major event, the Monaco Grand Prix, saw Allison crash badly during practice. Like most competitors of the day, Allison drove without a seat belt and was unceremoniously pitched out of his car when it slammed against some curbing. He suffered head and facial injuries and was out for the season. (Allison later told writer Alan Henry that he awoke in a French hospital speaking French. "That was strange, because I didn't know any French," mused Allison, who is now in retirement.)

The big news at Monaco was the long-awaited, hopelessly over-due debut of the Ferrari rear-engine 246, driven by Ginther. The car had been tested previously at Modena by Hill and test driver Martino Severi (a man who was unbeatable at the Modena autodrome, but slow everywhere else). The car handled reasonably well and worked creditably until its transmission acted up late in the race and Ginther straggled home sixth. Hill drove brilliantly in his antique and man-aged a third behind Moss's Lotus 18 and Bruce McLaren's Cooper.

Good news came from Le Mans a few weeks later, where Gende-bien and French journalist Paul Frère won the 24 Hours with a Tipo 250TR. NART's similar car was second. This victory helped to erase a defeat by Aston Martin the year earlier (driven by Roy Salvadori and Carroll Shelby—a man who was about to embark on his daring Cobra car-building effort and who was to become a major antagonist of the Scuderia).

The remainder of the Formula One season was an unmitigated disaster. The British cars were unbeatable, and the team didn't help itself by employing Mairesse for the Belgian Grand Prix at Spa—surely the most lethal of the major European circuits. For once Wil-ly's blowsy driving style was not at fault when he tangled with Englishman Chris Bristow as they wailed into the fearsome, twisting downhill right-hander at Burnenville. Bristow's Cooper slipped off line and sailed through a fence, sending the driver tumbling to his death. Near the end of the race another fine British prospect, Alan Stacey, was killed on the Masta straight when his car veered out of control, perhaps because a bird flew into his face. The best finish for Ferrari was Phil Hill's fourth, a long lap down.

All three Ferrari cars—driven by Hill, Trips and Mairesse—failed in the French Grand Prix. Then Hill and Trips dribbled home a distant fifth and sixth at the British Grand Prix. There was no success in Portugal, but relief came at the German Grand Prix when the race was set for Formula Two cars. Porsche was doing well in the class, and the Germans arbitrarily excluded the all-winning British Formula One teams. The Italians decided to follow suit, but in a slightly more subtle way. The ACI (Automobile Club of Italy) announced that the Italian Grand Prix at Monza would be contested on the combination oval-and-road course—a decision loudly denounced by the Cooper and Lotus contingent. They objected to the banking, which was bumpy as a cob and could tear up the suspensions of their fragile machines. The Italians' ploy worked. Knowing the rugged old Ferraris were capable of withstanding the abuse of the lumpy concrete banking, they had hoped that the British would stay home, thereby giving the hometown favorites an undisputed win. Then the Ferraris almost withdrew. The Commendatore had an argument with the Monza management over team arrangements and threatened to pull out. But his ruffled feathers were smoothed and the aging 246s, which were about to run their last race, qualified for the first three places in the hands of Hill, Ginther and Mairesse. During the race they circulated in an orderly formation unchallenged by the few independents who had the temerity to enter the contest. Hill won easily. It was a hollow victory, but one that the ecstatic *tifosi* would accept without the slightest regret.

While the record of the Scuderia Ferrari in Formula One competition is rosily recalled as one of unvarnished triumph, the fact remains that since the Ascari years of 1952–53 the team had been in a terrible slump. Excluding 1956, when Ferrari competed with Fangio and the Jano-designed Lancia D50s—which had been bestowed on him like pennies from heaven—his own cars had won but *eight* races in seven seasons (two in 1954, one in 1955, none in 1957, two in 1958 and 1959 and the single, hollow victory in 1960). Since 1954 he had won Le Mans and Sebring three times, the Mille Miglia twice, the 1,000-kilometer of Buenos Aires five times and numerous other major sports-car events, which enhanced the image of the marque.

But in many cases these victories had come against weak opposition, and more often than not, when faced with major factory teams like Mercedes-Benz, Jaguar, Lancia and Aston Martin, the scarlet machines from Maranello were hard-pressed to stay in the hunt.

The source of much of Ferrari's success over the years was not technological brilliance or tactical cleverness, but dogged, gritty, unfailing persistence in competing—a willingness to appear at the line no matter what the odds and run as hard as possible. Some of this must be attributed to his unflinching pride, but his stubbornness did not always serve him well. Surely his refusal to accede to the realities of the new wave of British machines, with their rear engines, light weight, disc brakes, coil springs, magnesium wheels and fiberglass bodywork, cost him dearly. Had he listened to men like Hawthorn, Chiti, Hill, Rocchi and others, all of whom were urging him to give up his antediluvian prejudices, he might have been far more successful. As it was, it took the badgering of his senior staff, plus shameful defeats on the racetrack for three long seasons, to force the change. This was hardly the response of an enlightened, daring, visionary engineer. Racing historian Mike Lawrence passes this rather harsh judgment on the situation: "Apologists have suggested that Ferrari was resistant to any ideas which his firm did not originate, and the late use of disc brakes is cited as an instance. If Ferrari was so insistent on original ideas, the marque would not have gotten off the ground, for it has never, never introduced a new idea into motor racing." If one discounts the accidental Ginther-Chiti discovery of the rear-deck spoiler, it is impossible to dispute Lawrence's stark claim.

As 1960 drew to a close, Ferrari had discarded the oft-used title of Commendatore and preferred to be called Ingegnere (Engineer) by those not daring to refer to him simply as Ferrari or—rarely—Enzo. The new appellation was based on an honor conferred in July by the University of Bologna—an honorary degree in engineering, of which he was justifiably proud, although in truth he was technically ungifted and never, in the parlance of men like Chiti and Rocchi, "put down a line."

His friend and informal biographer Gino Rancati recounts a story that offers some insight into a man supposed to make decisions as a matter of course. According to Rancati, he was invited to make the short trip to Bologna with Ferrari to witness the presentation. They had met for breakfast at the Cavallino, where a discussion ensued about how to travel the twenty-five-odd miles down the Via Emilia. At first it was decided to take one of the company's grand touring machines, but Ferrari demurred, saying that might be too ostentatious. After driving a series of modest Fiats, Alfa Romeos and Lancias over the years, Ferrari had finally begun to travel in one of his own cars, but decided against it on this day. Pepino Verdelli suggested Laura's brother's old Peugeot 404, which was currently in the household. Ferrari rejected it as inappropriate. After all, what sort of impression would be made if Italy's premier maker of exotic cars pulled up in a French sedan? A train was rejected for no good reason, and it was known that Ferrari refused to fly. The discussion dragged on. Finally a solution was reached. The Peugeot was employed, although it was parked a kilometer away and the two men trudged through the thick Bolognese heat to the university!

Slowly, cautiously, fifteen-year-old Piero Lardi was being edged out of the shadows. The young man, with his father's firm jaw and height, was now known to men like Tavoni, Gerolamo Gardini (the company's commercial director), Bazzi and Chinetti, all of whom had been invited to Lina Lardi's home in Castelvetro. "Let's take the long way home" would be Ferrari's signal that a diversion would be made before returning to Modena. Soon the tight retinue of men around Ferrari knew that a second son was in the drama, although to what extent he would finally influence the affairs of the factory was a mystery. Others were not apprised of the new presence. At the 1961 Italian Grand Prix, Gaetano Florini, who was in charge of customer sales of racing cars, spotted Piero in the clutch of hangers-on around his father. Mistaking him for a *tifosi* elbowing his way in for an autograph, Florini gave the boy a swift kick in the trousers. Panicked, poor Pepino Verdelli grabbed him and revealed the shocking truth. In such odd ways the reality of Piero Lardi spread through the staff and by the early 1960s he was a known, notorious meteor entering the Ferrari firmament.

As the prototypes of the new Chiti designed, rear-engine Tipo 156 Grand Prix cars were being completed for 1961, Laura Ferrari began to exert a new, strangely disturbing influence on the proceedings. She remains an enigma, in terms of both her background and the sudden increase in her activity during the early 1960s. For years she had remained in the shadows, an apparently strong influence over Ferrari's private and business life, but completely separated socially as well as sexually. The relationship is believed to have deteriorated following Dino's death, although it was an unsteady union almost from the beginning. She and Ferrari, along with Adalgisa, were sometimes seen in public, dining at the Fini or the Oreste restaurant as well as at the Cavallino. The table would also be arranged so that Enzo Ferrari sat between Laura and Mama, and witnesses recall that the meals were often punctuated by loud arguments.

During this period, Laura began to attend races, with either Tavoni or Chiti enlisted as her chauffeur. This was not a happy time for either man. Laura was a sightseer, insisting on stops at various churches, cathedrals, monasteries and other points of interest while they were trying to arrive at the various racetracks on schedule. Laura carried a briefcase loaded with money—uncounted millions of lire—although she never paid for anything and was inclined to sweep items off shopkeepers' shelves and let those in her retinue take care of payment.

Once at the races she would sit quietly in the corner of the pits, dressed in classic Italian Madonna black.

She came to America at least once, attending the Sebring 12 Hour race, where she incongruously wore a floppy white sun hat, and to the fashionable, but less serious sports-car frolic at Nassau in the Bahamas. There she stayed with the Chinetti family, no doubt in part because Luigi Sr. was a close friend and supporter of hers in the internecine factory warfare. It is recalled that she arrived in Nassau toting only a tiny suitcase (no briefcase, apparently) and dressed with shocking elegance for evening meals. Witnesses marveled at how she was able to produce such a diverse and impressive wardrobe from such a tiny valise.

Some claim that Ferrari induced Laura to travel so that he might carry on his affairs with Fiamma and others—who also included an eager lady from Paris who visited him frequently and the wife of a wealthy customer from Chartres who lived in an immense château. This may be the case, although it seems improbable. For years he had philandered under her nose and it seems unlikely that he would suddenly have sent her off to distant lands simply so that he could chase women. Laura Garello Ferrari was, if nothing else, a wary, battle-scarred woman and it is doubtful that she would have fallen for such a blatant charade.

But surely Laura, who was about to celebrate her sixtieth birthday and who had evidenced no intense interest in automobiles since her courtship with Ferrari forty years before, had a compelling reason to arise from her comfortable, sedentary life at 11 Largo Garibaldi and begin dragging around the world in company with the uninterested and decidedly hostile racing team. No one, including the men who would finally tire of her meddling, has ever been able to supply a reasonable explanation for this odd behavior.

There is a body of opinion that holds that Laura was operating as the eyes and ears of her husband, whose empire was burgeoning to a point where even his finely tuned intelligence network may have been transmitting confusing signals. There is no question, despite their deteriorating marital relationship, that Enzo and Laura Ferrari were linked in a firm partnership to preserve the company (which, in 1960, had been made into a formal corporation—Società Esercizio Fabbriche Automobili e Corse Ferrari, or SEFAC—with them as the principal stockholders). This financial interest on her part may have led her to observe how the operation functioned beyond the factory gates, something her husband was hardly prone to undertake. It is known that she was suspicious of the toadies and flunkies who hovered around Ferrari, and she acted as a trenchant critic of the Byzantine politics within the Maranello factory. A number of new men, not the least of which was the rotund, ebullient, opinionated Tuscan, Chiti (called a *Toscanaccio*, or outspoken Tuscan), were beginning to edge into the circles of power and it is possible this breed of modernists alarmed both Enzo and Laura, although this is pure speculation. Either way, Laura Ferrari exerted a strong and lasting influence over her husband. A longtime neighbor of the family who lived next to the Ferraris' summer villa at Viserba on the Adriatic recalls that Enzo often called Laura as many as five times a day during her long stays on the Adriatic shore to escape the Po Valley's summer heat. These calls were strictly business, implying a deep and abiding link between the two in terms of the company.

Laura was not the only female member of Ferrari's family to have an involvement in his daily life, although the role of his mother, Adalgisa, was not commercial. He visited her each morning and acted like a solicitous son in the presence of this stubby, inflexible woman who often dressed him down as though he were a runny-nosed boy. During moments of rage she was heard to shout, "The wrong son died young!" Whether or not this was a brand of mordant, octogenarian humor is unfathomable, but the fact remains that she stayed close to her son up to the day she died. She was like her daughter-in-law, a gritty, tough-minded woman who remained, despite her tirades, devoted to her son. There were occasions on sunny mornings when she would take short walks around the square and burst into his office at the Scuderia and, in the midst of a business meeting, inquire, "How's my little boy?" One can only imagine the humiliation of the great *ingegnere*, posturing as usual in his doge-like public trappings, being referred to as someone's "little boy."

Still, it is Laura Ferrari who was the enigma in these years. The recollections of witnesses are conflicting. Some say she behaved irrationally and was prone to scooping the tips off restaurant tables and snitching food from shopkeepers' shelves. There appears to be no question that as she moved into her seventies Laura's mental faculties were severely diminished, as was her ability to walk. On the other hand, many people argue that she has been slandered by Ferrari's allies and that she was an intelligent, well-mannered, attentive, if martyred wife who was a great help in operating the business. Luigi Chinetti, Sr., for example, is adamant that she was the victim of a cruel campaign of gossip and that she even remained devoted to her husband despite his maltreatment. "I still love Ferrari," she supposedly lamented in the early 1960s, when their private life appeared to be at its most turbulent.

Whatever troubles were afflicting his wife, Enzo Ferrari's social calendar was hardly empty. While he never, with the possible exception of a few nights spent during tests at Monza, slept out of his own bed for the final sixty years of his life, he managed constant dalliances with bevies of interesting women—including the lady from Paris who apparently roared south and across the Alps the moment he hung up the phone. However, he was not a lavish entertainer. One Modena restaurateur recalls an evening when, following a romantic meal, Ferrari's date regally called for champagne to cap the interlude. Ferrari quickly responded to the waiter, "Bring her *acqua minerale, gassata.*"

Ferrari seemed to like children. He expressed concern about the health of senior employees' children, and in the early 1960s he organized a youth club, which still meets occasionally. It was composed of a small number of enthusiasts who met on Mondays to discuss automotive subjects with the great man. As the years passed, Ferrari also donated heavily to the Modenese Hospital and the Mario Negri Institute in Milan, concentrating on gifts that would aid in the treatment of muscular dystrophy.

While the turbulence in his private life showed no signs of subsiding (nor did he want it to), Ferrari was at least able to look forward to the return of a modicum of stability on the Formula One front. Chiti, Rocchi and a new engineer named Mauro Forghieri were making great strides with the new Tipo 156 Grand Prix car. It would finally be a rear-engine machine. The chassis was a rather crude derivation of the lightweight tubular structures developed by the British, as was the independent coil-spring suspension. The advantage, as with so many classic Ferraris, lay with the powerful engine. Chiti and his staff had in fact created two engines, both V6s but with different cylinder-head angles. The earlier one was a version of the old 2.5-liter 65-degree block, while Chiti and Rocchi developed a new 120-degree V6 that afforded a lower center of gravity and was better suited to the new chassis. Both engines produced ample power to outrun the British, who, because of modest funding and a late start, could not count on Coventry-Climax to have its new 1.5-liter V8 running until the following season. Better yet, the Porsche Formula One car was overweight and somewhat underpowered, so that the Prancing Horse contingent appeared to have a clear field ahead for 1961. What a happy prospect considering the dismal results of recent seasons!

The new machines were breathtaking. Their rakish bodywork featured a daring, twin-nostril front end (first seen earlier on a special 250F Maserati built for the American sportsman Temple Buell) and Ferrari's single refusal to part with tradition: Borrani wire wheels.

Initially he decided to enter the lists with a completely non-Italian team. Again, no Number One was selected, although Phil Hill and Taffy von Trips were considered the senior members. Richie Ginther, who had been well briefed about the factory politics by his pal Hill, was also on board, as was Wild Willy Mairesse. Alarmed at the absence of an Italian, a group of wealthy sportsmen formed a consortium called the Scuderia Sant' Ambrocius to develop young talent and to essentially "rent" a 156 for twenty-five-year-old Milanese Giancarlo Baghetti for selected events. Nationalism was still rampant in Grand Prix racing, and the British, French and Germans, in addition to the Italians, constantly fretted that their national honor was not being upheld in Formula One competition. Ferrari received considerable criticism for not manning his team exclusively with Italians, but he considered the issue to be inconsequential compared with the uproar over the deaths of Castellotti and Musso.

The 156 made its maiden lap at the Syracuse Grand Prix, a minor event in the Sicilian seacoast city. Baghetti and Ginther were sent along as token entries, although Ginther, because of car trouble, did not actually compete. But the rookie Baghetti, relying on the lusty horsepower of the V6, managed to fend off the repeated assaults of Dan Gurney's Porsche and win. It was a surprisingly successful debut for both car and driver and was to serve as a harbinger for the season to come.

Until its tragic finale, 1961 was a year of almost constant triumph for Ferrari. Hill and Gendebien won at Le Mans in splendid style, and English gentleman driver/engineer Mike Parkes (who was beginning to make his presence felt at Maranello in sports cars), along with Mairesse, took second place. Laura was on hand to witness the great moment, which came against no serious factory competition other than the small-displacement Porsches. The same open field awaited the team in Formula One, as the English flailed away in the 156's wake with their feeble 1.5-liter Climax 4s. Still, Moss put up one of his masterful drives at Monaco to win with Rob Walker's tiny Lotus 18, while Ginther showed surprising skill in hectoring him all the way in second place.

Trips won the Dutch Grand Prix at Zandvoort, while Hill had to battle with Scotsman Jim Clark and his spindly Lotus 21 to finally secure second. The Ferrari's monster 30-hp advantage came into play at Spa, where Hill bested Trips, with Ginther third and Gendebien—aboard a Belgian yellow 156—in fourth. Baghetti won on the flat-out triangular road course at Rheims by outpowering Gurney to the flag after his team's leaders fell back. The top Ferraris rebounded at the British Grand Prix at Aintree, with Trips leading Hill and Ginther to a one-two-three finish. The struggle for the World Championship had boiled down to a *mano a mano* duel between the courtly, easygoing German and the testy, emotional, utterly committed American. A victory by either one at the Italian Grand Prix, to be held, as usual, at Monza, would clinch the title. Enzo Ferrari was in an expansive mood when he made his traditional appearance on the Saturday prior to the race, but he would return home without actually witnessing the appalling event that was about to unfold.

The team entered five cars, the three front-line machines for Trips, Hill and Ginther, with extras for Baghetti and Ricardo Rodriguez, now nineteen and throbbing with competitive zeal. Facing them was the talented but outgunned English contingent and their little 4-cylinder Lotuses and Coopers. It promised to be yet another Ferrari benefit run, highlighted by an intramural duel between the two contenders.

In order to conserve their engines on the long track, which employed both the road course and the bumpy oval banking, the Ferraris had been fitted with rather tall final drive ratios, which made them sluggish off the starting line. This permitted the plucky Clark to poke the lime-green nose of his Lotus up among the Ferraris at the start. On the second lap Trips and Clark streaked down the straight behind the pits and toward the right-hand 180-degree corner called the *Parabolica*. Under hard braking at perhaps 120 mph, Trips got the sharp snout of his Ferrari inside the Scotsman's Lotus and their wheels scuffed together. The impact sent the Ferrari pinwheeling across the track, up an embankment and into a clutch of spectators standing behind a wire fence. The car pounded into the mob full bore, pitching its hapless driver to earth before thumping onto its back at trackside.

The disaster was instantaneous and unspeakable. While the race droned on a few feet away, fifteen people, including Trips, lay dead, with dozens more injured. Once again a Ferrari had acted as the scythe of death.

Phil Hill drove onward to a grim victory and a tragically marred championship. Clark, in accordance with the vagaries of Italian justice, was charged with manslaughter. Like Ferrari and Ascari before him, he was the chosen scapegoat for unconscionable behavior on the part of the race organizers, who permitted innocent citizens to view the proceedings from totally vulnerable positions on the verge of the circuit. If anyone should have been charged with criminal acts, it was the officials who ran the race, but in the end it was all a charade. As in the prior cases, Clark was cleared of all charges after sufficient legal posturing and a proper passage of time.

Once again the press was at full cry, as was the Vatican. Ferrari fled for cover, feigning grief over his fallen driver. Legend has it that

Ferrari liked Trips more than Hill, and that is perhaps the case, but his affection was at best measured in tiny degrees. In fact, he cared little, if at all, for the men who drove his cars. A case in point: Shortly after poor von Trips was laid to rest at the family castle (his casket hauled to his grave by a badly overheating Ferrari GT) Ferrari mused to a priest with whom he maintained a close friendship, "I think I did a good job faking my sadness for the death of von Trips."

What had started as a season of soaring hopes had now plunged Ferrari into yet another black period of his life. It was at this point that the notorious "palace revolt" took place, an internal schism that removed no fewer than eight key men, including racing manager Romolo Tavoni, chief designer Carlo Chiti, financial wizard Ermanno Della Casa, plus several other players like Federico Giberti and Giotto Bizzarrini. The exact reasons for the breakup remain unclear, despite a number of public explanations by the protagonists. It is commonly believed that Laura's interference in the racing operation was a component of the dispute, but this seems a bit simplistic on the surface, *unless* her behavior was more bizarre than her supporters would have us believe. To be sure, she was nettlesome to Chiti and Tavoni, both of whom were charged with her welfare while on the road, but unless she began to meddle extensively in the financial affairs of the company, which was Della Casa's domain, or in the car sales for which Giberti was responsible, it would appear that the real reason for the upset was more complex.

It may have been a simple battle for control of the company, which was rapidly expanding in scope and influence. At the time employment was edging toward 500 and in 1961 as many as 441 passenger cars would be built, plus two dozen or more Formula One, Formula Two, and sports-racing cars for major endurance races and hill climbs. Ferrari was himself a workaholic, with no time taken off for weekends or holidays (he is recalled to have worked on both Easter Sunday and Christmas), and there was extreme pressure on his senior staffers to maintain the same regimen.

When it came to his factory, Enzo Ferrari's megalomania knew few boundaries. He was jealous of anything or anyone that might impede his power or intrude on his position in the limelight. He was devoted to men like Bazzi and Rocchi, who were quiet loyalists, content to work in the shadows of the great man. But visible men like Tavoni and Chiti, as well as all the drivers, were a different matter. They tended to diffuse his image and to focus attention away from him and his automobiles. Yet he needed brilliant managers and designers and brave men to extend the reach and the reputation of himself and his operation. Therefore anyone who became too successful posed a threat and therefore was expendable. Ferrari was prepared to share the spotlight with no one, and he chose to further illuminate himself through a series of informal memoirs begun in 1961.

Ferrari enthusiasts have made much of his eloquence as a writer and journalist. They point to his publication in 1962 of *Le Mie Gioie Terribili (My Terrible Joys)* as a prime example of his skills with the pen. Surely Ferrari had a flair for expression in the classic, rather pompous prose of nineteenth-century Italian letters, but it is doubtful that he had either the time or the inclination to write a complete book. *Le Mie Gioie Terribili* (updated two years later as *Due Anni Dopo*) was artfully self-serving and often, by omission, revealed much about the writer. He chose to almost totally ignore Chinetti—a man who was responsible for nearly half his total passenger-car sales—as well as the Maserati brothers, the Orsis and other rivals. Colombo was more favorably treated, but his dislike of old antagonists like Ricart and Fangio was blatantly apparent. He glossed over his modest career as a racing driver while attempting to create the impression of a docile, determined, slightly martyred and misunderstood artisan making his way in a hostile, insensitive world.

All this was the work not of Ferrari but rather of a ghostwriter named Gianni Roghi. A successful Milanese journalist, Roghi spent a year at Ferrari's side, taking notes and surely recording incidents in the exact words of his employer. He was paid for his work not in lire but with a spanking-new Ferrari 250GT coupe. As luck would have it, Roghi had little chance to drive his new prize. Two years later he went to Africa to do a story on wildlife and was killed by a rogue elephant. It is interesting to note that all of the drivers who won the

World Championship for Ferrari—Ascari, Fangio, Hawthorn, Hill and later Surtees, Lauda and Scheckter—left shortly after winning the title. Only Hawthorn and Scheckter, both of whom retired, departed on anything that could remotely be described as pleasant terms. By late 1961 Phil Hill's relationship was souring at the factory, and not a few people noted that the Californian never received a word of thanks, much less formal acknowledgment of his achievement from Ferrari after winning the World Championship.

Therefore it must be assumed that the Great Defection that removed Tavoni, Chiti, Della Casa and Giberti (the latter two later returned) from the staff was a battle of egos, perhaps induced in part by Laura's unsettling presence but also by Ferrari's decision to centralize and consolidate power within his rapidly growing operation. In any case, the blowup was, in typically Italian fashion, loud, profane and grandly operatic in scope. The dissidents met with Ferrari to pose a series of demands, with the threat of quitting employed as a lever. Despite his bombast and imposing persona, Ferrari could sometimes be bluffed and forced to back down by determined adversaries, but that was not the case in this instance. Ferrari would not budge in the face of the billowing rhetoric and the widening breach. Finally the dissidents had no choice but to walk out, shocking the Italian racing community and leaving Ferrari badly shaken but unbowed.

He quickly reached into the substantial talent pool that swirled around the factory. He called a meeting of his junior staff members. "We got rid of the generals," he said. "Now you corporals must take charge." He quickly elevated two young men, Mauro Forghieri and Angelo Bellei, both trained engineers, to take over for Chiti. Bellei's responsibilities would ultimately be directed toward creating real road cars, as opposed to the lightly modified, detuned racing machines that heretofore had been represented as passenger cars, while Forghieri would carry on Chiti's designs for the racing machinery. They would initially be assigned to work alongside veterans like Franco Rocchi, Walter Salvarani and of course Bazzi before establishing their own domains. While both Forghieri and Bellei would remain with the company for years and become devoted and valued employees, it was the former who would become the most visible and therefore threaten Ferrari himself with his notoriety.

Forghieri was only twenty-six years old at the time of his promotion, but his connection with the Ferrari operation went back to his childhood in Modena. His father, Reclus (a French name, because Mauro's grandfather was something of a socialist dissident and raised his family along the Côte d'Azur), was a pattern maker and had worked with Ferrari in the old Scuderia when the Alfa Romeo 158s were being created in the late 1930s. Now he was back at the factory and recommended that his son Mauro be hired in 1960 following his graduation from the University of Bologna with an advanced degree in engineering. Initially Mauro Forghieri had planned to immigrate to California and enter the aircraft industry, but a call from the Commendatore persuaded him to stay at home. Intelligent, willful and full of energy, young Forghieri was to have an impact on Ferrari's fortunes almost from the day he arrived in Maranello.

With the promotion of Forghieri and the hiring of a new team manager, wealthy, headstrong Eugenio Dragoni, whose real business was perfume manufacturing in Milan, Ferrari was able to regroup his forces with amazing success following the so-called revolution. But once again his racing team was decimated. Trips was gone, and Ferrari, utterly without justification, had decided that Phil Hill had lost his will to win. Then Richie Ginther quit. The Californian, who had exhibited real talent in his few Grand Prix races and was a genuinely perceptive test driver, had been, like so many others, duped by his own enthusiasm into driving for the team for ridiculously meager remuneration. He was essentially a part-timer, with no formal position on the Formula One team. According to his contract, Ginther was a utility man, on call for all sorts of duty but without a real role within the operation.

Ferrari invited him back for 1962. The meeting took place in Ferrari's Maranello office, a dingy, stark, blue-walled room that had taken on the trappings of a shrine. It was virtually empty, having only a small conference table and a large photograph of a smiling Dino facing the *ingegnere*'s immense, totally bare desk. Beneath the portrait of his dead son was a vase filled with fresh flowers. These were strange, somewhat sinister surroundings in which Ferrari enjoyed a powerful psychological advantage. He handed Ginther a contract. It was more or less the same as the one he had offered the year before. Ginther scanned it, then refused to sign it. "Sign it or you'll never drive in Formula One again," said Ferrari darkly. Ginther wadded up the paper and tossed it in Ferrari's lap.

Ferrari said nothing, then buzzed for one of his assistants. "Take the key to Signor Ginther's car and check in the trunk to be sure the jack is still there," he said imperiously. And so ended, rather inelegantly, Richie Ginther's short but illustrious career in Maranello— although he did go on to compete in Formula One for four more years.

Once again it was time to regroup the Formula One team—those temperamental, demanding egomaniacs who were essential to the success of the operation. No doubt, if Ferrari could have figured out a way to use simple faceless peasants to drive his automobiles, he would have done so, thereby relieving himself of the endless negotiations, massaging and discourses with men who for the most part cared only for their personal gain and glory, not that of the marque or the man behind it.

The 1962 team would be led by Phil Hill, who gained the honor more because of seniority than because of any enthusiasm on the part of Ferrari. Willy Mairesse, Olivier Gendebien, Giancarlo Baghetti and Ricardo Rodriguez were back, along with the twenty-six-year-old Florentine garage owner and Formula Junior star Lorenzo Bandini. He had joined the team at the urging of Dragoni, who had taken it upon himself to groom the fiery young man as Italy's successor to Ascari, Castellotti and Musso. Although Hill's skills have sometimes been underestimated, it is generally agreed that this was not a particularly powerful squad of drivers. Hill and Gendebien were, of course, fine endurance men, especially skilled at Le Mans, while Rodriguez and Mairesse were known to be capable of maniacally fast laps, but limited in the critical area of prudence. Baghetti's long-term commitment to the sport was suspect, although he had surprised the experts with his performance during the previous season.

Worse yet, Forghieri found the Chiti chassis design to be sadly lacking. The 156s were ancient machines under their rakish skins, at least in terms of the spindly, hopelessly flexible tubular frames that were little unchanged from what Ascari and Villoresi had sat in a decade earlier. Except for their powerful engines, which had enjoyed a 20 percent horsepower advantage over the outdated British Climaxes the year before, the cars were decidedly crude, and Forghieri reckoned they needed major modifications in order to face the British challenge.

A wave of English talent was about to wash over the venerable craftsmen from Maranello. Not only were talented drivers coming from all corners of the Commonwealth, but both Coventry-Climax and BRM had just completed tiny, highly efficient new V8s that were rumored to rival the Ferrari V6 in terms of horsepower and torque and were significantly lighter. Tucked into sophisticated monocoque chassis (aircraft-type quasi-fuselages in which the skin was not hung over the frame but was actually a stressed structural member) that Colin Chapman and others like Lola's Eric Broadley had on the drawing boards, the Ferraris that Forghieri and Company were frantically trying to update suddenly appeared to be as antediluvian as oxcarts. Ferrari was implored to spend more time and money updating the chassis, suspension and aerodynamics of his cars, but he remained convinced that superior power was the single important component of a racing car and demanded that development be concentrated on getting more horses from his aging V6s.

An updated version of the 156 was shown to the motoring press in late December 1961. This was one of Ferrari's carefully choreographed public appearances. They were pure theater. The Commendatore would regally preside and answer questions about the new product line before permitting pictures.

There is no question that the man Enzo Ferrari wanted most on his team was the superstar Stirling Moss. The two men had grown somewhat close during the 1961 season, when Moss had driven sponsor Rob Walker's 250GT short-wheelbase Berlinetta coupe (painted in Walker's dark blue livery) in a number of sports-car events. Moss's pledge never to race for the Scuderia after the decade-old Bari fiasco was softening and he had gained considerable respect for Ferrari, based on his tenacity and single-minded devotion to the sport. Stirling Moss was a pure racer, and this quality, which Ferrari found to be lacking in many of his drivers, was the basis of a natural attraction. He was openly covetous of the Englishman, which did little to increase morale on his own team. This was further exacerbated when Moss visited Maranello and the two men had a long conversation over lunch at Ferrari's table in the Cavallino. Rumors swirled through the factory that a 156, with the fresh 185-hp 120-degree V6, was being prepared for shipment to England, painted in the blue Walker colors. The driver would surely be Moss. Unfortunately, this union never took place. Stirling Moss was gravely injured when his Lotus-Climax V8 crashed at Goodwood on Easter Monday 1962 and his brilliant Formula One career was terminated due to near-fatal head injuries. As it turned out, even his towering talents would probably have been unable to keep Ferrari in the hunt. The new British cars were dazzling, thanks to the new Climax V8s. The vaunted team of the Prancing Horse was dealt repeated knockout blows on the 1962 Grand Prix circuit and failed to win a single race, although Hill and Gendebien repeated at Le Mans against feeble competition. But Ferrari's passion was Formula One. The sports cars remained important as sales tools, and participation in endurance contests like Le Mans were to bring the factory repeated Manufacturers Championships. But they paled in comparison to the glories of Grand Prix competition.

The disappointment of 1962 was overwhelming, especially coming on the heels of the previous year's domination. Ferrari refused to accept responsibility for fielding obsolete cars and blamed his hapless crew of drivers. Dragoni, never tactful, was the point man in the assault, directing his insults at Hill and Baghetti while continually feathering Bandini's nest. Ferrari grumbled that Hill had lost heart after the Trips crash, although if he was losing enthusiasm, it was over the endless backbiting and intrigues that swirled through the factory. Exhausted and frustrated, he walked out at the end of the 1962 season to join the ill-fated ATS team that had been formed by twenty-four-year-old Count Giovanni Volpi and industrialists Jaime Ortiz Patino and Georgio Billi.

The ATS operation (Società per Azioni Automobili Turismo e Sport Serenissima, as it was initially known) was built around the two major Ferrari émigrés, Tavoni and Chiti, and caused a great furor at its announcement. But the project was doomed from the start. Volpi soon argued with his partners and dropped out. Chiti's fuel-injected V8 was powerful enough, but his chassis was a warmed-over version of his mediocre creations at Ferrari. The team was hopelessly under-funded after Volpi's departure and proved to be a terrible embarrassment for all concerned.

Ironically, the woman who may have triggered the great schism of 1961, Laura Garello Ferrari, was no longer a factor—if in fact her meddling had been the prime cause for the defection of Tavoni, Chiti et al. and the subsequent formation of the doomed ATS team. She dropped out of the scene as suddenly as she had appeared. After two seasons of reasonably regular attendance at the races, Laura retired from the sport, as it were, and began to spend her summers at the family's villa at Viserba on the Adriatic.

But in her absence other hostile forces were beginning to exert themselves on Enzo Ferrari. The Italian economy was on a seesaw ride, with the government in turmoil. Since De Gaspari's death there had been twelve governments in ten years. The centrist Christian Democrats held a slight majority but could not maintain power without endless, shifting alliances with the socialists, rightists and smaller splinter parties. The Communists, although they had moved out of Moscow's orbit after the Hungary invasion, were a massive, unruly force in the Po Valley. They maintained enormous power within the labor unions. The Ferrari factory, like all industries in Modena, was subjected to frequent strikes and work stoppages. Total walkouts were rare, but specific groups of employees were continually walking off the job, sometimes only for hours at a time but nonetheless disrupting work schedules and the cadence of the factory.

Whether it was the unsettling political situation in Italy, or the frustrations of the racing season, or his advancing age, or a combination of all three is not known, but in 1962 Enzo Ferrari began to entertain thoughts of selling his company. The first serious bidders arrived from faraway Texas. The fabulously wealthy oilmen John Mecom, Sr., and John Mecom, Jr., had been regular customers of Ferrari since 1957. They had vast oil interests in the Middle East and spent a great deal of time in Europe. Both men made numerous visits to Maranello, although John Jr. was far more enthralled with racing than his father. Because they were so wealthy and powerful, Ferrari saw them regularly when they were in town. In 1962 John Jr. honeymooned in Europe and drove around in a Corvette he had airfreighted from the United States. On his stop in Modena he presented the car to Ferrari. Shortly thereafter discussions were begun between Ferrari and the two Mecoms about purchasing the factory. The price, never settled, ranged between $20 and $25 million. It was a deal that was never consummated, much less mentioned publicly, but John Jr. recalls that the discussions reached a very serious level until another suitor entered the bidding—a suitor with so much money that the Mecoms decided it was futile to continue and dropped out. (John Jr. then formed his own racing team, won the Indianapolis 500 with Graham Hill in 1966 and left active participation in the sport to buy the New Orleans Saints franchise in the National Football League.)

The new player that shoved the Mecoms and their millions aside was not another millionaire sportsman but the second-largest automobile company in the world. In June 1963 the Ford Motor Company renounced the long-standing Detroit-based Automobile Manufacturers Association agreement not to participate in motorsports and announced it was entering into a massive automobile-racing campaign. Ford had correctly identified a growing American youth market and a strong potential demand for high-performance road cars that would spawn such successes as the legendary Mustang. The new promotion theme would be called "Total Performance." While no one at Ford was actively seeking to purchase Ferrari, word was received in February 1963 through its German subsidiary, Ford-Werke AG in Cologne, that the German consul in Milan had noted that a small specialty car builder in Italy was seeking some sort of alliance with a major automaker. Subsequent investigation revealed that the company in question was Ferrari. (Ferrari's version claims that Ford contacted him first, but this is only partially true; the call was made *only* after he made it known that the company was on the block.)

Lee Iacocca, then the brightest star in the Ford Motor Company, recognized the potential for such a purchase. He was seeking prestige, pure and simple. Earlier he had toyed with the notion of trying to buy Rolls-Royce to counter the then-large General Motors lead in the upscale domestic market. The Ferrari buyout, which could be made with pocket change, was an ideal substitute.

In mid-April 1963 a legion of engineers, accountants and management types arrived in Maranello. They conducted a massive inventory of the factory, down to the last bin of bolts and the last ingots of aluminum alloy in the foundry. What they found was a marvelous little car-building operation tucked at the base of the Apennine foothills. Along the Abetone road were the offices. Ranging off to the east was a long, two-story open gallery. Here the racing cars were lined up diagonally in open work bays, with the most talented mechanics in the place tending them. The flooring was red tile and, like all the factory work spaces, was spotlessly clean. At the rear of the cathedral-like race shop lay the engineering design and drafting rooms. The back side of the triangular formation of structures was by far the largest and housed the foundry and automobile assembly lines. Across the road was the Cavallino restaurant, while up the road toward Formigine sat the large, three-story villa used both by Ferrari and by important guests. In the city of Modena was the old Scuderia, now faced with an ugly brick addition, which served as the customer sales and delivery area, as well as Ferrari's in-town office.

According to Leo Levine, who recounted this attempted acquisition in his extensive history of Ford racing, *The Dust and the Glory,* this was probably the only complete and accurate inventory of the factory equipment ever made and remains in the Ford Motor Company archives. The accountants poked through the books to discover that the company was marginally profitable and that rumored subsidies from Fiat and the Italian Automobile Club were no longer necessary to keep the operation in the black.

The price would be about $18 million (surprisingly less than the Mecoms were apparently prepared to pay), for which Ford would receive full rights to the Ferrari name and all trademarks, all patents and subsequent technical developments and 90 percent of the stock of SEFAC, which was owned by Ferrari and his direct family. The company name would be changed to Ford-Ferrari, with the Prancing Horse emblem retained as part of the new company logo. Ferrari's title would be vice president. On the other hand, the racing effort, which was Ferrari's first and only love, would be retained by him. He would hold 90 percent of the stock, with Ford owning the remaining 10 percent. The caveat was that Ferrari would have to construct cars for competition in the areas that Dearborn chose, not exclusively in Formula One and endurance racing, as had been the traditional practice.

When the Ford Division assistant general manager Donald Frey arrived in Modena, negotiations were proceeding without serious problems. It was early May, and Modena was enjoying one of the few periods of the year when the weather was warm, clear and relatively free of humidity. Frey was a bespectacled man in his early forties whose cherubic face concealed a tough, laser-sharp mind. He could speak broken Italian, which impressed Ferrari, and the two men got along rather well, at least on a superficial basis. Numerous evenings were spent together, either at the opera next to Ferrari's home on Largo Garibaldi or at one of the restaurants in the area. One evening Ferrari gathered up a fast grand touring car and took Frey into the Apennines for dinner. He drove as hard as he could, skimming the curves that edged along 500-foot drop-offs. He was clearly trying to scare the Detroiter, although Frey stoically sat in silence as Ferrari's driving became more frantic and ragged. Following their meal in the mountains, and well fueled with champagne, Ferrari drove equally hard on the way back to Modena—until he was stopped by a local police patrol. Furious, Ferrari regally announced who he was and demanded to be set free. But he was far from his home, where the constabulary was more cooperative, and was given a ticket for speeding. The contrite Commendatore proceeded the rest of the way at a considerably reduced velocity.

The meetings dragged on, with lawyers from Italy, America and Switzerland (where Ferrari had a complex of holding companies and bank accounts) amassing piles of papers for the principals' signatures. Frey, who later joined the faculty at Northwestern University, recalls that Ferrari spent hours doodling with the Ford and Ferrari logos, trying to effectively integrate them into one official trademark. He was disposed to sell the passenger-car facility without argument. The price had been settled and only the routine legalities had to be completed. But the racing operation was clearly another matter. The competition department had been discussed informally over dinner at the Tucano restaurant—on the ground floor of Ferrari's house across the street from the Real-Fini hotel—but no firm conclusions had been reached.

By the third week in May the possible sale of Ferrari to Ford had become a national issue. The press was bellowing about the loss of a treasure somehow equal to the Sistine Chapel, and representatives of Fiat, Lancia and Alfa Romeo were hovering in the background. Word was leaked that Ferrari was being required by the agreement to get clearance from Dearborn for any expenditures over $10,000 (which Frey says is utter nonsense). An image was quickly created of an American Goliath about to stomp the gallant David of Maranello into oblivion. But the controversy was swirling exclusively outside the factory gates. Inside there was relative calm, simply because none of the really salient issues regarding the racing team had been addressed. And clearly that was where Ferrari's priorities were centered.

It was a sunny Saturday morning when Frey arrived in Maranello. He had long since tired of the local accommodations and the dreary, heavy sameness of the Modena diet and had begun commuting from the elegant Principe di Savoia Hotel in Milan each day. As usual, the meetings were held in Ferrari's bizarre office with the eerie shrine to Dino looming behind Frey's back. On this day the fate of the Scuderia and its motorsports program would be dealt with one on one. There were no lawyers or advisers present, only the two men who would decide the fate of the sale.

Ferrari opened the conversation with a question that addressed the entire philosophy of how the new operation would be run. "Dottore Ingegnere," as he liked to call Frey, "if I wish to enter cars at Indianapolis and you do not wish me to enter cars at Indianapolis, do we go or do we not go?"

"You do not go," responded Frey without hesitation.

Ferrari stiffened in his chair and said nothing for a moment. Then he stood up and gave Frey an icy glare. "It was nice to know you," he said, and Frey understood instantly that the negotiations were over. Under no circumstances would Ferrari release control of his racing operation to the Ford Motor Company. They could do anything they wanted with the road cars, but he would maintain total command of the racing department or no deal would be possible. Conversely, Ford would not accept an independent operator working within the fabric of their organization. Thus an unbridgeable rift opened up. A few quick goodbyes were made, and Frey returned to Dearborn the next day with nothing to show for the negotiations except an autographed copy of Ferrari's informal autobiography, *My Terrible Joys*, and a few extra, tortellini-created inches around the waistline. When he reported the news to Henry Ford II—a man as iron-willed as Ferrari and with considerably more power—the head of one of the world's largest automobile empires noted, "Okay then, we'll kick his ass." At that moment one of the most expensive and elaborate motorsports programs in history was begun. The object was to win Le Mans and other major endurance races and, as a bonus, to crush the man in Maranello who had resisted their advances.

At this point a small group of wealthy California businessmen and sports-car enthusiasts briefly considered trying to buy the Ferrari operation—which was still most decidedly for sale. One of the principals was Chick Vandagriff, the much-respected owner of Hollywood Sports Cars and one of America's most prominent Ferrari dealers. He recalls that the price had by then dropped to about $7 million, but the preliminary negotiations were cut off when Fiat began to become involved with Ferrari through increased financial support. This does, however, establish that in the middle 1960s Enzo Ferrari definitely had the company on the block—and with a steadily declining price tag.

One of the first men to hear the news of the Ford withdrawal from the negotiations was Ferrari's rival Carroll Shelby. The tall, laconic Texan, whose racing career had been cut short by heart trouble after his 1959 Le Mans victory, had disliked Ferrari from the 1950s and the feeling was mutual. Now they were about to butt heads on the race-track.

In late 1961 Shelby had approached General Motors with an idea. AC Cars Ltd. of Surrey, England, was producing a neat, aluminum-bodied roadster called the ACE into which a small-block Chevrolet V8 would fit perfectly. Shelby proposed the hybrid for the growing American sports-car market, but the General Motors officials turned him down. He then talked to Ford, which began to supply him with similar-sized 289-cubic-inch V8 engines from Dearborn. Thus was the Cobra born, and with infusions of Ford capital, Shelby-American set out to win races not only in domestic competition but also on the continent, where his machines would face the best Ferrari grand touring cars.

In 1962 the FIA had created a Manufacturers Championship for grand touring cars—a class for automobiles produced in quantities of at least 100. On February 24, 1962, Ferrari had introduced the famed 250GTO. This rakish machine, of which 39 examples would be built, was supposed to be an extension of the earlier 250GT and therefore, claimed Ferrari, met the quantity requirement. The GTO was the last front-engine racing machine built by the factory and would come to stand as the ultimate Ferrari road car. But when it was introduced on that chill day in 1962, its purpose was simply to win the new Grand Touring Championship and to dispose of the upstart

Cobras in the process. This it would do in the near term, but by 1964 the lovely coupes from California, in the hands of experts like Dan Gurney, were considerably faster and more than capable of wresting the championship away from the red machines. That the Ford Motor Company, along with its ally, Goodyear Tire & Rubber, was pouring millions into the project did not hurt, although the fact that an American production-based, pushrod type of engine mounted in a sleek, lightweight body could run away from the finest Italian exotica did not escape enthusiasts around the world. The GTO was a sensuous, sexy, aesthetically riveting machine, but as with so many Ferraris, its success was based on mystique and a lack of serious competition.

Meanwhile the personal dramas surrounding Enzo Ferrari were hardly diminishing. Young Piero Lardi was enrolled in Modena's European school, where he was being taught English by a fair-haired British schoolteacher named Brenda Vernor, who had come to the city on vacation and was never to leave. She fell in love with the dashing Mike Parkes, who was steadily rising in the ranks of Ferrari endurance drivers while gaining a reputation as a development engineer. Vernor took a job at the European school to earn her keep. In later years she would become one of Ferrari's personal secretaries and a noted personality on her own within the factory. When Laura Ferrari learned of Piero's existence is a matter of conjecture, but Ferrari's mother knew of the boy at the time of her death in 1965 and it must be presumed that Laura did as well. However, in 1963 it was reliably reported that Piero blundered into his father's office, only to find Laura sitting across from his desk. He quickly fled the scene, although Laura did not recognize him.

By this time Fiamma Breschi had moved away from Bologna to set up a boutique in Florence, where she was to remain for years. Despite her absence, Ferrari never lacked for female companionship and his dalliances at Castelvetro and at the Grand and Real-Fini hotels continued for many years to come. Women and their conquest was to remain a central theme in his life. (He financed a bar on Via Ciro Menotti near the Maserati works for a girlfriend in the early

1980s.) Many close to him puzzle over which came first—automobiles or sex.

The 1963 season started well enough with yet another walkover win at Le Mans against practically nonexistent opposition (none of the major factories sent teams, other than Porsche, which competed in the smaller-displacement classes). The win by Lorenzo Bandini and Lodovico Scarfiotti in one of the new mid-engine V12 250Ps made Italy delirious with joy. It was a total victory for the nation, Italian drivers in Italian automobiles bringing home Ferraris in the first six places. The other major threat came from team member Mairesse and his new co-driver, Englishman John Surtees. Their sister 250P had set the lap record (with Surtees up) after leading for fifteen hours and having put two full laps on the eventual winners. A sloppy pit stop—which was not uncommon for the SEFAC operatives—caused the car to catch fire on the racetrack and Mairesse was forced to evacuate while it was still rolling. His arm injury was to put him out for much of the season.

The arrival of Surtees was welcomed by everyone at Maranello, save for Dragoni, who disliked most foreigners, and Parkes, who snubbed his working-class countryman. But Surtees was a giant in the sport. He had already won the World Motorcycle Championship for MV-Agusta and was much loved in Italy, a favor he returned in kind. He was brave to a fault and a skilled car and chassis tuner. His effortless transfer to four-wheeled racing—an entirely foreign discipline—had been accomplished by only a few men, including Nuvolari, Rosemeyer and Varzi, and was a major accomplishment in itself. That Surtees was instantly a contender against the best in the business bordered on the miraculous.

The Formula One team was supposed to be formed around two men, Surtees and the wounded Mairesse. The cars would once again be the aging 156s, updated slightly by Forghieri while he and Bellei were laying down drawings for a pair of new engines: a four-cam V8 and a flat-12-cylinder. The 1963 cars carried stiffer frames, a conventional single-scoop nose and magnesium knockoff wheels in place of the pretty but heavier Borrani wires. No matter, the engine, even with Bosch fuel injection, was feeble in comparison with the newer, more potent Climax and BRM V8s. The chassis was out of date, in stark contrast to the lighter, more rigid monocoques from England. It promised to be a long season.

But thanks to the man they were beginning to call "Big John" in deference not to his physique but to his heart, new life was slowly infused into the old V6 engines. Working in concert with Forghieri and with the Bosch fuel-injection people, Surtees steadily gained speed on the fleeing British, led by the dazzling Clark and his new pencil-slim Lotus 25. Bandini was gone for a year to the Italian Centro Sud team, thereby removing Dragoni's incessant chauvinism from the atmosphere. The ailing Mairesse was temporarily replaced by sports-car specialist Scarfiotti until he was able to return for the German Grand Prix at the Nürburgring. There Surtees scored the only Formula One victory for the team in 1963, setting a new lap record along the way and generating genuine joy in Maranello, which had not celebrated a victory since the tragedy-marred Italian Grand Prix two seasons earlier. Sadly, Mairesse ended his career as Big John was on his way to the checkered flag. Wild Willy overcooked it entering the lethal *Flugplatz* hump and spun off the road, killing an ambulance attendant and shattering his arm to such an extent that he never drove a Ferrari Grand Prix car again.

A few weeks later Surtees won a nonchampionship race at Enna in Italy, where Bandini pressed him so hard in his independently entered BRM that he was returned to the team. These back-to-back victories helped to compensate for a string of dismal bad luck that befell the team for the remainder of the schedule—including mechanical failures for both Surtees and Bandini at the all-important Italian Grand Prix. But with Bellei's new four-cam V8 almost complete, prospects for the 1964 Formula One calendar looked more than optimistic.

So too did the booming car market. Thanks in no small way to the enthusiasm of the Americans, passenger-car production was just two shy of six hundred units in 1963 and projections for 1964 went well beyond that number. But while the creations of Pininfarina and Sca-glietti had been magnificent hunks of kinetic sculpture in the late 1950s and early 1960s, the stylists at Pininfarina were beginning to jump the tracks as the middle of the decade approached. The bodies became bulbous and in some cases absurdly long-nosed. Such aesthetic barbarisms as dual headlights and busily sculptured roofs and side panels began to appear, and some of the resultant automobiles— namely, the 330GTs, 365GT 2-plus-2s and 275GTBs of the late 1960s and early 1970s—were so ugly that Chinetti and other distributors found them almost unsellable. Moreover, the cars began to arrive with bizarre color combinations. Chinetti recalls reeling with horror as a shipment was unloaded in New York that included a 330GT with owlish twin headlights, canary-yellow paint and kelly-green leather. Another appeared with chestnut-brown paint and light blue leather seats! Moreover, the cars were badly made. Beneath the flash and glamour of the engines and the Prancing Horse label lay machines made up of simple welded tube chassis and components often cheaply or badly fabricated. The bodies were inclined to rust and leak, and no one, not even the best mechanics, could figure out the electrical wiring that had apparently been improvised on each automobile. The clutches were the true Achilles' heel, and the Chinettis were lucky to escape a major lawsuit when one exploded and tore off a woman's foot. Somehow they avoided litigation by replacing the automobile and selling the damaged machine to another unwitting customer.

The automobiles were sufficiently unpleasant to operate (other than for short, noisy, thoroughly delightful bursts down open stretches of road) so that a Bolognese tractor manufacturer, Ferruccio Lamborghini, had begun in 1962 to design a grand touring machine of his own. The story is probably apocryphal, but it is known that Lamborghini told people that he had decided to build his own car after driving to Maranello to complain about a Ferrari he had purchased and being forced to steam in isolation outside the Commendatore's office. This may or may not be true, but he did hire an engineering team headed by Ferrari expatriate Giotto Bizzarrini to create a V12 super-car of his own—presumably with better quality and reliability than his Modenese rival. The result was a magnificent 24-valve, 3.5-liter 12-cylinder from the pen of the man who had been the project engineer on Ferrari's vaunted GTO. Much to Ferrari's displeasure, Lamborghini showed his first complete car at the Geneva auto show in 1964. While the frog-eyed 350GT was less than an aesthetic triumph, it was the beginning of an honored line of two-place grand touring cars that were to rival Ferrari in terms of workmanship and performance. However, Lamborghini was not interested in racing and his machines never contested Ferrari for mastery on the track.

But Enzo Ferrari had considerably more to worry about in terms of corporate rivalry than Ferruccio Lamborghini. After being rebuffed, the Ford Motor Company was coming at him with a fury. Word was filtering into Modena that Iacocca and Company had marshaled a staff of brilliant engineers and experienced racing men in both the United States and Great Britain to do battle against his best sports-racing cars. Their mission was to win the 24 Hours of Le Mans, along with as many major international road races as possible. Not since the massive assault by Mercedes-Benz in 1954–55 had he faced such opposition. Surely the aged GTO, which was now being beaten by the Cobra coupes, would have to be updated if the World Championship for grand touring cars was to be retained (although the factory's hold on the overall Manufacturers Championship, which was held for out-and-out sports-racing cars, was not in immediate jeopardy).

While plans for the 1964 Formula One car were moving along, with Bellei's little V8 to be the main weapon in the coming war with the British, a new weapon would have to be created to fend off the barbarians from America and their crude but effective passenger-car V8s. The solution was to build a mid-engine grand touring car powered by one of the proven Colombo V12s, now radically updated but similar to that which the fabled designer laid down twenty years earlier. The car would be called the 250LM (for Le Mans). International rules required that one hundred be made before it could be sanctioned for grand touring competition. This was a hopeless endeavor, considering that the hand-built cars would be priced at $22,000 apiece—a princely sum indeed in 1964.

Ferrari was in an expansive mood. The years of legal wrangling over the Mille Miglia accident of 1957 were finally coming to an end, with a predictable and thoroughly proper not-guilty verdict assured. Informal contact had been made with the Agnellis at Fiat, and several joint projects that would produce much-needed capital were in the works. In April he made an official application on behalf of the 250LM to the sanctions committee of the FIA's Commission Sportive Internationale. But after he was forced to admit that only ten cars had been completed, Ferrari was turned down and told to reapply in July. With time running out, he asked his friends in the Automobile Club of Italy to exert pressure on the authorities. It was known that sanctioning was often a farce and in many cases requirements for production numbers were bent (as in the case of the GTO) or ignored altogether. Ferrari applied again in July, and this time a delegation drove down from Paris to have a look for themselves. What they found were seven 250LMs ready for delivery, seven more being assembled, four semi-complete cars awaiting engines and transmissions and nine bare chassis. Four more 250LMs were under construction and six bodies were on hand from Scaglietti. A total of thirty-seven machines. Again he was turned down flat. Enzo Ferrari went berserk. Screaming treachery, he announced once more that he was quitting competition and would never carry the flag of his ingrate countrymen. Moreover, in a theatrical gesture, he turned in his competition license. The wrangling was to ebb and flow through the summer until Ferrari was forced to accept defeat. Furious, he announced that his Grand Prix team would race at the United States Grand Prix and at the final race in Mexico under the blue and white colors of the United States.

Ironically, this would mean that his cars would operate, albeit briefly, in the livery of the nation whose most fabled car company was embarking on a full-blown campaign to dislodge him from the pinnacle of motorsports. Not only were Shelby's Cobras gaining strength by the day, but now the Ford Motor Company was coming at him with a hot new prototype to compete with the potent mid-engine 275Ps that were incontestably the fastest sports-racing cars in the world. Working with a small crew at the recently formed Ford Advanced Vehicles facility in Slough, England, Shelby, former Aston Martin team manager John Wyer, designers Roy Lunn and Eric Broadley and the brilliant mechanic Phil Remington managed by early spring 1964 to complete tests on a car called the GT40 (simply because it was a mere 40 inches high). The sleek, flat-nosed mid-engine coupe was to be powered by a modified version of the company's 289-cubic-inch pushrod production-car engine—a far cry technically from the overhead-camshaft masterpieces from Maranello.

After middling results in tests and mechanical failure at the 1,000-kilometer race at the Nürburgring, Ford entered three of the rakish little machines at Le Mans. Ferrari countered with no fewer than six 275Ps (four from the factory and one each from Chinetti's NART and the British importers, Maranello Concessionaires). Surtees, who was teamed with Bandini, set the fastest time in practice, averaging a prodigious 135.6 mph around the narrow, abundantly dangerous 8.36-mile network of otherwise public highways. But the Fords, driven by two-man teams—Phil Hill and Bruce McLaren, Frenchman Jo Schlesser and Englishman "Dickie" Atwood and Americans Masten Gregory and Richie Ginther—were in the hunt. When qualifying ended the fastest thirteen cars were Ferraris and Fords.

In those days the 24 Hour race began with the traditional "Le Mans start," in which the contestants sprinted across the track to their cars, leapt aboard, started the engines and raced off. The marathon began at four o'clock on a warm June afternoon and, as expected, the first lap was completed with three red Ferraris leading the pack. In fourth place was the blue-and-white GT40 with Ginther at the wheel. As the field streamed out onto the 3.5-mile Mulsanne straight, Ginther shot past the three Ferraris of Rodriguez, Jo Bonnier and Surtees to take the lead. Ginther said later that as the roadside trees melted into a green smear on the windshield, he felt the steering get light. He dared a quick glance at the tachometer. It read 7,200 rpm. Ginther quickly calculated that he was running close to 210 mph!

He kept the car in front for the first hour and a half before giving up the lead to Surtees during a slow pit stop. Eventually the Colotti gearboxes (made, ironically, in Modena by a former Ferrari employee) broke on both the Ginther-Gregory car and on the Hill-McLaren machine, but only after Phil Hill had set the fastest lap in the race and the GT40s had revealed themselves to have enormous potential. Surely the gauntlet had been dropped.

The race ultimately went to Ferrari, with the 275P of Frenchman Jean Guichet and Nino Vaccarella leading a three-car parade at the finish. But Ford salvaged some honor when Dan Gurney and Bob Bondurant brought their Shelby Cobra home in fourth to win the grand touring category and soundly defeat the GTOs.

In an interesting footnote to the first Le Mans confrontation between the two giants, Shelby and his crew took one of the GT40s back to Dearborn for wind-tunnel tests. Working with aerodynamic experts from the aircraft industry, they discovered that over 75 hp was being gobbled up by poor ducting to the engine and other inefficiencies in the bodywork. They could only puzzle how badly the Ferraris worked at speed, considering that they were lighter and had more horsepower. Yet the GT40s were significantly faster in a straight line! While the Ferrari racing-car bodies were always aesthetically pleasing and appeared to be streamlined, very little emphasis was placed on that critical component of performance called aerodynamics. Maranello had but a tiny wind tunnel, where only scale models of the cars would be tested. Despite the urging of his engineering staff, Ferrari refused to spend funds on a proper facility right to the end of his life. He remained true to his origins, believing until his death that superior horsepower was the key to victory at the expense of all else.

Despite the looming Ford threat, the Formula One wars went surprisingly well through the summer. The determined Surtees won the Grand Prix of Germany, Austria and Italy aboard the new V8 car. But sure enough, Ferrari was faithful to his threat, and when the team appeared at the United States Grand Prix at Watkins Glen in October, the cars were painted blue and white and had been entered by Chinetti's NART. His efforts to dragoon the FIA and the ACI into sanctioning the 250LM as a grand touring car had failed, although he gained a modicum of success by convincing the Monza management to cancel their late-season endurance race, in which the Cobras were likely to beat up on the GTOs in front of a loyalist home crowd.

The final two Grand Prix races of the season in North America were to produce a happy ending. While the Lotus of Jim Clark and the BRM of Graham Hill were the class of the field, Surtees's pluck and guile had kept him in contention throughout the long season and a second-place finish at Watkins Glen behind Graham Hill left the two men with a clear shot at the World Championship as the circus re-formed at Mexico City on October 25. Surtees had his familiar V8, while Bandini had the much-improved 12. Pedro Rodriguez, driving in memory of his brother Ricardo, who had been killed at the same track two years earlier, was given one of the aged V6s.

At the beginning of the race, Clark, as he had done so often during the season, simply fled from the rest of the field. Gurney rode in second in his Brabham, while Graham Hill muscled past Bandini into third. Surtees, who had gotten a bad start, began pushing through the field in pursuit of the leaders. On the thirty-first lap Bandini overcooked his Ferrari as he sailed into the track's only hairpin and bumped Hill's BRM. The hit damaged the Englishman's exhaust system and forced him into the pits for repairs. Clark appeared a sure winner but on the final lap his oil pressure failed and the singing little Climax clanked to a halt. That put Gurney into first, with Bandini behind him. But suddenly the Ferrari pit crew realized that if Surtees, who was now lying third, could finish second, he would edge Graham Hill for the World Championship by one point! Frantic signaling to Bandini prompted him to slow sufficiently so that Big John nipped into second place and the title.

It was a lucky victory (although John Surtees was thoroughly qualified to be World Champion). Over the long season Graham Hill had enjoyed higher overall finishes, but rules were rules and Ferrari had the laurels. At first the British press grumped that Bandini had been assigned to bump Hill into the weeds so that his teammate could win the championship. But no less an authority than Hill denied this. Pulling himself up to his considerable height, the regal, mustachioed Englishman, whose blue helmet carried the stripes of the London Rowing Club, snorted, "Of course he didn't mean to do it. It was just bloody bad driving."

No matter, the World Championship was back in Maranello after three years and John Surtees came as close to being a national hero as any non-Italian could be. Ferrari seemed to genuinely like this graying, rather moody, but intensely competitive thirty-year-old. Like himself, Surtees was self-made, a school dropout who had learned mechanics before rising to stardom on motorcycles and then in automobiles. Both men were utterly single-minded when it came to racing, and unlike many of the dilettantes who had driven for the team but never got their hands dirty, Surtees was prepared to work endless hours with Forghieri in the dyno rooms and engine shops of the factory seeking more power and reliability. He was a "racer" to the core, and Ferrari pulled him as close, relatively speaking, as any of his drivers.

The two men spent considerable time together away from the shop. Surtees was one of the rare few to be invited into the private precincts of the ocean villa at Viserba, where tactics and future car designs were discussed far into the night. They had long lunches at the Cavallino, where Ferrari delighted in serving drinks called "Formula One, Two and Three," based on their alcoholic content. Surtees, a fit man and light drinker, always chose the mildest Formula Three concoction, much to Ferrari's amusement. This tight association was a frustration to Dragoni, who continued to advance the fortunes of his protégé, Bandini, at court. But full attention was being paid to the new champion, and Bandini (who was embarrassed over Dragoni's incessant, strident advocacy of his cause) was forced into the shadows.

Not surprisingly, the contretemps over the 250LM sanctioning was quickly forgotten in the blush of Surtees's victory, and as preparations were begun for 1965, it was clearly understood that the Ferraris would once again campaign under the national colors. All seemed to bode well for the future, although the massive forces being marshaled against SEFAC in faraway places like Dearborn, Michigan, and in English towns like Cheshunt, Hertfordshire, where Lotus Cars Ltd. was located, were about to pose a challenge to the Ferrari reputation unlike anything seen before and, in a broad sense, were to remove forever the legend of invincibility—justified or not—that had been so carefully nurtured by Ferrari and his ever-expanding legion of true believers.

16

The linkage between Enzo Ferrari and Sergio Scaglietti was forged by a common heritage and a passion for fast, sensuous automobiles. A man in his mid-forties, Scaglietti had known Ferrari since the 1930s, when as a teenager he had repaired bent Alfa Romeo fenders at the old Scuderia workshop. Scaglietti was also a Modenese, and in private the two men conversed only in the thick local dialect that on occasion could be incomprehensible to upper-class Tuscans and Romans. After the war Scaglietti had established his small *carrozzeria* near the main Modena railroad station before moving to more elaborate quarters along the Via Emilia on the eastern outskirts of the city. Despite his lack of formal training, he was a pure artisan possessing an instinctive sense of proportion and scale. From his shops had come such classically shaped automobiles as the Testarossa, the 250GT Spider California and his two masterpieces: the 250GT short-wheelbase Berlinetta and the GTO.

While much of the formal design work for the Ferrari bodywork was done at Pininfarina in faraway Turin, it was this man with the gnarled craftsman's hands and the commoner's sense of reality who was closest to Ferrari in terms of an overall approach to life. Each Saturday, Scaglietti would receive a phone call from the man he always referred to as Ingegnere. "Well, Scaglietti, where are you going now?" would be the standard question. The reply was rote. Cast in stone. He had no plans, Scaglietti would respond, and a lunch would be arranged. It would seem impromptu, but in fact this same schedule was maintained for over twenty years, until Ferrari's health failed in the mid-1980s. In the early years, Scaglietti recalled, the talk centered on automobiles and the racing wars that were being waged at any given moment on any number of major circuits around the world. In later years, the cars slipped into the background and women took their place in the dialogue. Ferrari would lapse into recollections of his youth, repeating, hundreds of times, the story of his painful recovery in the Army hospital near Bologna and the sound of coffin nails being struck home on the floor below. The lunch, heavy Modenese fare in cream sauces, amply lubricated with champagne, was classic Ferrari—habitual, regulated, predictable and Modenese to the core. This was a simple man, seeking simple pleasure, and he found it only within the confines of his own familiar territory.

Several major distractions faced Ferrari as winter broke and the 1965 season was upon him. The V8 and the flat-12 Tipo 158s would have to suffice for the final year of the 1.5-liter formula. In 1966 the rules would require 3-liter unblown engines, with 1.5-liter supercharged power plants permitted. This meant that a whole new family of engines and chassis would have to be created. To complicate matters, the Ford challenge in endurance racing would only amplify, and new sports cars would have to be built if the Le Mans dominance and the Manufacturers Championship were to be preserved.

Moreover, there was new competition on the production-car scene. Porsche was becoming a serious contender in the upscale grand touring market with the introduction of its 911 series. Jaguar was scoring well with its rakish XKE coupe and roadster. Even Lamborghini and crippled old Maserati were nibbling away at the Prancing Horse's upscale position. With products from this segment of the business so critical to underwriting the financial burdens of the racing program, any slump in sales could have serious consequences.

Ford's influence on Ferrari was far from restricted to the major endurance races. With the new 3-liter formula about to be adopted, Ford money was about to be poured into the firm of Cosworth Engineering Ltd.—an engine-building operation formed by the talented engineering duo of Frank Costin and Keith Duckworth. They were known to be creating a radical, compact four-camshaft V8. BRM was working on a fiendishly complicated but powerful 16-cylinder which would also be a threat. Overall, the intensity of competition, which had been escalating over the past four years, was now about to increase to almost intolerable levels.

In the roller-coaster ride of fortunes that composed the Ferrari racing history, 1965 was a period of near-total disaster. Even Le Mans was nearly a debacle, with the factory cars breaking early in brutal, flat-out combat that also removed the best of the Ford contingent. The Dearborn forces had arrived with improved machines, but once more their novice status at the big track penalized them and for the second year in a row the duel went to a Ferrari. But it was not an official SEFAC entry, but rather a tired 250LM entered by Chinetti's NART that was never expected to be a factor. The NART weak link was supposed to be its driver combination, a pair of men noted for wild sprints but lacking the discipline to be successful in long events like Le Mans.

Masten Gregory was more adept at endurance driving than the young Austrian hot shoe Jochen Rindt, who was gaining a reputation in open-wheel Formula Two and Formula Three cars and short, fierce fifty-mile races. Gregory had been around big-time European road racing for over a decade without accumulating much in the way of important finishes, much less victories. But at least he had considerable experience in long events such as Le Mans, which was more than could be said of his teammate. Considering their reputations, and their aged war-horse of an automobile, they were given no chance for victory.

But after the early-race Ford attrition Gregory found himself sitting in fourth place when he pulled into the pits for his first refueling and driver change. No one, including Rindt, had expected the old car to last that long, much less be among the leaders. Gregory glanced across the pit counter to find Rindt in his street clothes! A mechanic sauntered over to the car, raised the hood and routinely signaled that something had broken and the race was ended. Everyone was thinking of a good meal and early to bed. Except Gregory. Furious, he leapt out of the car and dragooned Rindt into donning his driving suit and carrying on. This the Austrian did, with Gregory fully expecting that he'd break the engine out of sheer spite. But Rindt rolled on, getting ever closer to the faltering leaders. Surtees, who was teamed with Scarfiotti, was struggling with failing brakes as the night gathered around the immense track. Only the garish carnival lights and the masses of campfires in the infield served as beacons for the drivers who ripped through the darkness at 200 mph. Bandini's 275P2 was crippled, as was the Rodriguez-Vaccarella 365P2. In the blackness Gregory overtook the Cobra coupe of Dan Gurney and the pair battled for a few laps before the Cobra, which was lying third, broke under the strain. By then Rindt and Gregory were breezing along, convinced the old machine wouldn't go the distance but was well enough placed to go out in a blaze of glory.

As dawn broke they were hovering near the lead. Only a similar 250LM entered by the Belgian amateurs Pierre Dumay and Taf Gosselin was able to run with them. By noon, with only four hours remaining, Gregory and Rindt now could see victory in sight, but were haunted by the abuse they had heaped on the car early in the race. Their tired brains were filled with imagined clankings and rattles inside the leaking, frazzled engine. Surely, in retribution for their maltreatment, the car would stop within sight of the finish.

The NART entry was running on Goodyear tires, while the Belgians' car was shod with Dunlops, the tire company to which Ferrari was contracted for Formula One. At one o'clock in the afternoon the Belgian Ferrari burst a rear tire on the Mulsanne straight, destroying the rear-deck spoiler with flying rubber. Crucial minutes were lost as the car limped around the track to the pits. Victory seemed assured for Rindt and Gregory. At this point Eugenio Dragoni had a deal to offer. Luigi Chinetti, Jr., recalls that Dragoni noted that because Dunlop was associated with the factory, it would be in everybody's best interest if the Belgians won, rather than the Goodyear-equipped NART car. Dragoni's proposal was simple: Let the Belgians take first and lavish discounts would be offered to Chinetti on a batch of new customer cars. The deal was tempting. Thousands of dollars could be scooped from the extra profits. But there was the prestige of winning Le Mans as a counterbalance. Chinetti rejected Dragoni's offer, thereby creating yet another rift in his relationship with the marque he had served so loyally. Therefore the wrong car, with the wrong drivers, riding on the wrong tires, rolled into the victory circle at the world's greatest endurance race. It was a Ferrari win, but one that generated limited joy in Maranello.

If Enzo Ferrari was disappointed with the result, Henry Ford II was furious. When Don Frey reported the results to the boss, Ford shrugged and said, "Well, you got your ass whipped." Frey observed that a win was possible, but not with the current budget constraints. "Did I ever say anything about money?" countered Ford. This instantly unlocked the millions needed to overpower Ferrari. It was also to mark the end of Ferrari's dominance at Le Mans. When Gregory and Rindt stuttered across the line in the early summer of 1965, it marked the last time a Ferrari was to win at Le Mans and signaled the end of the Prancing Horse's longtime domination of international endurance racing.

Adalgisa Bisbini Ferrari died in 1965, the victim of a choking accident while eating a hard-boiled egg. She had been a major part of Enzo Ferrari's life, and her legacy was to force him to make a major decision. She had been fully aware of Piero Lardi's existence, although at what point she discovered the secret is unknown. What is known is her insistence that the boy be legitimized, backed up by the threat that she would directly include him in her will unless Enzo agreed. A bargain was struck that would legalize Piero as a son, but only following Laura's death. There is no question that Piero was by then a notorious presence in Modena and that Laura Ferrari was extremely angry and hurt. Any sightings of the tall, square young man around the factory would elicit her screams of "Bastard!" and Piero made a point of avoiding her at all costs as he drifted through a variety of ill-defined jobs. But regardless of Laura's ire, Ferrari was to stand by his bargain after he laid his mother to rest in the family tomb at San Cataldo.

Despite the breakdown in negotiations between himself and the Ford Motor Company, Ferrari's connection with the United States became irresistibly stronger. Thanks to a solid sales network that included men like Chinetti, Nevada's gaming lord and car collector par excellence, William Harrah (who now had the West Coast distributorship), and California racer/dealer Johnny Von Neumann, Ferrari's stock in America was climbing. But times were changing for Chinetti. In the 1950s he had enjoyed exclusivity with the marque on American shores, but now his franchise was dwindling by the day. It was apparent that the old bonds—which, for all the acrimonious stress they absorbed, seemed unbreakable—were now about to shred. Slowly the Chinettis were being eased out of the picture, and Enzo Ferrari was now treating his old friend as little more than a nettlesome dealer from a faraway land. The Ferrari road cars of the mid-1960s were far from aesthetic triumphs, but the appetite for European status in the booming American economy was insatiable. By now the image of Enzo Ferrari had been inflated to mystical dimensions. His adoring customers accepted him as an imperious but lovable artisan laboring with monklike devotion over his beloved machines. Little did the provincials know that he had been prepared to unceremoniously dump the road-car business not once but three times within the past few years and was now engaged in a serious attempt to establish strong financial relations with Fiat.

By 1965 a joint effort had been established with the massive Turin conglomerate to produce a four-cam 2-liter V6 engine for production cars. Designed by Franco Rocchi, the power plant was to be fitted into a pair of "Dino" automobiles—a mid-engine Pininfarina-designed coupe that remains one of the most beautiful of all time and a less successful front-engine two-seater. The 206 Dino, with its engine mounted behind the driver, was to be sold as a Ferrari, although Fiat was to manufacture the engine and Pininfarina the bodywork. The front-engine car, called the Fiat Dino, was to be sold without the

Ferrari name, although the engine would be billed as a Ferrari design. The project was primarily Fiat's, with Ferrari himself advocating that the engine be a 3-liter flat-12 of the type that was being developed for Formula One competition. But this would have been expensive to manufacture and would have created a much faster, more exotic machine than Fiat's marketing philosophy warranted. That Fiat's plans for the car, rather than Ferrari's, were implemented is proof that the Agnellis and their monster firm were about to exert enormous influence that would ultimately lead to a complete takeover. This, claim some close to Ferrari, was exactly what he had desired all along, and he had simply used the Mecoms and Ford as bait to attract Fiat into the negotiations.

John Frankenheimer, at thirty-five years of age, was at the height of his directorial powers in Hollywood, having created such commercial and critical successes as *The Manchurian Candidate*, *Seven Days in May* and *Birdman of Alcatraz* within the past three years. A devoted automobile enthusiast, Frankenheimer was embarking on an ambitious project to produce a big-budget film on Formula One racing called *Grand Prix*. It was to be shot at a variety of locations across Europe during the 1965 and 1966 racing seasons, and Frankenheimer traveled to Maranello to solicit Ferrari's cooperation. The veteran Italian actor Adolfo Celi had been selected to play a role unabashedly based on Ferrari, and the *ingegnere*'s blessing was desired, if not essential.

Refusal was immediate. Ferrari would under no circumstances aid in the production, despite Frankenheimer's forceful urgings. It was not until Frankenheimer returned in the fall of the same year to screen for Ferrari forty-five minutes of sample footage shot at the Monaco Grand Prix that the position was modified. Ferrari could see that Frankenheimer had an incisive and sympathetic eye for the sport, and only then did he indicate that help might be forthcoming. Frankenheimer spoke fluent French, which helped to convince Ferrari during a series of long lunches at the Cavallino that his cars and his sport would be well treated on the big screen. Frankenheimer recalls one lengthy meal during which talk turned to the road cars and the irritating problem of air-conditioning. Ferrari lamented that his American customers wanted the pesky, expensive, power-robbing units, while his European loyalists rejected them. What to do? He groaned. It was too expensive to equip a small batch of cars with the air-conditioning and omit it from the rest. Frankenheimer had a solution. Why not, he suggested, put air-conditioning in all the cars and merely give customers the option of using it or not? Ferrari was delighted with the solution and implemented it immediately on several car lines. He also relented with regard to the picture and ultimately permitted Frankenheimer to shoot several scenes inside his factory. In the epic final race of the film one of the stars, Yves Montand, is killed at Monza in what is clearly a Ferrari. Ferrari had no objection to a driver being killed during a race, but was adamant that none of his cars lose to any rivals on the track. Ferrari not only sanctioned the film but traveled to Monza for several days while the final scenes were being shot. He was clearly intrigued with the flash and glamour surrounding the multimillion-dollar production.

Grand Prix was also influential in aligning Ferrari with the Firestone Tire & Rubber Company, which was being challenged for supremacy in American motorsports by its crosstown Akron archrival, Goodyear. The notorious "tire wars" of the 1960s consumed hundreds of millions of dollars and spread across four continents as the two giants went head to head on the world's racetracks. At first the rivalry was confined to places like Indianapolis and Daytona Beach, but it soon spread to the Formula One scene. Thanks to Carroll Shelby, Goodyear had gained a foothold in Europe first, both by signing several racing teams to lucrative contracts and by gaining exclusive usage in Frankenheimer's film. Firestone responded by contracting with several other Formula One teams, including Ferrari. The 1966 season would see the SEFAC entries using Firestones manufactured in the company's Brentford, England, facility. But because rain tires were not used in American competition and Firestone had none in its inventory, Dunlop would continue to supply that type of competition tire until the Akron firm could develop one of its own.

One would expect that Ferrari would have extracted a king's

ransom from Firestone to use their tires. But that was not the case. H. A. "Humpy" Wheeler, who currently manages the elegant Charlotte Motor Speedway in Charlotte, North Carolina, but was at the time with Firestone's racing division, recalls that Ferrari cared only about quality. "His sponsorship fee was quite reasonable, considering the astronomical numbers that were being tossed around in those days. Ferrari wanted the best tires available. That's all that mattered. Firestone spent millions developing tires for his race cars and passenger cars. Raymond Firestone was irritated that Goodyear had an exclusive presence in Frankenheimer's *Grand Prix* and he wanted a European presence at all costs. Ferrari was the most prestigious name in the business and the deal was done. The exact cost is impossible to determine because it came from a number of internal budgets, but compared to the millions being spent at Indianapolis and in Southern stock-car racing, it was not an outrageous figure."

The image—assiduously formed and polished by Ferrari himself—of a lone visionary, battered but unbowed, struggling against giant forces like the Ford Motor Company, generated waves of sympathy. He had successfully fostered the notion that his factory was a tiny clump of buildings housing devoted craftsmen hand-forming exquisite machines from exotic metals. Visitors arrived in Maranello expecting to find dirt-floored sheds packed with sweating Modenese laboring like Renaissance bronze sculptors. Instead they discovered a thoroughly modern factory, lavishly equipped and manned by a work force of nearly a thousand men who were capable of mass-producing over 650 road cars a year while maintaining a vast motorsports operation. Compared with an adversary like Ford, Ferrari could be considered an underdog, but contrasted with the British *garagistas* who had been regularly beating him for the past five years, he was a veritable Gulliver being brought down by a gang of Lilliputians.

It would take more than mythic images to save the team in 1965. Jimmy Clark, the taciturn Scot, dominated the Formula One schedule in his lithe Lotus 33 and had clinched the World Championship by mid-season. The best that Big John—or "Shirtsleeves," as his British mates called him—could do was to labor along in the popular

Jim's wake. After he finished second in the first race of the year in South Africa, Surtees's fortunes went into steady decline. Mechanical failures, poor chassis setups and low power outputs dogged him for the entire year. Worse yet, he faced the strident jingoism of Dragoni, who was currying favor with Ferrari and the press in order to advance the fortunes of his favorite, Bandini. On the other side was Michael Parkes, who disdained Surtees's engineering skills, not to mention his working-class origins, and was openly covetous of his place on the Formula One team. Again Ferrari did nothing to bring peace to the camp. Discord, in his mind, was the mother of fast cars, and this internecine struggle was only viewed as the source of more victories on the racetrack. But this clearly was not the case. Surtees was far and away the most talented driver on the team, and despite Dragoni's efforts to get Bandini the best cars and Parkes's incessant maneuvering in the background, Big John was the only man capable of staying near the lighter, more agile British machines from Lotus, BRM and Brabham. Other than a hard-earned third place at the British Grand Prix, the season was a string of dismal failures for the Scuderia's top driver. It ended for him in late September when he had a massive crash at Mosport Park outside Toronto. He was at the wheel of his own Lola-Chevrolet Can-Am sports car and not a Ferrari. He had been racing independently in Can-Am competition with an English-built machine since Ferrari chose not to field an entry in the popular new North American series. The crash left him seriously injured in a Toronto hospital. At that point his relations with Enzo Ferrari were still close. Not only did the factory cover his medical bills, even though he was not racing one of their cars, but during his recovery the boss telephoned him to inquire about the healing of his badly fractured left leg. Ferrari kiddingly offered to build a car with an automatic transmission for his star when he returned the following season.

Major changes were in the works for 1966. The Grand Prix formula was increased from 1.5 liters to 3 liters, unsupercharged, which required new automobiles from the competing teams. Because Ferrari had been building powerful 3-liter sports-car engines for nearly fifteen years, it was assumed that the Scuderia would produce winners right out of the box—especially when it was considered that the new Ford-Cosworth V8 was a year away and teams like Cooper were forced to rely on Maserati's ten-year-old V12. Jack Brabham's operation was reduced to modifying an obsolete Oldsmobile V8 passenger-car engine, while the BRM engineers fiddled with an insanely complex H16 layout with nearly as many moving parts as a Swiss chronograph. Yes, the field seemed clear for Ferrari, although the health of their star driver remained a question mark.

Angelo Bellei was moved out of the racing department and assigned exclusively to the road-car engineering operation. This left the brilliant if mercurial Forghieri to create the new cars for the 3-liter formula. Because massive funds were being thrown into the development of the new 4-liter P3 sports car with which to battle Ford at Le Mans and elsewhere, Forghieri had no choice but to stint on the Formula One car. Lacking the money and manpower to create a new engine, he gamely modified the aged, heavy, bulky 3.3-liter V12 developed for the sports cars—a power plant with direct lineage to the Colombo design which had started the business twenty years earlier.

An early prototype of the Tipo 312 was hauled out to the Modena autodrome in the spring of 1966. Surtees, still hobbling on his game leg, was there to test the new machine. Also present was a smaller, lighter 2.5-liter V6 single-seater intended for the Tasman series in New Zealand and Australia. The session provided a terrible shock. The small, underpowered Tasman car was 2.5 seconds a lap faster around the truncated Modena circuit. The bigger V12 was a lumpy, cumbersome, tail-heavy, sluggish mongrel of a machine that Surtees instantly judged would be outclassed even by the ragtag opposition from Great Britain. This was to be the beginning of the end of John Surtees's connection with Ferrari and was to trigger a disastrous slump in the team's fortune.

After the little V6 proved to be the faster of the two cars, an internal struggle developed over whether Bandini or Surtees was to be cursed with the creaking V12. As the team leader, Surtees got the assignment, simply because it was the main-line automobile for the

Formula and team honor demanded that their Number One be at the wheel. Moreover, the Ferrari road cars were powered by V12s, and it was deemed important that Ferrari's long-term philosophical linkage to that type of engine be exploited as much as possible. Hence Surtees was doomed to forced labor with the ungainly machine while Bandini was left to dance lightly around the circuits in a car that was theoretically uncompetitive—although in truth it had nearly as much power as the 312 and handled effortlessly. This only strengthened Dragoni's bleatings about his man's ability to outrun Surtees in a smaller automobile, while Parkes kept badgering Ferrari and Dragoni for a place on the Formula One team. Outside the gates, the Italian press, with whom Dragoni had excellent rapport, was screeching for Bandini to be elevated to the Number One position in place of the wounded, faltering Surtees. While this was going on, Forghieri and the engineering staff fed Ferrari wildly optimistic horsepower numbers and test results. (It was not uncommon for Ferrari team managers to "edit" practice times at Grand Prix races when reporting to the boss in order to make the operation look better than it actually was.) In all, it was a period of Medici-like internal politics at Ferrari. Pressures were building to intolerable levels, and someone's temper had to rupture. It was to be John Surtees's.

He had won in gallant style on the rain-swept Spa circuit, where no fewer than seven cars crashed on the opening lap—including all three BRMs when they plunged into a rain squall on the back half of the monster circuit. Still, Surtees had to battle with the plucky Rindt and his hot-rodded, overweight Cooper-Maserati before finally prevailing. Dragoni twitted Surtees for not leading the entire distance against what he described as meager opposition. This only deepened the rift between the two men. The entire situation unraveled a week later at Le Mans, where the new 330P3 sports cars faced a blitzkrieg of ultra-powerful 7-liter Mk II Ford GTs handled by an elite Anglo-American team of drivers, engineers, team managers and mechanics. Ferrari sent a trio of the potent P3s with their magnificently rounded coachwork from Piero Drogo's Modena body shop. Surtees was teamed with Parkes in the lead car, while Bandini and French long-distance ace Jean Guichet were in the second. The third machine was entered by Chinetti's NART operation and was to carry Rodriguez and Richie Ginther (who after leaving Ferrari had risen back to prominence by winning the Mexican Grand Prix at the end of the 1965 season).

When Surtees arrived at Le Mans he was informed that Lodovico Scarfiotti had been signed on as a third, "reserve" driver with him and Parkes—a tacit pronouncement that Maranello did not consider Surtees fit to run the entire distance with a single co-driver. This slight was the triggering mechanism that broke up the relationship. Surtees demanded that Scarfiotti be removed, but Dragoni—then Ferrari by telephone—flatly refused. In a huff Big John packed his bags and left for England. It was over, and yet another World Champion had left the Ferrari realm in anger. He was immediately hired on by the Cooper-Maserati operation and went on to organize his own Formula One team, which operated with middling results.

The finely tuned network of toadies and apologists for the Scuderia immediately began passing the word that Surtees had been fired for a variety of offenses, including his complaints to the press, insubordination and an alleged diminishment of his skills following his Mosport crash. It was said that Franco Gozzi, Ferrari's aide, had been assigned to fire him following the Belgian Grand Prix, but his victory had delayed the decision. Then came the blowup at Le Mans practice, which, claimed the Ferrari claque, led to his sacking. It is certain that Surtees quit, but corporate pride in Maranello would not tolerate the humiliation of the former World Champion walking off the team. It was stridently claimed that he had been dismissed.

Despite the internal contretemps and the company-sanctioned misinformation campaign, Ferrari issued his typically syrupy, ingenuous tribute to Surtees, expressing deep regret at the loss and wishing his former driver the best. On the surface it appeared that the Commendatore practically sent him on his way with a first-class plane ticket to London. Not quite. Later in the season John Frankenheimer was filming *Grand Prix* during a private session at Monza. Ferrari had driven up for the day to watch the shooting. Surtees appeared with his

Cooper-Maserati and asked permission to take a few test laps. Frankenheimer was willing, but Ferrari not only refused to speak with his former star but flatly objected to his running any test laps while he was present.

John Surtees's departure from Modena also created a blowup in the Ferrari household. At that point Enzo Ferrari was still juggling his relationships with Laura and his mistress, Lina Lardi. This sensitive situation always bordered on the chaotic. Sometimes it broke into open warfare, as when Big John left town. He had been living in an apartment on the lovely tree-lined Viale Vittorio Veneto which was being paid for by Ferrari. On the day he returned to England, Ferrari called an old lady friend—the wife of a journalist whom he had known for years—and instructed her to take the apartment key from Surtees and give it to, she thought he said, Laura. This she did, accepting the key from Surtees and dutifully driving to Maranello, where she found Laura having lunch at the Cavallino. She handed over the key, only to watch Laura's face begin to percolate with rage. A few hours later the woman received a phone call from Ferrari, who was screeching a chain of profanities.

The woman, a feisty Modenese who was aware of Ferrari's foibles and brooked no nonsense from him, screamed back. Finally, in the hysterical garble boiling through the phone, it became clear: Ferrari had wanted the woman to give the key to Lina, not Laura. The plan was for her and Piero to move into the quarters vacated by Surtees. When Laura received the key she understood exactly what was up and landed on Ferrari like a wounded Bengal tiger. Such were the risks of trying to maintain a pair of households cheek by jowl in Modena.

While the loss of Big John Surtees was a serious blow, it is doubtful that his talents could have forestalled the battering the team took at Le Mans. The monster Fords dominated the race, easily capturing the first three places in front of a delighted Henry Ford II. The first Ferrari to finish was the British-entered 275GTB of Piers Courage and Roy Pike, hundreds of miles back in eighth place. The Fords, with their massive engines developed for Southern-style stock-car racing, were clearly better machines than the 330P3s in every respect, and apologists for Ferrari began to whine that the little Maranello operation was being overwhelmed by the giants from America. This conveniently ignored the fact that the giant from Maranello, which had been the 800-pound gorilla in Formula One for years, was being humbled by Jack Brabham and his Repco V8—a mini-budget racing conversion that used the aforementioned Oldsmobile engine block designed six years earlier by General Motors.

There was simply no excuse for a massive racing operation like Ferrari's, manned with legions of capable, committed engineers, to be bettered at its own game by the small British Formula One teams unless the latent conservatism of the boss, coupled with the debilitating political atmosphere fostered by his leadership, was a major factor. As has been said, much of Ferrari's early success came against feeble competition. When faced with serious, committed opposition, the factory was more often than not beaten on the racetrack.

The Formula One team won but one more Grand Prix race during the 1966 season. Scarfiotti was victorious at the Italian Grand Prix, the benefactor of widespread mechanical failures on the part of the enemy. This sent the *tifosi* into paroxysms of joy—a routine state prompted by every Ferrari victory at Monza, but doubled by the knowledge that Scarfiotti was the first Italian in an Italian car to win the event since the late Alberto Ascari in 1952. Such moments were treated as quasi-religious experiences by the motorsports addicts of the nation. Still, it was clear to everyone else—perhaps excluding those seized with self-delusionary pride inside the Maranello factory walls—that the joy would be short-lived. The British were coming, in the form of the heavily financed Ford-Lotus effort and the new, brilliant DFV Ford-Cosworth V8—an engine that was to win more major Formula One races (as well as Indianapolis car competitions in America) than any other racing power plant in history. Using this minutely packaged engine, which made the Ferrari V12s look like truck diesels by comparison, men like Jackie Stewart, Jim Clark, Graham Hill and Denis Hulme were to dominate the sport for years to come, despite the best efforts of Ferrari.

Dragoni's impassioned politicking was even beginning to affect Ferrari. Clearly the team had suffered with the departure of Surtees. He had been replaced by Parkes, who had deluded himself about his capacities as a Formula One driver. He was a competent journeyman at the wheel of a sports car, but the twitchy, demanding single-seaters were simply more than he could handle. Scarfiotti was also well suited to sports cars, but lacked the scalpel-sharp skill to run in Grand Prix competition. Only Lorenzo Bandini was a legitimate first-rank driver, and he was so possessed by the burden of being Italy's torch bearer that this otherwise decent, uncomplicated man was being forced to levels of performance well beyond his emotional and physical capacities. Much of the blame lay at the feet of Dragoni, whose fevered nationalism had contributed to Surtees's defection and had left the team with a retinue of highly motivated but modestly talented drivers.

Part of the responsibility had to be borne by Ferrari himself, whose manipulation of the press over the years had created a manic loyalty on the part of Italian motorsports enthusiasts. The Scuderia had become a national passion, in no small part because Ferrari had choreographed the press into treating him and his operation like an icon upon which Italian honor rested. This attention was turning out to be a curse as well as a blessing. The press was becoming more fickle in proportion to the rising stature of the company. When the Scuderia had been small and impecunious, Ferrari had been able to elicit sympathy for his occasional absences from the victory circle. But now that he had so carefully groomed himself as the patriarch of an all-winning operation whose machines were by their very nature superior to all others, his pleas for forbearance in thin times generated less and less tolerance. Having bought the Ferrari claim of invincibility based on superior technology, the Italian media began to evidence decreasing levels of patience when Ferrari experienced a debacle of the magnitude of the 1966 season.

Dragoni had to go. His abrasive personality had caused turmoil throughout the organization, and by November a move was on to replace him. This move was in part prompted by Enzo Ferrari's obsession with the press. His morning newspaper readings were becoming longer and more detailed. Slights and overt criticisms directed at the company were carefully noted and retribution was exacted. More serious offenders received loud, threatening phone calls or cryptic notes containing dark threats. Lesser offenses were duly noted, to be dredged up at later press conferences, in which Ferrari would appear with a massive dossier on each journalist present. Clearly, a way had to be found to neutralize the frustrated yelps from the media. After all, they had been groomed to expect that Ferrari's vaunted V12s would walk away with the 1966 championship and that the crude interlopers from Ford would be turned back at Le Mans. When both campaigns disintegrated into bitter, humiliating defeats, the press rose to full cry. Each morning Gozzi and Ferrari would wear red pencils down to stubs recording the angry assaults coming from the ingrates manning the nation's major sports papers and magazines.

An inspiration: If one of the opposition in the media could be enlisted as team manager, it might silence the critics. After all, if this vocal group persisted in critiquing every new engineering change and second-guessing every strategic decision, why not let one of them bear the heat for a while?

Franco Lini had known Enzo Ferrari since, as a young journalist, he had covered the team for *Auto Italiana* in 1949. Since then he had risen to become the nation's leading auto journalist. In November 1966 Lini was in Portugal covering the Rally Costa del Sol when a phone message was received at his hotel. It stated simply, "Call Gozzi." Franco Lini had received dozens like them before. They usually involved a complaint about something—often trivial beyond belief—Lini had written about Ferrari. Considering it to be yet another gripe, Lini ignored the message until he returned to Rome. Another call. This time it was Ferrari himself. "You must come to Modena immediately. Take the night train," he demanded. Curious but unintimidated, Lini traveled to Modena and arrived at Ferrari's office at the old Scuderia early in the morning. Clearly something was wrong. Ferrari was unshaven. His hair was disordered. It appeared that he had not slept well.

Once inside the office, Ferrari locked the door. Franco Lini was not a large man, and it ran through his mind that perhaps Ferrari had come unglued over the unremitting assaults of the press and was about to kill him. Ferrari walked behind his desk and slumped in the chair. "How much do you make each month?" he asked. Lini told him. It was a good salary. A comfortable living. "Dragoni is gone. You must become the team manager," said Ferrari bluntly. He offered him a 50 percent increase in wages and other perks. At first Lini refused, citing a total lack of experience in such matters. Ferrari insisted. They spent the afternoon locked in negotiation at the Cavallino. Still no decision. Franco Lini returned home. Finally the deal was settled by telex. The appointment was announced on national television. It would be a two-year contract, with Lini expected to regroup what was clearly a team in chaos. Forghieri was laboring day and night with Rocchi and his staff to improve the lumbering 312s. The new 3-liter formula had energized the sport. Six major teams now faced Ferrari, all with excellent machinery. The sweet-handling Brabhams, which had carried Jack Brabham to his third World Championship the year before, would be back in improved form. Dan Gurney, funded by Goodyear, had a new Eagle, with a powerful V12, ready for battle, as did Honda—which had retained John Surtees as their driver. BRM had a new V12 on the drawing board to replace its quirky H16, while the Cooper-Maserati could be counted on for reliability if not great speed. In the wings was the much-rumored Lotus 49 with the new Ford-Cosworth V8, although it was not due to appear until mid-season.

Clearly, Lini needed another front-line driver. Bandini was trying hard and on his best days was capable of running at the front of the pack. But Parkes was simply not up to the task, despite his soaring motivation. Worse yet, he was a tall man, well over six feet, which meant that his rangy frame required a special long-wheelbase chassis. That involved extra expense during a period when the onslaught of competition on the Grand Prix front and from Ford in sports cars was stretching budgets to the maximum. The Italian economy was not good. The north was being swept by strikes by an increasingly churlish work force. Ferrari had been disabled by repeated stoppages in 1966 and more unrest was expected during the upcoming summer months, when the humidity frayed even the most tranquil of tempers.

Lodovico Scarfiotti was available, as was Nino Vaccarella and Giancarlo Baghetti. They were gentlemen drivers, skilled in sports cars and on occasion capable of fast drives in Grand Prix machines. But Ferrari knew all too well that wealthy amateurs were a different breed, devoid of that cruel emptiness in the pit of their stomachs that drives the true professional. He wanted a man like a Nuvolari or, God forbid, a Fangio, who would compete with that brutal, heartless, unremitting determination to win at all costs. They were difficult, impudent, egocentric, selfish, detached men, but they could win. Their loyalties were directed only to their own often monastic needs, but when called upon to face the gruesome odds of a rain-swept, oil-soaked racetrack, or a mean, ill-handling brute of a race car, they would not falter.

At that point the best professionals—Clark, Gurney, Hill, Surtees, Stewart, Hulme and Brabham—were contracted to other teams and not available. Franco Lini had his eye on a twenty-three-year-old New Zealander named Chris Amon. The son of a prosperous sheep farmer, Amon had knocked around the European scene for four seasons, gaining a reputation as something of a party boy as well as a quick driver in outclassed machinery. At the time he was racing in the North American Can-Am series with Bruce McLaren and, through his contacts at Firestone and Shell, learned that Ferrari was interested in his talents. A slight, sleepy-eyed young man, Chris Amon was hardly the prototype of a strutting, macho race driver, but he was abundantly talented and was a desirable prospect for the depleted Maranello contingent. After flying to Modena and meeting with Ferrari, he was hired for the 1967 season. Naïve about the politics and the niggardly wage scales of the Scuderia, Amon agreed to drive for no salary and 50 percent of the purse. Once again the Commendatore had bluffed a zealous youth into risking his life for the privilege of driving for the last "grand constructor" in the field. It was to be a dubious honor.

The season started with a stunning success at the Daytona 24 Hour sports-car race. An early struggle developed between the Fords and the General Motors–financed Chaparral of Texan Jim Hall (teamed with a resurgent Phil Hill), and the ensuing attrition left the field open to a major win for Ferrari. Amon and Bandini drove their new 330P4 to an easy win, finishing ahead of Mike Parkes and Scarfiotti in an identical machine. Pedro Rodriguez and Jean Guichet brought home their older NART P3/4 in third to complete the domination. Once again the major test came at Le Mans, where Ferrari entered the same pair of P4s that had performed so effectively at Daytona six months earlier. But this time they faced the full might of Ford, headed by a super-team of Dan Gurney and his partner A. J. Foyt—who only two weeks earlier had won the Indianapolis 500. Gurney was by far the more accomplished road-racing driver of the two and set a judicious pace for the race that kept the Ferrari at bay, yet preserved the rather fragile drive train of the big new Mk IV Ford GT. This was to be the final major confrontation of the two teams, and the Parkes-Scarfiotti car made a valiant battle of it until an hour and a half from the finish. The towel was then tossed in and they cruised home second behind Gurney and Foyt. Not only would this mark the end of the classic Ford-Ferrari duel at Le Mans, but it would spell *finito* to the monster 250-mph machines that had acted as the primary firepower in the confrontation. The FIA, alarmed at the escalating speeds, mandated that in 1968 the endurance-racing rules would permit displacement limits of 3 liters for prototype racing engines and 5 liters for limited-production units. While the Fords would compete for a number of years under the banner of John Wyer's Mirage team (with 5-liter engines and later with 3-liter Ford-Cosworths) and Ferrari was to develop 3-liter machines, the epic struggle that had begun when Donald Frey left Maranello five years earlier had ended.

Following two nonchampionship races in England early in the season, the first serious event on the Formula One calendar took place at Monaco on May 7. Both Amon and Bandini appeared with 312s equipped with 36-valve cylinder heads developed by Franco Rocchi. The bodies of the cars were leaner and more streamlined, and Forghieri had done considerable work on the suspension and brakes. (This young engineer also had the overall responsibility of simultaneously developing the 330P4 sports car—a backbreaking effort for his small team of engineers.) Bandini was recognized as the team leader and certainly felt the same oppressive pressure as had Luigi Musso and Eugenio Castellotti. It was one thing for an Italian driver to race for the Scuderia in a secondary role. It was entirely different to wear the mantle of Number One. Suddenly all of Italy seemed to expect colossal efforts, as if the best Italian car and the best Italian driver possessed superhuman powers in the face of the infidels. How much this burden affected Bandini is unknown, except that he drove with exceptional determination in practice at Monaco and qualified in second place—while Amon could do no better than thirteenth. The Italian made a magnificent start and led the first lap ahead of Denis Hulme's Brabham, then fell back to fourth after Stewart, Surtees and Gurney slipped past. Then Gurney's Eagle broke its fuel-pump drive and Stewart's ring and pinion failed. Bandini repassed Surtees into second behind Hulme and was driving the race of his life. As the contest droned into its final stages, Bandini held second. As he whisked through the same chicane that had sent Ascari tumbling into the harbor in 1955, Bandini bobbled. The Ferrari nicked a barrier, vaulted some trackside hay bales and flopped upside down onto the middle of the racetrack. The ruptured fuel tanks spilled gasoline onto the red-hot exhausts and the car erupted in flames. Before beleaguered rescue crews (who lacked fireproof clothing) could douse the flames, Bandini was horribly burned. Hulme drove through the smoking wreckage to take a joyless win while Amon pressed on to finish third.

Bandini was rushed to the hospital with terminal burns. He lingered a few miserable days before passing away. Once again Italy was plunged into mourning. Once again the assaults began on Ferrari and the sport. Accusations were made that Bandini's car was unsafe—a veritable coffin of fuel. That was nonsense. While the modern technology of packaging volatile fuel in racing cars was undeveloped, there is no question that the 312 of Bandini represented the best available packaging for the day. If Enzo Ferrari did nothing else, he built racing cars that were stronger and safer than they needed to be.

Bandini's funeral was held in Milan, where thousands turned out to mourn the fallen hero. Even Enzo Ferrari was sufficiently saddened to make the trip to the funeral and to the post-service wake at Bandini's apartment. As luck would have it, Bandini's quarters were on the seventh floor, so that the sixty-nine-year-old Commendatore, who was fearful of elevators, had to climb the stairs to reach the gathering.

Exactly a month later disaster of another kind struck the team. Jim Clark and Graham Hill arrived at the Dutch Grand Prix with the long-rumored, much-feared Lotus 49 and its potent new Ford-Cosworth V8. Clark, who was living in Paris at the time to avoid England's tax laws, had never seen the silky green-and-yellow machine before he drove into the Zandvoort paddock. Hill, his dashing teammate, had already done extensive testing with the powerful if rather twitchy machine. Ford of England had invested over $250,000 with Cosworth Engineering to create the engine and perhaps an equal amount with Colin Chapman's Lotus Cars to develop the advanced chassis. The machines were shockingly fast from the very moment they fired up along pit row. Hill won the pole position with relative ease, while Clark's lack of familiarity with the high-revving engine forced him back to eighth on the starting grid. But after Hill's engine broke ten laps into the race, Clark slowly got comfortable in the saddle of his new mount and surged into the lead. He won with ease and was mobbed in victory lane by the dozens of Ford executives who had apprehensively crossed the Channel to witness the debut of their new charge. The Ferraris of Amon, Parkes and Scarfiotti finished in fourth, fifth and sixth, respectively, and were relegated to the role of bit players in the entire proceedings.

It was clear to Franco Lini that, aside from the newcomer Amon, his team lacked first-rank driving talent. This he needed to correct, although conventional wisdom in Maranello deemed that the cars were so capable that anyone could win with them. This ridiculous notion ebbed and flowed with the success of the team; when it was winning, it was boasted that drivers were pawns in the game, but losses were generally attributed to the sloth, cowardice and ineptitude of the men in the seats. Lini was a more practical man. He recognized that Parkes and Scarfiotti were not up to the task of handling the powerful but cranky-handling 312s and began to poke around for new talent. Three men caught his eye. Jacky Ickx, the son of a prominent Belgian automobile journalist, was driving brilliantly in smaller-formula cars and showed great potential. Jackie Stewart was a young Scotsman who appeared, as a driver, to be a near-clone of the great Clark. The third was a brash dirt-tracker and Indianapolis star from America named Mario Andretti. The trio—in particular Andretti—evidenced the kind of bold, tigerlike qualities that Ferrari coveted in his drivers. Andretti was an Italian by birth, having been raised near Trieste before immigrating with his family to the United States. At twenty-seven, he was a rising star in American racing and seemed a natural candidate for Ferrari. But at the time his commitment to Indianapolis-type competition was total and his interest in Formula One was limited by his schedule in America. Ferrari fantasized about this Italian youth returning to his homeland to win for the Prancing Horse, although he later would be angered to learn that Andretti often said that the greatest day of his life was the one on which he received his American citizenship. Both Ickx and Andretti would later drive for Ferrari, and would gain important victories, but neither would turn out to be the reincarnation of Tazio Nuvolari that consumed the old man's dreams. Stewart, a canny man with no illusions, visited Maranello and did not like what he saw. He chose to remain with Englishman Ken Tyrrell, whose small team was to produce a variety of excellent machines and three World Championships for its star driver.

Try as he might, Chris Amon could not buy a Grand Prix victory with Ferrari, or with any other team for that matter. No top-flight driver in the history of the sport was more star-crossed. Acknowledged by his peers to be in the very top rank, the likable New Zealander led race after race only to experience mechanical failures—broken throttle cables, ruptured water pipes or fractured fuel pumps. He was to labor mightily for Ferrari for three seasons and never to gain a victory.

Moreover, the entire responsibility for the Ferrari Grand Prix campaign would soon rest on his shoulders. In only his second race for the team, at the Belgian Grand Prix, misfortune struck again. On the first lap of the race around the fearsome eight-mile circuit, Parkes slid on oil spilling from Jackie Stewart's H16 BRM and sailed off the 150-mph left-hander at Blanchimont. Amon, who was immediately behind, watched in horror as Parkes's machine snap-rolled up an embankment and tossed its husky driver to the turf like a rag doll. Amon drove on, convinced that Parkes was dead. He gamely pressed ahead to finish third. Scarfiotti also witnessed the crash and immediately lost heart. He dazedly cruised home to finish a distant eleventh. Mike Parkes survived but with horrific leg injuries that ended his brief, ill-advised Formula One racing career. He would continue to compete in sports cars after recovering and remained with Ferrari as a valued engineer and development driver until dying in a road accident on the way to Turin. The Spa experience effectively neutralized Scarfiotti as well in Grand Prix racing, leaving only Amon to carry the colors. (Sadly, poor Scarfiotti was to be killed a year later while competing in a minor hill climb event.)

In the middle of the disastrous season Amon was joined by an Englishman named Roger Bailey, who had been lured away from the Ford grand touring racing program. Bailey was an inspired racing technician and was hired to act as Amon's personal mechanic and compatriot among the Italians. A crusty man with a unique instinct for high-performance machinery, Bailey quickly assimilated into the Maranello scene even though he couldn't speak a word of Italian. He learned that the factory's coveralls came in one size only, with smaller men simply rolling up the sleeves and pant legs, and that the no-smoking rule in the shops was totally ignored. As in the past, whenever the Commendatore appeared, cigarettes would simply be stuffed inside toolboxes, creating the bizarre impression that hundreds of wrenches and screwdrivers were smoldering from overuse. Ferrari, of course, understood what was going on but chose to ignore the reality in favor of the illusion.

Bailey, now settled in America and a prominent member of the Indianapolis racing community, recalls those days at Ferrari with amusement and great fondness. He remembers riding to work down the Abetone road each morning on his 50-cc motorbike, with the acrid fumes of insecticide from the bordering vineyards and orchards filling his nostrils, and the long lunches at the Cavallino with Piero Lardi—who always kept a wary eye out for Laura, lest he accidentally confront her. In those days the racing mechanics were often paid in the currency of the prize money from the last race. "Gozzi would hand you this batch of Argentine pesos or Dutch guilders or Austrian schillings and by the time you got it changed into lire or pounds at the local bank, you'd have nothing left," he muses. Franco Gozzi was a regular at the races in those days and handled all of the non-racing details, even cooking up batches of pasta for Bailey and the rest of the crew in the paddock areas. Amidst the pressure and the shouting matches and the insane rush to constantly modify the cars in search of more speed, Bailey recalls a simple, easygoing charm to motorsports that has long since been replaced by steely, humorless professionalism.

"Mike Parkes had an apartment on the park behind the Old Man's house. It was on the Martini della Liberia, I think, and Lina and Piero lived in the same building. On warm summer nights we'd see the Old Man come in. He'd just stroll over from the Garibaldi Square for a few hours. Everybody knew about the arrangement, but never said a word. On the other hand I got along quite well with Laura. She seemed normal to me, except when she spotted poor old Piero. I'd talk to her on occasion at the Cavallino. She loved After Eight mints from England and I'd bring them to her whenever I got back from London."

It was Bailey who agonized with Amon over his misfortunes in Formula One and who traveled the world with him seeking success. They found it in the winter of 1968–69 in New Zealand and Australia. The Tasman series was organized for 2.5-liter racing cars with production-based engines. Ferrari created specials for Amon and newcomer Derek Bell, a bright English prospect who would later gain fame with Porsche as an endurance driver. Chris won the Tasman—which attracted most of the world's Grand Prix stars—with ease. "It was a prestigious win for the factory in a pretty lean period," says Bailey. "They had a celebration for us when we got back to Modena. Ferrari gave me a watch with the Prancing Horse on the face. Chris didn't get a thing!"

It was during this period that Ferrari became famous for its Keystone Kop pit work. The chaos of a Ferrari tire change or simple fuel fill-up became a worldwide attraction. No one, before or since, has been able to duplicate such utter madness. "It was beyond belief," says Bailey. "I remember the first sports-car race I crewed in. We had a meeting. Each mechanic was given a specific assignment—right front wheel, gas fill-up, left front wheel, et cetera. It was all very specific. Forghieri had everything worked out to perfection. I was assigned the job of changing the right rear wheel. During the race the car came screeching in. I leapt over the wall only to find *five* other mechanics trying to remove the right rear!" It was at Sebring in 1969 that Bailey was attempting to fuel Chris Amon and Mario Andretti's 312P when Mario opened the car door during a hurried driver change, displacing the hose and sending several gallons of gasoline into the cockpit. Amon was forced to drive his shift with high-test sloshing around his feet as if he was aboard a leaky rowboat. "Never, ever will we see the likes of those Ferrari pit stops again," says Bailey.

Part of the problem, surely, was Ferrari's insistence on being everywhere at the same time—at least in spirit. During that period the Scuderia was competing in Formula One and Formula Two, the Tasman series, the Can-Am, major sports-car endurance races and the European hill-climb championships. Franco Lini tried to convince Ferrari that the demands on Forghieri and his engineering staff were too great. He insisted that the resources of the team, both financially and physically, were insufficient for such a broad-based campaign. But he was a voice in the wilderness. Too much pride, too much tradition, too much bravado stood in the way of practical thinking. Many of those around the Commendatore had been with him since the beginning, when a single engine and simple chassis could be cobbled and converted for a variety of purposes. Surely the vaunted foundry and ingenious men like Rocchi and Forghieri could innovate and modify sufficiently to compete against the lowly *garagistas*.

It was obvious by the end of the 1968 season that the old V12 was hopelessly outclassed by the DFV "Cossie." Not only was it more powerful but it was lighter and more compact, which aided handling. Forghieri struggled on, adding a movable airfoil for the Belgian Grand Prix, thereby becoming the first Formula One team to install such a device (although the American Chaparral had been employing one for several seasons). But it was simply a finger in the dike.

Ickx was brought onto the team in 1968 and managed to win the French Grand Prix while Amon's wretched luck continued. Moreover, friction developed between the two because the rather aloof Belgian disliked testing, which left Amon to do a majority of the development driving at Modena and Monza. The team was clearly handicapped by the absence of a full-scale wind tunnel and a legitimate test track. The autodrome on the edge of town was now obsolete as a site for testing the ultra-fast racing cars of the day, and Monza was a hundred miles down the road and a logistical nightmare. Plans were underway to update the old circuit at nearby Imola, but completion was several years away. Ferrari owned plenty of orchard property across the Abetone road on which to construct a private test facility, but in the fluid, rapidly changing world of automotive technology in the late 1960s, it was simply one of hundreds of priorities within a limited budget.

From 1968 to 1969 car sales slumped by over 100 units, from 729 to 619, adding more red ink to the ledger. Much of this decrease came in the American market, which now accounted for almost half the passenger-car sales. The U.S. Congress was on the verge of setting strict exhaust-emission and safety standards, which would boost manufacturing costs. At that point the company was a middling-sized operation, capable of generating profit in a stable economy and constant market, but lacking funds for much-needed expansion. Complicating matters was Ferrari's insistence on competing across a number of fronts, while men like Franco Lini urged him to limit his efforts to Formula One, where the full force of his small but devoted engineering staff could be concentrated.

Enzo Ferrari had by now become a prisoner of his own carefully wrought public image. He had portrayed himself as the standard bearer of Italy's motorsports fortunes—a sort of motorized knight-errant who was prepared to defend national honor at all costs against the foreign interlopers. His adoring public had bought the imagery in toto and paid dearly in a kind of mindless worship unprecedented in a major sport. Enzo Ferrari had become an icon, a paterfamilias for a nation starved for respect in the international community. Italy simply was not taken seriously as a European power. Its chaotic governments, unstable economy, the blight of the Mafia, its increasingly unruly political left and the gripping poverty of the south had long since erased memories of the postwar "economic miracle." Small triumphs like the Scuderia Ferrari had become focal points of passion for a population starved for victory and notoriety. But as with all lovers of the frivolous, devotion was cruelly fickle. For the chanting mobs who appeared each year at the Italian Grand Prix, waving giant yellow-and-red Ferrari flags like French revolutionaries storming the Bastille, victory for the Scuderia would generate insane celebrations. But defeat would prompt anger and frustration. With the slump in fortunes came brutal broadsides from the press. The major media—*Gazzetta dello Sport, L'Unità, Il Giorno*—were staffed with automotive journalists whose primary mission was to generate daily stories on the fortunes of the Scuderia. During successful seasons, as in 1964, the stories were upbeat, filled with bravado and hyperbolic boasting about the team's success. But during dry spells like the three years of 1967, 1968 and 1969, when but one Grand Prix victory was recorded, the press was transformed into packs of wolves baying at the factory gates. Endless rumors were circulated, nonsensical stories were printed as truth and the activities inside Scuderia degenerated into an utterly bogus, sensationalized soap opera.

Ferrari, now seventy years old and obsessed with the role he had created for himself, was consumed by these attacks. Each story was examined in detail, and more attention was paid to rebuttals and threats of revenge than to the labors of Forghieri and his engineering staff. More and more time was spent with Gozzi preparing countering press releases, with their copy edited, refined, edited again, analyzed and reviewed as if they were papal bulls or declarations of war. Gozzi would write the copy, then present it to Ferrari, who would make careful notations with his customary purple-inked pen. It would then be returned for typing by Valerio Stradi, Ferrari's longtime personal secretary. Other staffers would be consulted, and the release would be edited again and again. A routine announcement of a schedule change in the introduction of a new model sometimes consumed days before it was given to the press. This preoccupation with the opinions of the national media began to take up more and more of Ferrari's time, as his concerns about his personal image transcended the very business itself.

One component of the lurid Ferrari drama that remained unknown to the public was Piero. By this time the press fully understood the situation but remained silent, either out of fear or in deference to the age-old custom of keeping family matters on what was politely called the "private side." Piero was by then twenty-three years old and had grown into a tall young man with dark features and large brooding eyes. He was well supplied with money by his father and was able to maintain a reasonably normal life away from the factory. On February 10, 1968, he traveled to Cesena, outside Forli, and married the former Floriana Nalin. They would return to Modena to set up housekeeping and a year later their only child, a daughter named Antonella, would be born. This was to be the only grandchild of Enzo Ferrari.

While revelations about Piero would have infuriated and embarrassed him during the Fiat negotiations, he was slowly gaining control in his perpetual jousting with the press. Whether he was the butt of criticism or the hero of the day, Ferrari always tried to maintain control of the game. Victories would bring eloquent testimonies to the courage of his drivers and the talent of his engineers, all working in concert in behalf of the regal house of Ferrari. In public, he was gracious in moments of triumph and defeat, although his private reactions were usually overblown. During seasons in which wins were as rare as a mosquito-free summer day in Modena, he played the press like a maestro, moaning about the international rule makers, the unfairness of the competition, the lack of money, the grinding labor problems and his overworked staff, while keeping in reserve the ultimate threat of retirement from the sport. This had by now become the shoe that never fell; he had declared so many times over the years that the Scuderia could no longer continue that it had become a joke within the circle of journalists who covered motorsports. Most viewed such decrees as blatant ploys to generate more financial support or as sulky recriminations to a nation that was offering insufficient appreciation and had to be startled into a more attentive and grateful state. Those who knew him best understood that Enzo Ferrari would never retire. There was little else in his life besides automobile racing. It was that simple.

Although he eschewed any role in politics and in fact refused to travel to Rome—a city he held in contempt—Ferrari had become a major public figure and a man with considerable power within the environs of Modena. Now that the Orsis had ceased to wield much influence with Maserati, which was now aligned with Citroën, Enzo Ferrari remained the king of automobiles in the city. If he had a rival for the title of major impresario of Modena, it was probably Giorgio Fini. The Fini family had expanded their tiny sausage shop into a massive food-service business, which included a chain of autostrada restaurants and food-processing operations, as well as Modena's most prestigious restaurant and the old Real hotel (renamed the Real-Fini) across the street from Ferrari's home. Another Modenese name was spreading beyond the perimeters of the ancient city—that of a dazzling prodigy whose riveting operatic presence would soon gain him worldwide fame and adoration. His name was Luciano Pavarotti.

Enzo Ferrari maintained a cordial relationship with both men as they grew in fame and stature to rival his own. Both he and Giorgio Fini had been named Knights of Industry, an honor accorded to commercially successful Italians, and Ferrari tended to treat the younger Fini—who had waited on the Commendatore as a young man in the family restaurant—as an equal. When Dottore Fini was recovering from major heart surgery in a Houston, Texas, hospital, he received a one-word telegram from Ferrari. It read: *"Forza!"* (fight).

As Ferrari gained in prominence, he maintained his regular schedule, although now he and Pepino drove to work on back roads, avoiding the small clusters of adoring fans who waited for him outside the main factory gates. He ate only in the private room at the back of the Cavallino or in equally secluded places at his favorite in-town haunts like the Fini or Orestes or Bianca, where he often dined with non-racing associates like Benzi or the brilliant Giacomo Cuoghi, who handled his financial and legal affairs. Women remained a part of his life on a regular basis beyond the immediate demands of Laura and Lina. Fiamma Breschi had by then drifted out of his life. Still, females obsessed him, and as he grew older, private conversations circled more around them than automobiles. But both served as the trappings of success rather than success itself. Had it been otherwise, why would he have treated both with such disdain? To be sure, he was a prodigious lover, but his affection for females was superficial at best. His relationship with cars, thought by his adoring public to be a deep, abiding passion, is equally suspect. They were a means to an end, instruments of personal aggrandizement and reinforcement of what was probably a rather feeble ego beneath all the posturing and bluster. Said one of his closest personal associates, "Only three things truly mattered to Enzo Ferrari. They were prominence, property and prestige. All the rest was window dressing."

Audiences were granted to the favored few, although even eminences like the Shah of Iran, a regular customer, were required to wait before entering the dreary office-shrine at the factory. There he told witty stories and offered murky, florid philosophical observations to a string of rich customers, businessmen and journalists—some of whom were summoned to Maranello to be dressed down for printing an affront to the company. His desk was enormous and always bare of paperwork. The drawers were a veritable souvenir vendor's warehouse. Their contents would be dispensed according to the status of the visitor. Low-level types would receive key chains or pins bearing the Prancing Horse (a symbol Ferrari had tried to trademark, until it was discovered that the West German city of Stuttgart held proprietary rights). Middle-level customers received lovely silk scarves, also emblazoned with the *Cavallino Rampante*, while important dignitaries would be presented with autographed copies of the latest edition of the often revised, reprinted and updated autobiography that had begun life as *My Terrible Joys*.

One of Europe's most prominent Ferrari importers recalls visiting the factory with a wealthy count who was taken with the Ferrari mystique. The importer managed to gain an audience with the Commendatore and announced that the count had purchased six Ferraris and was an excellent customer. It was his one great wish, said the importer, to receive an autographed copy of The Book.

"Impossible," grunted Ferrari. He then reached into a drawer and took out a scarf. He stared at the count and asked, "Do you have a mistress?" The count was too taken aback to answer. Ferrari repeated the question: "Do you have a mistress or not?" Finally the count blurted, "Yes." "All right, then," said Ferrari. He then reached into the drawer and took out a second scarf.

The collection was also employed for small bribes. One morning Roger Bailey was clocked speeding through the village of Formigine a few kilometers to the north on the Abetone road. He was driving Chris Amon's 500-cc Fiat to work. Bailey was later summoned into Ferrari's office, where he was confronted by two angry Formigine policemen who had pursued him to the factory. After much yelling, Ferrari brought the confrontation to a close by saying, "Roger Bailey is a valued mechanic for our great driver, Chris Amon. If you involve him, you involve Chris Amon, which in turn involves me and the Scuderia Ferrari. That cannot be." Ferrari then handed out a collection of key rings, pins and other small items that could probably be sold for substantial sums on the street. The incident was over.

Frustrated by his inability to alter the inertia of ego and tradition, Franco Lini left at the end of 1968 and returned to the field of journalism. As the black years of the 1960s wound down, Enzo Ferrari, at an age when most men have long since eased into graceful retirement, was about to begin a major renaissance for himself and his beleaguered company.

It was clear from the 1965 agreement between Ferrari and Fiat to build the Dino engine and its various automotive permutations that the two firms were moving closer to an alliance. It was reasonable to assume that SEFAC was at an impasse. Massive infusions of capital were necessary if it was to remain in the field of manufacturing luxury cars, what with Porsche, Mercedes-Benz, Jaguar, Lamborghini and BMW all evidencing newfound strength in the market. All were backed with considerably more financial strength than the relatively tiny Maranello operation. Whether he liked it or not, the road cars were the heart and soul of his beloved racing operation. Without the profits accruing from their sale (the net profit was an estimated 200 percent for the flashy but relatively cheaply built customer cars) the Formula One operation—not to mention the lesser sports-car competition—simply could not be funded.

At that point Fiat was the most powerful industrial conglomerate in Italy and a major force on the European economic scene. Thanks to a cozy relationship with the government, the Agnelli family had created a web of tariffs that made imported cars outlandishly expensive in Italy. This gave Fiat what amounted to a national monopoly in the automobile business. But without competition at home, quality and engineering stimuli were lacking and Fiat products were notably weak in the export market. Repeated attempts were made to sell Fiats in the United States, for example, but all failed due to a shoddy product line. Clearly, Fiat had an international image problem that had to be corrected if sales were to increase in the competitive arena of the European Common Market. What better way than to embrace the prestige and mystique of the famed Prancing Horse of Maranello?

Conversely, what better way for Enzo Ferrari to regain his competitive edge than to dip into the bottomless pockets of the regal, white-maned Giovanni Agnelli and his mega-rich Fiat S.p.A.?

When and how the negotiations began is unknown, but it is presumed that Ferrari and Agnelli began informal talks about an alliance in 1967–68, with the legal details finally resolved by the early summer of 1969. On the basis of the Ford experience and subsequent comments by Agnelli, it can be assumed that Enzo Ferrari did not give up control without a struggle. Here was a self-made man, bloated with pride, who was not about to be swallowed up like some helpless Jonah. While a tiny entity in the industrial world, his company was a giant in terms of prestige — an automotive boutique more comparable to Coco Chanel or Cartier or Gucci or the House of Christian Dior than to the fuming, steaming, grinding industrial monolith that was Fiat.

It would be an alliance based on opposing needs. For Ferrari it was a simple solution to capitalize his sagging business, but for Agnelli, the patrician Torinese, it was to venture into the dusty hinterlands seeking from a bumptious commoner those most precious of all ingredients, status and respect. He would find them in Modena, for an estimated 7 billion lire. This translated to about $11 million — a substantially smaller sum than either the Mecoms or Ford had offered, although Fiat gained less than half the company in the short term. The deal was so attractive to Ferrari that he even traveled to Turin to sign the final papers — a city he had not visited in decades and a classic example of the mountain visiting Mohammed.

The alliance was announced on June 21, 1969, with a cryptic statement from Fiat: "Following a meeting of the president of Fiat, Dr. Giovanni Agnelli, with the engineer Enzo Ferrari, the prominent intent of it being to ensure the continuity and development of Ferrari automobiles, it has been decided that the past technical collaboration and support will be transformed within the year to a joint participation." This came on the heels of yet another of Ferrari's tiresome pledges that he would retire from competition following the September running of the Italian Grand Prix. Clearly the Fiat connection rendered obsolete that already meaningless bleat and opened vast new vistas of opportunities. It was also a measure of revenge for the old man. In his book Gino Rancati gives a hint of Ferrari's mind-set when he comments, "The process that was set in motion was to wipe out the shame of 1918 when Ferrari was not welcomed by Fiat." Surely Enzo Ferrari had long since forgotten that as a youth he had been rejected for employment by Fiat some *fifty* years earlier. Surely he hadn't nursed that tiny carbuncle of bitterness for half a century and now, in his own convoluted way, considered the account settled. Or had he?

Whatever the case, it was a marvelous deal. For its billions of lire, Fiat received 40 percent of the existing Ferrari stock. Enzo Ferrari retained 49 percent—to be ceded to Fiat upon his death. His old colleague Pininfarina received 1 percent, and Piero Lardi, the young man still in the family shadows, got 10 percent. The arrangement was in a sense similar to the one almost worked out with Ford. Ferrari was to have total control of the racing operation, while Fiat was to operate the passenger-car side.

Regardless of the myths built up around Enzo Ferrari and his later road cars—such as the 308GTB/GTS, 328GTB/GTS Boxer, 512BB Boxer, 400i Automatic—he had almost nothing whatsoever to do with their creation. The last road car that might legitimately be considered a true Ferrari was the 365GTB4 Daytona. Subsequent automobiles were really more like limited-production Fiats (or mass-produced Ferraris) than like the classic quasi-racing cars of yore. In fact, during the early planning stages after the Fiat-Ferrari alliance, the Agnelli management team seriously considered expanding the Ferrari production to Jaguar-like levels of 20,000 cars a year. But for reasons of investment, marketing, and so forth, it was decided to escalate the Maranello output only slightly. By 1971 production had been boosted beyond the 1,000 mark and by the end of the decade production surpassed 2,000. In 1988, the year Enzo Ferrari died, production totaled 4,001 cars—a far cry from the paltry 619 machines created the year the Fiat deal was consummated.

Surely a new age was about to dawn over beleaguered little Maranello. Fiat could do what they chose with the road cars; Enzo Ferrari couldn't care less. In private he remained openly disdainful of the status slaves who bought them. His passion for racing was undiminished, and with Fiat's bank account behind him, he was ready to launch a counterattack against the likes of Lotus and Ford and Porsche and BRM and all the other upstarts, as well as the jackals in the Italian press. Now that would be true revenge!

17

With his coffers amply filled with Fiat lire, Enzo Ferrari was now ready to exact some tribute from the scoffers who had derided him for the past three years. New engines and chassis were on the way for Formula One and the sports cars. Better yet, Fiat was moving ahead to fund a new line of ultra-powerful road cars that would help to recover that market. In all, he could look forward to 1970—and his seventy-second birthday—with considerable hope for renewal.

Chris Amon, star-crossed as usual, chose to depart on the eve of the renaissance. Forghieri had developed a potent new 3-liter flat-12, which Chris had begun testing at Modena in the summer of 1969. Early examples were failures, breaking important parts and leaking oil like soaked sponges. Discouraged that he faced yet another binge of bad luck, Amon moved to the British March team, which had been supplied with the all-conquering Cosworth engines. Had he remained in Maranello it is entirely possible he would have won a World Championship and certainly would have erased his reputation as a talented but luckless also-ran.

Replacing him, after a year's absence, was the aloof Ickx, a pouty, dark-haired youth possessed of almost feminine good looks and the heart of a lion. With him would be a rough-riding Swiss named Gian-claudio "Clay" Regazzoni, an Italian-speaking Swiss from the border canton of Ticino. Regazzoni had starred in Formula Two, where he had gained a reputation for being something of a bully on the race-track. He had been blamed for the death of Englishman Chris Lambert, plus any number of other "shunts," and was viewed as a bad actor, especially by the British motoring press. His heavy features and dark mustache only served to enhance this image of thuggery. Also joining the team was a twenty-eight-year-old Roman named Ignazio Giunti, a talented driver whom many believed to be a potential World Champion. Sadly, Giunti was to die a year later at the wheel of a Ferrari 312PB sports-racing car during the Buenos Aires 1,000-kilometer race. The likable young man never had a chance to exhibit the skills that everyone agreed he possessed.

Following Amon's departure, Forghieri and his engineers success-fully debugged the flat-12 and by the end of the year were able to ex-hibit the new 312B Formula One car—a rakish, long-nosed machine with a three-finned spoiler mounted over the rear wheels. The car was a striking amalgam of form and function, a prime example of the Italian flair for building a special dash of glamour into what was essentially a metal tube suspended on four fat racing tires. This auto-mobile, in various updated permutations, would carry Ickx and Regaz-zoni to a number of major victories, including the Italian Grand Prix at Monza in 1970.

With the Fiat personnel taking over the day-to-day functions of designing, building and selling the road cars, Ferrari was relieved of a task that had never interested him. The customer cars were simply a means to an end, and as he grew older he became openly disdainful of the slathering nouveaus and empty aristocrats who appeared with gaping wallets to receive their automotive benedictions.

Millions were poured into the Maranello factory to update the assembly lines, enlarge production areas, modernize the foundry and construct a massive test track called Fiorano on the old orchard across the Abetone road. The circuit, with television cameras and telemetry gear installed around its perimeter, was the most advanced of its kind in the world and relieved the racing team of its traditional, arduous trips to Monza or the Modena autodrome for special tests. Moreover, Carrozzeria Scaglietti was slowly being absorbed into the company, and by the mid-1970s had been situated on its own new, immense complex on the Via Emilia near the edge of the city. The quaint days of yore were gone forever at SEFAC. It was well on its way to becoming yet another modern appendage of Fiat merely glazed with the tradition and imagery of the old Scuderia.

But life had changed little within the racing operation. The old guard, led by Enzo Ferrari, operated with the same combination of panic, intimidation and fiendish devotion to the cause. Forghieri was becoming a star in his own right with the creation of the successful 312 flat-12 (which was also spun off into a successful 3-liter sports-racing car, the 312PB, although Ferrari dropped all endurance racing after the 1973 season and thereafter totally concentrated on Formula One). Under the pressurized canopy that was the Scuderia Ferrari, Forghieri could be mercurial in the extreme. He was thoroughly liked within the team, although Laura Ferrari suspected him of being a manipulative favor seeker. Emotional, intense and completely devoted to his job, Forghieri could lapse into near-madness in the heat of competition. Brian Redman, a superb English driver whose career was on the rise in the late 1960s, recalls his first encounter with Forghieri in the 1968 Eifelrennen at the Nürburgring. The event was for Formula Two cars, and Redman, at the time infused with enthusiasm, charged off around the twisting, tree-shrouded circuit among the leaders. On the first lap a flying stone shattered his goggles. Bleeding from the damage, he darted into the pits. There he found Forghieri in a state of apoplexy. He was furious that Redman had stopped because of such a minor problem. A mechanic was sent to rummage through the spares for another set of goggles. Finally a pair was found, although the lenses were deeply tinted for driving in bright sunshine. But the day was dull and bleak, with a roof of nimbus clouds clinging to the Eifel Mountains. Redman put on the new set and drove off, but he soon discovered he was nearly blind in the deep forest sections where the track wound in a series of looping, blind corners. Still, he

flailed on, driving, as he recalls, "like a madman, like a man possessed by demons." By the end of the race he had fought his way back
to fourth. It had been a titanic effort, with Redman risking his life
every few yards throughout the event. Yet his finish was greeted only
with sullen disregard by Forghieri. Clearly, Redman's stop for new
goggles had revealed him to be a man not prepared to offer up his life
and limb to the cause. This attitude reduced Redman's enthusiasm
for the operation, and a year later he declined an invitation to drive
for the Formula One team. Redman did, however, return to the
sports-car team in 1972, and won 1,000-kilometer races at Spa, Zelt-
weg, Dijon, Monza and the Nürburgring. By then two major changes
had occurred. Peter Schetty, a Swiss former hill-climb driver for the
team, had taken over as manager and had run the team with considerably less theatrics than Forghieri. Drivers were also being paid. Alfa
Romeo had reentered sports-car racing, and Ferrari had little choice
but to pay like his old rivals in Turin. Redman recalls that he received
20,000 English pounds plus expenses and a small percentage of the
prize money in 1972. At long last, men were being compensated for
the "privilege" of driving for the vaunted Scuderia.

By 1973 Ferrari had withdrawn from the sports-car races—hill
climbs, Tasman series, Can-Am, and the rest—which had for so long
diluted the Grand Prix racing operation. The graceful little 312PBs
and the larger, more potent 512 coupes were parked forever, either to
be scavenged for parts and broken up or to be sold to privateers. A
512S coupe was purchased by American entrepreneur Roger Penske
in 1971 for $28,000.

Penske was the ultimate racing businessman who used motorsports as a launching platform for a burgeoning empire of car dealerships, tire distributorships, truck rental firms and diesel engine
businesses. He purchased the Ferrari because he reckoned he and his
engineer-driver, Mark Donohue, could transform it into a winner.
He was the antithesis of the starry-eyed aficionado. In fact, after he
had finished his negotiations with Gaetano Florini he was invited to
an audience with Ferrari. Noting that he had a plane to catch in
Milan, Penske quickly departed, thereby stunning the entire factory

staff and infuriating the great man. Audiences with Ferrari were not granted easily, and a turndown was as rare as missing an appointment with the Pope. That Penske received no further cooperation from the factory is hardly a mystery. Like so many customers, Penske was shocked to discover that the allegedly new car was shabbily constructed and not remotely race-ready. (A wealthy privateer of the day reportedly bought a 512M for $25,000 only to discover that it was a badly misused former team car rather than the new model he had ordered. He returned the car to Maranello, demanding correction. He got it back, slightly repaired, with a bill for another $20,000!) Penske and Donohue tore their Ferrari apart, replacing and redesigning masses of suspension and chassis bits. The V12 engine was sent to the experts at Traco Engineering in California, where remachining of the rather casual factory tolerances produced another 40 horsepower. This car, probably the best-prepared Ferrari sports car ever to be fielded, was also the fastest of the lot. When Forghieri inspected the car at the Daytona 24 Hour race—where it was quicker than the lighter, nimbler factory-entered 312Ps—he proclaimed it the finest Ferrari racing car he had ever seen. It was raced but four times, winning the pole position on three occasions. But by then Porsche had assumed almost complete domination of major endurance competition (which would be a factor in Ferrari's departure), and Penske's 512M, for all its potency, could not keep up with the ultra-fast 917 coupes from Stuttgart that were shattering records at Le Mans, Daytona and almost everywhere they ran.

Despite the infusions of capital from Fiat and the reduction of the motorsports activity in order to concentrate on Formula One, life for Enzo Ferrari remained essentially unchanged. Each day began as the one before had, with a trip to the barber, a drive to Dino's grave and a morning session in the Modena office before leaving for the factory. Now he and Laura were alone in the immense house. They lived in only a few dark, rather shabby rooms, dominated by dingy green wallpaper and filled with dusty, faded furniture that had been unrepaired and unpolished for decades. The immense structure was dotted with empty rooms littered with effluvia from the past—masses of trophies (probably including the long-lost 1936 Vanderbilt Cup, a silver bowl created by Cartier large enough that Nuvolari could sit inside!), books, photographs and even, it was reported by one visitor, cases of Rolex watches. "It was like a grave in there," said one close associate of the family who was a constant visitor to 11 Largo Garibaldi.

As she entered her early seventies, Laura Ferrari's behavior became more bizarre. She had taken to snitching rolls and loaves of bread from tables in the Cavallino as she passed by and lifting tips left for the waiters. One day she encountered a longtime woman acquaintance on the Via Emilia. She demanded that she turn over her dress, wailing that Enzo gave her no money for clothes. This odd confrontation took place with a woman who was a head shorter and perhaps two dress sizes larger than Ferrari's wife. But a man who knew her as summer vacationer at Viserba claims that she behaved normally while at the Adriatic shore.

By the middle 1970s she was becoming physically infirm and spent considerable time in bed. Friendly biographers have, as noted, described her illness as a "dystrophy of the legs," which is an ailment without medical foundation.

There is a story, probably apocryphal, that has made the rounds of Modena that is revealing—if not in substance, at least in the way the relationship between Ferrari and his wife was perceived. It was said that Laura became despondent and attempted to commit suicide by leaping into the Panaro River. A group of Ferrari mechanics spotted her in the water and dove in to save her life. Once they returned to the factory they were summoned to Ferrari's office. There they were told, in angry tones, "If she jumps in again, leave her in there!" While there is no evidence that she ever attempted to drown herself, the fact that the story was concocted at all indicates that the hostile relationship between Ferrari and his wife was common knowledge.

For decades the standard litany of Enzo Ferrari and his growing legion of courtiers and sycophants held that the Scuderia's cars would not carry any product endorsements other than those closely associated with motorsports. Since 1966, when the Formula One cars displayed no decals on their riveted hulls other than the vaunted Prancing Horse of Maranello, the involvement of outside sponsors had grown steadily. By 1973 the 312B3 of Ickx carried prominent product decals from Goodyear (the switch had been made from a financially strapped Firestone the year before), Marelli magnetos, Ferodo brake linings, Champion spark plugs, Shell Oil, Heuer watches and several other suppliers. The sponsorship game had burgeoned into a mega-dollar business, and a massive bidding war was being fought by fuel, tire and ignition companies to be represented on major teams like Ferrari. Tobacco companies like Rothmans, R. J. Reynolds and Philip Morris had also discovered that automobile racing was an extremely effective advertising and promotion medium, and by the mid-1970s were spilling millions per year into the sport. In the face of this massive infusion of funds, Enzo Ferrari stated repeatedly and with increasing ire that the cars of his Scuderia would remain "pure" and unsullied by the ugly, foreign logotypes of cigarettes, toiletries, sexual aids or financial institutions that were beginning to be emblazoned on the flashy bodywork of the competition. But, as in so many cases, Enzo Ferrari had a public and a private position on the issue. And, as in so many cases, the debating point was not ethics but money. At the time Philip Morris Europe was making a major thrust into the sport with its Marlboro brand of cigarettes, and they entered into serious discussions with Ferrari about developing a full sponsorship package with the team. Had this happened, it would have been a total contradiction of Ferrari's bleating about sponsorship purity. And the deal came perilously close to becoming a reality. According to Aleardo Buzzi, currently the president of Marlboro Europe, he was present at a series of meetings in the early 1970s in which full Marlboro sponsorship of the Ferrari team was discussed. A deal was almost concluded when Buzzi's boss noted that because of certain financial requirements within the company, the sponsorship funds would be paid in lire. Ferrari demanded either dollars or Swiss francs. Surely the funds, had they been forthcoming, would have been deposited in banks in Monaco or Geneva or both. When the

Marlboro people would not budge from their offer of lire or nothing, Ferrari stood up from the table, cast a glance at his watch and said, "Gentlemen, my wife is very ill. I must visit her. Therefore the meeting is concluded." As with Ford's Donald Frey a decade earlier, he left the session, never to return. Marlboro was not to appear on the cowling of the Ferrari Grand Prix cars for another decade, and then only under the subterfuge that the sponsorship was on behalf of the drivers, not the cars.

Following the 1970 season, when Ickx had barely lost the World Championship to Jochen Rindt (who had died in a practice crash at Monza prior to the Italian Grand Prix but had gathered enough points to take the title posthumously), the fortunes of the drivers, the 312Bs and the man who created the cars went into a steady decline. Mauro Forghieri redesigned the rear suspension for 1971, but after some initial successes with Ickx, Regazzoni and Mario Andretti—who was an occasional third man when his American racing schedule permitted—the cars advanced steadily to the rear of the pack. Twitchy handling was the primary complaint, and the more Forghieri fiddled with the suspension, the weight distribution and the aerodynamics, the slower the cars went, the more frustrated the drivers became and the louder was the bellowing of the Italian fans and the press.

While the elaborate Fiorano track was opened in time for testing before the 1972 season, the second-generation car, the 312B2, was by all accounts a technological step backward. When given the choice, both Ickx and Andretti would opt for the older B1s, although good-natured Regazzoni, a *cabezón* of the old school, drove both types with equal vigor. (Sometimes this enthusiasm went beyond the bounds of propriety, as in the German Grand Prix, when he bulled Jackie Stewart off the track on the final lap to take second behind Ickx.) With the exception of that one-two team finish at the Nürburgring, the season was a dismal parade of mechanical failures, crashes and outright defeats. In frustration, Ferrari blamed Forghieri, who was nearly hyperventilating in an effort to correct the quirky machines. There was horsepower aplenty—as much as 30 more than the all-winning Cosworths—but the erratic behavior on the track nullified the advantage. By mid-season Forghieri was demoted to work on special engineering projects within the factory. The chore of correcting the cars was handed over to Sandro Colombo, a young engineer recently enlisted from Innocenti. It was this newcomer (unrelated to the great Gioachino) who tried radical measures, including commissioning John Thompson's English-based TC Prototypes to fabricate three special chassis. This was heresy to the Italian loyalists, who had comforted themselves with the myth that, for all the failings of Ferrari, at least it was *Italian*. But now the red cars were riding on American rubber, being fed fuel through German fuel-injection systems, with their Belgian, Swiss and American drivers seated in tubs of metal and composite materials made in England. Where would it all end?

Initially, at least, it would conclude with a chain of crushing defeats in 1973. Regazzoni had accepted what appeared to be a better offer from BRM, now run by the preposterous, Blimpian Lord Louis "Big Lou" Stanley and his fabulously wealthy wife. His teammate would be a scrawny, buck-toothed Austrian named Niki Lauda. The member of a moneyed Viennese family, Lauda had borrowed heavily to rent rides from a variety of racing operations to learn his trade. His progress had been slow, and he was taken onto the creaking, underfunded BRM operation as the third driver. Before the year was out, this steely, totally committed twenty-four-year-old would be driving considerably faster than his teammates and would catch the attention of Enzo Ferrari.

At this point Ferrari had a phenomenal intelligence network. Not only were men like Colombo and Forghieri feeding him information from trackside (often "edited" for favorable consumption back home) but literally dozens of suppliers, hangers-on, journalists and old Scuderia loyalists were transmitting juicy rumors, timely tips, sly innuendos, corporate leaks and internal secrets from rival teams to the Commendatore. Much of this flow of information was meant to flatter, to gain favors or to delude the ego-driven leader in order to advance some personal agenda. Never were the automobiles blamed for defeats. Invariably it was the inept drivers, the churlish race officials, the blundering accessory manufacturers, the incompetents who supplied the tires, the idiots who ran the timing and scoring, but *never* the cars—which, if given an equal chance against the conspiracies mounted against them, would have won every race. It was this fogbank of self-delusion and bloated expectations that rational men had to penetrate when working with the increasingly isolated and misinformed Ferrari during the final twenty years of his life.

But with defeat piling upon defeat, it was clear to the men in Turin and even to the Commendatore himself that radical measures would have to be taken if the team was to regain not only its honor but its commercial value to Fiat in the upcoming 1974 season. Ickx and his diminutive teammate Arturo Merzario had encountered nothing but bad luck and distant finishes in 1973. Both were to depart at season's end. Ickx was a fine driver, and surely raced with soldierly determination, but with the modern Grand Prix car, equipped as it was with tricky, infinitely adjustable suspensions and wings, no driver, no matter how brave or talented, could overcome a bad chassis. In the old days, a superstar like a Moss or a Fangio or a Nuvolari could simply overpower a second-rate car with sheer talent. But in the era of ultra-high technology, the driver had been reduced to perhaps 20 percent of the equation (the car being the other 80 percent), and no man, regardless of his skill or courage, could perform effectively in a mediocre automobile.

Colombo was removed from the racing department, and Forghieri was recalled from his purgatory and reinstated in his old job. He was to work nonstop, with the manic energy only he could sustain, to modify and improve the 312s—now in their B3 permutation. Other changes were also in the works for the team. Shell Oil, the Dutch petroleum giant that had backed Ferrari since the early days of the Scuderia, withdrew its support. There was no dissatisfaction with Ferrari, but rather with business in Italy as a whole. Shell Italia turned over its interests to the home-based operation, Agip. As part of the massive transfer of markets, Agip picked up the sponsorship of Italy's most honored racing team. But the most important changes were in personnel—not only in the retention of the talented young Niki Lauda but in the arrival of Luca di Montezemolo. A patrician lawyer and a member of the Agnelli family, Luca, as he quickly became known in racing circles, was ostensibly installed as the team manager in the mold of Tavoni, Ugolini and Dragoni. But he was considerably more powerful and in the end a more positive influence on the operation of the team. Montezemolo ultimately answered to higher powers. While he was totally deferential to the seventy-four-year-old titular leader, at least in public, there is no question that Luca di Montezemolo held very high cards and was quietly more influential within the racing organization than any man other than Ferrari. He was a stabilizing force when the team needed it most.

But the centerpiece of the Ferrari renaissance was Niki Lauda, a personality who would ignite Maranello more than anyone since John Surtees. Lauda was the first of the new breed of racing drivers to confront Enzo Ferrari. Heretofore he could count on men to drive for him out of sheer love of the sport, to risk their lives in the name of glory, fame and the inner gratification of ego. They were a flamboyant, devil-may-care lot, infused with the glamour and panache of the Formula One scene and energized by the rush of emotion produced by skimming on the ragged edge of control over the surface of the earth. But now a different type was entering the sport. The prototype had been Jackie Stewart, an analytical Scotsman who carefully weighed the risks against the rewards and drove accordingly. Recall that Stewart had been courted by Ferrari in 1968 but had rejected his advances, discarding the trappings of tradition and romance for what he believed to be a better automobile to carry him to the World Championship. Lauda was similarly inclined. While he was hardly poor (very few paupers ever made it to Formula One, since they couldn't afford to enter the expensive sport to begin with), Lauda was the consummate professional. He was a money driver, a man with an icy resolve to drive at the limit, but only when all the pieces were in place, when all the contracts were signed and when all the checks were properly deposited. Like Stewart, he was as devoted to the sport and as courageous as any man who ever planted himself in the cramped seat of a single-seater, but there was more to the equation. Niki Lauda was not about to risk his life for a chimerical mission, and especially not to gratify the ego of an old icon ensconced in a dusty town in northern Italy.

As with so many driving candidates, Lauda's arrival in Maranello was kept secret. The paparazzi of the Italian motoring press had taken up permanent station at the Cavallino and around the factory gates, and the presence of Niki Lauda would have been trumpeted around the nation within moments of his sighting. To avoid such a meeting, Lauda was met at the Modena West exit of the autostrada by Montezemolo and escorted directly to Fiorano. There the brash, self-confident Austrian took some test laps in the current 312 Formula One car. Later that day he was given an audience with the Old Man, as Ferrari was now called by everyone (behind his back, of course). Also present were Forghieri and Piero Lardi, who was acting as translator for Lauda, who spoke in English. Ferrari asked how he liked the car. Lauda bluntly replied that it was a mess. An undrivable machine with severe understeer. Piero breathed hard, and a look of panic crossed his face. "You can't say that," he said to Lauda. After all, no man dared to criticize the automobiles to Ferrari's face. Over the years he had created for himself the grand delusion that his cars were above reproach and that their losses were attributable only to bad driving, bad tactics and the disgraceful behavior of his rivals.

Lauda told Piero Lardi to rephrase the answer, noting that the front suspension needed work. Ferrari then turned to Forghieri and asked him how long Lauda's recommended changes would take. The work could be done in a week, said the engineer. Ferrari turned again to Lauda and said, "If you are not one full second faster by this time next week you're out." This incident, recounted by Lauda in his candid autobiography, *Meine Story*, had a happy ending. One week later Forghieri's fixes permitted Lauda to run more than a second quicker at Fiorano and the job was his. He was to contrast sharply with his teammate Regazzoni (who had returned after a sojourn with BRM), who was a driver of the old school—a lusty man who embodied the image of the wild-living, extroverted, hard-driving international racing star. Whereas Lauda was cool and reserved, Regazzoni was openly emotional and therefore much loved by the *tifosi*. Lauda stayed fit and lived simply, while Regazzoni basked in his celebrity through exhausting bouts of socializing. They even contrasted physically. Niki was lean and drawn, with a crooked ironic grin and frigid blue eyes. Clay Regazzoni was a stubby, blunt-faced man with a broad black mustache arcing over a perpetual toothy smile. They were never close, but got along well as teammates, perhaps because their personal styles were so contrasting and their roles—with Lauda openly designated as Number One—never in doubt.

With a revived Forghieri back at the drawing board and the brilliant Lauda at the wheel of his improved creations and with Luca di Montezemolo operating as a stabilizing influence among the Byzantine politicians infesting the factory, the situation was bound to improve. Better yet, all the distractions of attempting to run sports-car programs were gone, and these three talented men were free to bear down on the single goal of dominating Formula One. One might conclude that Enzo Ferrari was shunted into the background, but that was hardly the case. He remained deeply involved and was surely the arbiter in all disputes, and still had the final say in driver selections, sponsor negotiations, dealing with the FIA and public policy statements. His annual press conferences, traditionally held to introduce the new Formula One cars, were developing into mob scenes. Hundreds of journalists from all over the world converged on Maranello to witness the theatrics—the long, Castro-like orations dealing with current issues, celebrations or past glories and often elegant, eloquent denunciations of enemies, past and present. While a growing number of capable professionals were slowly replacing the old court jesters, Ferrari remained the Emperor, clothes or no, who happily received the homage of the masses. Yet, as it turned out, the flunkies who had for so long influenced the fortunes of the Scuderia and its traditions were far from defeated and would soon make a comeback.

But for the upcoming seasons at least, the team would operate like a contemporary motorsports operation. Upon arriving in Maranello (although he would continue to commute by private aircraft from his home outside Salzburg), Lauda was overwhelmed by the scope and elaborateness of the Ferrari racing *équipe*. The mobs of mechanics, the foundry—which had turned out literally hundreds of prototype engines during its twenty years of operation—the teams of engineers, the test circuit and the dynamometer rooms impressed him no end and led him to wonder how, with these monumental advantages, Ferrari *ever* lost a race.

Lauda had a fine season in 1974, winning the Spanish Grand Prix at Jamara near Madrid and the Dutch Grand Prix at Zandvoort. He was leading the British Grand Prix when he cut a tire and then crashed twice—once on the opening lap of the German Grand Prix at the Nürburgring (which his teammate Regazzoni won) and again without serious damage at Mosport in Canada. Overall it was a transitional year filled with high hopes for the future. Luca di Montezemolo had served as a marvelous tempering influence, tamping down the rivalries inside the organization and for once dealing with Ferrari himself objectively. This alone permitted the Old Man, whose instincts were still very keen, to make sensible judgments based on fact, not on the bloated fantasies of cronies fearful for their employment. Montezemolo was hardly haunted by that threat and was therefore able to quietly and reasonably induce the chorus to at least sing from the same page of the songbook, if not always in the same key.

It was during this period that Enzo Ferrari lost his longtime friend, confidant and chauffeur, Pepino Verdelli. The little man took sick and died in 1975, leaving a void in Ferrari's life that was never to be filled. Pepino had been at his side since the mid-1920s, literally from dawn to dark. For years he had picked up Ferrari in the early morning and had remained on call until the boss's social adventures were completed late that night. No human knew more about the personal habits of Enzo Ferrari than Pepino Verdelli, and those confidences were never broken. He remained silent to the end, always content to remain in the shadows. But for all his loyalty, he was not always appreciated. A longtime Ferrari acquaintance recalls a late dinner at Orestes, a restaurant in the center of Modena's ancient city that was a favorite haunt of Ferrari's. As usual, Verdelli had taken his meal alone at a separate table, then had gone outside to await his master beside their Fiat sedan. Well lubricated with wine, Ferrari lurched out of the restaurant only to find that Pepino had briefly left the scene to relieve himself. "Damn that Pepe," Ferrari growled. "He's never around when you want him."

During the winter of 1974–75 Forghieri and his staff labored hard over what was to be his masterpiece, and what may be the high-water mark of Ferrari Formula One designs. His lusty flat-12 had now been improved to a point where it pumped out a reliable 440 hp—nearly 100 more than when the first 12-cylinder 3-liter Formula One engine had been introduced ten years earlier. Better yet, the handling was radically improved thanks to suspension updates and the development of a new five-speed gearbox mounted transversely behind the engine. This layout not only was more compact but lowered the center of gravity, which eliminated the nasty understeer of the earlier 312s and dovetailed perfectly with Lauda's precise driving style. The new car was called the 312T (for its *transversal* gearbox), and with its rounded, rather muscular bodywork it presented a vivid, unique shape among its more spindly English rivals. For the first time the cars were not all red. For 1975 the towering air intakes behind the drivers' heads (all the rage until the FIA banned them in 1977), which gave the cars a faint resemblance to midget submarines, were painted white with red and green stripes. These Italian national colors were something of a shock to the press and the aficionados, who thought the all-red Ferrari livery was perpetual. There is no question that the change was instituted at the behest of Fiat, which was interested in exploiting the image of Italian technology beyond its own borders.

After some early-season teething problems, Lauda got the 312T working perfectly at Monaco, where he dominated the race flag to flag. Thanks in part to Fiat's funding and Montezemolo's discipline, the testing procedures had been intensified, which in turn led to vastly better-prepared cars being sent into combat. Overall, it was a leaner, meaner racing team that set out from Maranello in the late spring of 1975 and the results were almost instantly apparent. Lauda won again at Zolder in Belgium a week later (the magnificently lethal Spa finally having been removed from the schedule in 1970 until it could be made safer), then at Andersdorp and again at the French Grand Prix at Paul Picard. He finally met defeat at the Nürburgring when he cut a tire. He clinched the World Championship—the first for Ferrari since 1964—when he finished third at Monza while the beloved Clay was delighting his fans with a solid win. Lauda capped the season with a final victory at Watkins Glen, but only after Regazzoni was black-flagged for blatantly blocking defending World Champion Emerson Fittipaldi in the closing stages of the race.

It had been an awesome season of domination for the team. Not since the 1961 campaign, when Phil Hill and Taffy von Trips had crushed the weakened British contingent, had the victories come with such satisfying regularity. Italy celebrated. Maranello was back on the map. Enzo Ferrari was restored to sainthood. All was well with motorsports and the national honor.

Overall the prospects for 1976 seemed even brighter. The 312Ts, thanks to Forghieri's constant fiddling, now boasted about 500 hp, which gave them nearly 20 hp over the best-tuned Cosworths. Although slightly heavier than their opposition, the Ferraris' tidy manners in the cool hands of Lauda appeared more than capable of outrunning the best of their rivals. Better yet, Emerson Fittipaldi, the most consistent driver of the current lot and a two-time champion, had left the McLaren operation to form his own team backed by an immense Brazilian coffee cartel. His place was taken by James Hunt, a brash English schoolboy racer who had shed his early nickname, "Hunt the Shunt," and was about to become a world-class competitor at the wheel of the new McLaren M23. The only negative aspect of the upcoming season involved the departure of Montezemolo. This steadying influence on the operation returned to senior management at Fiat (although he would remain closely linked to the Ferrari operation), and his place was taken by the tall, gentlemanly Daniele Audetto, the former manager of the Lancia rally team and yet another man closely connected to the Agnelli dynasty. Like Montezemolo, Audetto was operating as an official delegate from Fiat, although his management style would be more casual and better attuned to the factory's internal political intrigues than his predecessor.

There was a clash of styles between Audetto and Lauda. The former was a socialite who enjoyed the glamour of Formula One, whereas Lauda's lifestyle bordered on the monastic during the racing season. Compared with Montezemolo, Lauda viewed the newcomer as frivolous and disorganized. This was hardly the case, but the differences between the two men were to destroy the fine balance that Montezemolo had tried to create in Maranello and send the operation into another decline.

Montezemolo's candor was replaced by Audetto's courtly manners and circumspect reporting. Ferrari once again began to delude himself that anyone could win with one of his cars and that Lauda and Regazzoni were only two out of dozens of skilled drivers who could get the job done. This cant was echoed by the Italian press, who, spoiled by Lauda's excellent work the year before, had come to expect victory in every race. In this bizarre chain of self-delusion, the press would parrot the sentiments of the Old Man, who in turn would reinforce his prejudices by what they wrote. Without Montezemolo to inject some reality into this scene, it was certain that Ferrari's reasoning would once again jump the tracks.

It all started well enough, with Lauda winning at Interlagos in the Brazilian Grand Prix and in South Africa. Regazzoni was in form at Long Beach and won over Lauda, although by then it was clear that only Hunt in the new McLaren had a clear shot at preventing Lauda from winning a second consecutive title.

In late April, Lauda returned to his sprawling new lakeside home at Hof, outside Salzburg, for a few days' relaxation prior to the Spanish Grand Prix at Jamara. While he was driving a heavy garden tractor, the machine toppled over on him and cracked two ribs. He drove courageously a few days later and gave way to Hunt only after the pain in his chest became unbearable. It was clear that Niki Lauda was a man of immense determination and endurance, but the hierarchy at Maranello pointed out to him that such diversions as tractor riding were not acceptable during the racing season.

Lauda bounced back by winning at Zolder and Monte Carlo, but still the brash Briton stayed close in points. Funded with Marlboro money that might have found its way to Maranello, the McLaren team was a top-line professional operation by now, with the Old Man's *garagista* description an obsolete canard. Hunt scored twice, at the French Grand Prix and at Brands Hatch, before the Grand Prix circus traveled to the notorious Nürburgring for the August running of the German Grand Prix. Lauda had never enjoyed good luck at the Ring, and in fact had broken a wrist there in a 1973 crash at the wheel of a BRM. Lauda was in the vanguard of Nürburgring critics. Like the now-retired Jackie Stewart, Lauda was vocal in his protests that rescue crews could not effectively cover the entire fourteen-mile length of the old track and thus the drivers were in added peril. While the whole route had been lined with steel barriers and there were numerous fire and rescue crews, it was clear that the new breed of drivers, in their ultra-fast racing cars, were rapidly reaching a point where they would refuse to run on this most demanding of all circuits. Other problems would help to drive the old track off the international calendar, not the least of which was the expense and complication of lining the entire track with television cameras—an aspect of the sport that was to dominate its financial structure.

Race day dawned with the ugly threat of rain. Team managers and drivers puzzled over the choice of regular dry-weather tires or the specially grooved and compounded versions for running on wet surfaces. Lauda, like many of his competitors, started the race on rain tires, only to discover on the opening lap that the back section of the immense track was dry and that the sky was clearing. He stopped at the pits for a tire change, then charged back into the competition, far behind. He was taking a gamble, because the asphalt was still dotted with puddles and damp areas, but this did not deter the gritty Austrian from running flat out in a frantic game of catch-up. As he barreled at full speed into a sweeping left-hander leading into the tight *Bergwerk* right-hander, the Ferrari yawed left, then shot across the track and through a barrier into a stone wall. The impact of the crash shattered the fuel tanks and tore Lauda's helmet from his head. After pinwheeling crazily down the track, the 312T lurched to a halt engulfed in flames. The burning car was hit by the American Brett Lunger's Surtees. Three drivers whom Lauda had just passed, Lunger, the Englishman Guy Edwards and Arturo Merzario, skidded to a stop at the scene and rushed to the rescue along with a track marshal. Thanks to the heroic efforts of these four, Lauda was saved from burning to death. Finally the World Champion was lifted, badly burned and unconscious, from the smoldering wreck.

Lauda had inhaled enormous quantities of smoke and toxic fumes from the burning fiberglass. His face was badly burned, and it was feared that his lungs had been fatally scorched. He awoke in a hospital to hear a priest administering the last rites. For three days the racing world braced itself for news that the champion was dead. It was at this point that the unusually long period of tranquility at Maranello fell apart. The press was filled with reports that Lauda had crashed due to a mechanical failure—a contention that tended to be confirmed by blurry home movies shot by a spectator. They seemed to indicate that the Ferrari veered to the *left* before crashing. Logically, if Lauda had simply lost control, his car would have aimed to the right in a left-hand corner. But Lauda, upon viewing the films when he recovered, could ascertain little from the film, only that the rear end snapped around before he arrowed into an unyielding wall at about 120 mph.

Ferrari was furious over the suggestions that his car had somehow caused the wreck or that Lauda had been burned due to its frail construction (again, Ferraris were—and remain—as stout and sturdy as technology permitted). In a fit of pique he made one of his periodic withdrawals from the sport. Like the rest, this one was not taken seriously, although no Ferraris appeared at the Austrian Grand Prix. But Regazzoni was present at the start of the Dutch race two weeks later.

By then the European sporting world was swirling with insane, gruesome rumors about Lauda. More sensational journals carried lurid tales of how his entire face had been burned away and that years of plastic surgery would be required before the hapless driver could even appear in public, much less think about driving. Even the most rational accounts cast doubt on Lauda's ever competing again, much less by the end of the season. Imagine the shock that swept through the press when he, by dint of his iron will, had sufficiently recovered to tell Ferrari he wanted to enter the Italian Grand Prix on September 12—a mere six weeks following his near-fatal crash. It was an act to prove that the mind could dominate all weakness of the flesh. If Enzo Ferrari voiced any reservations about the Austrian's hasty reappearance, they were beyond Lauda's earshot (although Ferrari would later imply that he had stated that he felt Lauda's return was premature).

Whatever the case, Niki Lauda drove at Monza, his head swathed in bandages and his wounds still oozing blood. He used special padding to protect his healing, ultrasensitive skin. Most observers felt this was a brazen act of showmanship that would lead nowhere. It was hardly that. Lauda was a competitive professional who was going to the wall to protect his championship from the challenger, James Hunt. For this he was prepared to face supreme levels of pain and the risk of more injury. There was no exhibitionism in the act other than the tiny drama being played out between Lauda's brain and the fried skin that encased it.

A third driver was also entered for the Italian Grand Prix—a dour, rather suspicious Argentinian named Carlos Reutemann. "Lele," as he was called by his peers, had recently left the Brabham team and, through secret negotiations, was being groomed to replace Regazzoni at the end of the season. It was also widely believed inside the factory that Lauda's career was over as well, so Reutemann was virtually assured of a place on the team for 1977, although neither of the two contracted drivers was aware of the arrangement. The deft hand of

Luca di Montezemolo was present in the acquisition of Reutemann, and from that day on, Regazzoni, who had raced loyally for the Scuderia for six seasons, was doomed. He had been a good second man, capable of inspired wins on occasion and always ready to drive at the limit to aid Lauda in a team effort. But someone—be it Ferrari or Audetto or Montezemolo—believed that the gregarious Swiss was past his prime, and he was being ushered out of Maranello long before he realized he faced the door. Meetings between Regazzoni and Ferrari indicated that there was no problem, which was surely not the case. As in many instances, talk at Ferrari meant nothing; only action counted.

Hunt added to his point total at Monza with a convincing victory, while Lauda, in great pain, struggled home eighth with an ailing car. Regazzoni, driving with his usual verve, muscled his way into sixth. At Watkins Glen, New York, a month later, Hunt won again, although a much stronger Lauda managed third. The World Championship would be decided between the two men at the final race of the season—the Japanese Grand Prix to be held on the sweeping Mount Fuji circuit. The Italian press was again at full cry. They were yelping that Lauda had come back too soon and his presence was an unsettling force on the team. This in turn had led to a series of bad finishes, they said, while conveniently ignoring the rise of the powerful McLaren effort and a less effective Maranello test program following the departure of Montezemolo.

Lauda was hardly in the bloom of health when he traveled to Japan. Awaiting him during the off-season was extensive plastic surgery to repair his ruined face. But the immediate challenge was the ebullient Hunt, who was only three points behind in the race for the championship. The event was run in appalling weather. Low clouds hovered over the track, spilling down torrents of rain from their dark bellies. Few veterans of the Grand Prix wars could recall a worse day upon which a race had been started. Three men would contest the race: Hunt was on the pole, Mario Andretti sat in second position and Lauda a wink of the clock back in third. The flag fell and the field skittered off into the murk, leaving nothing more than a rooster tail of spray as they groped their way toward the first corner. There Lauda nearly spun into the fence. Two laps later he stopped. It was too much for him. There were limits even to his towering courage. Frazzled by the endless nattering and rumormongering of the press, the internal politics and the painful attentions of his doctors, he simply parked the car. Never one to mince words, Lauda refused Forghieri's offer to blame the retirement on car failure. Blunt and honest to a fault, Lauda told the press that he quit because the conditions were too much for him. If they could not accept that explanation in view of his injuries and his subsequent comeback, it was hardly his problem.

Hunt stayed in the soaking struggle and managed to slide his way into third place at the finish. That permitted him to win the World Championship by one point. Italy was again sent into a tizzy. Lauda called Ferrari from the Tokyo airport and told him what had happened. The Old Man said little and Lauda knew that he was displeased. No driver, regardless of his situation, was ever supposed to park a perfectly operable Ferrari, and the Austrian understood that from that moment on his Maranello inquisition had begun.

In public Ferrari tended to damn Lauda with faint praise, defending him for his decision at Fuji, but questioning his judgment about returning to competition at Monza so soon after his crash. This struck the theme for the Italian press, which began to caterwaul that Lauda had ruined the rhythm, destroyed the cadence, as it were, of the team by insisting on driving in the Italian Grand Prix. It was an act of supreme selfishness, they screamed, and resulted in Ferrari and the nation losing the World Championship. (The factory won the so-called Manufacturers Championship for compiling more points than any other team, but that title meant little compared with the highly publicized driver's crown.)

A week after the Japanese race Ferrari organized one of his famous press conferences to deal with the situation. It was mobbed with journalists from all over Europe, all eager to take part in the comic theater that such affairs had become. As usual, the Old Man dominated the proceedings, alternately funny, angry, pugnacious and humble. At one point a writer addressed him as Commendatore to which Ferrari growled, "Look, I'm no Commendatore—I prefer to be called Ferrari. When I go to the barber shop and am called Commendatore I don't object because there are many Commendatori, but if they call me Ferrari that is another matter, understand? Therefore if you want to call me Enzo then I shall close my eyes, imagine you are a beautiful girl and I will be even happier." This was the kind of opaque garble for which such sessions had become famous. Early in the conference he was asked about the Lauda retirement at Fuji and if he knew of anything like it. "Yes, I know of a driver," he mused. "His name was Enzo Ferrari. You don't believe me because you are young men. I was to go to the Lyon Grand Prix, where I was to drive the fourth car. At the time I suffered a nervous breakdown and I had the guts to say to myself, 'I have tried the circuit, have come home and have to go back—but I can't make it.'

"As soon as I got over the breakdown, which I suffered in 1924, I went back to racing. Do not forget that I was born in 1898 and that in 1918 I underwent two chest operations. I have suffered a lot, but God has not abandoned me. Like Lauda, I had to ask myself, when my Dino was born, should I race again? I decided to stop. Lauda decided to go on. I had to keep my word to him even if this should damage Ferrari."

From such jumbles of sentimentalities, distortions and vague inferences, the gathered press were supposed to infer that Ferrari supported Lauda's decision to park the car, but in private—within the inner circle of decision makers composed of Audetto, Piero Lardi, Forghieri, Franco Rocchi, Franco Gozzi and others—it was understood that in the eyes of their leader Niki Lauda had committed an unpardonable sin. Lauda knew this as well, and being the clear-thinking pragmatist that he was, he understood full well that at Maranello total absolution of his sins would be impossible.

As the disappointments of 1976 drifted away, the new golden boy at Ferrari was Carlos Reutemann, although it was obvious to all concerned that Lauda could outdrive him at will. Regazzoni, the loyal spear carrier, was gone, expelled without ever having received a proper explanation of his transgressions—other than that he tended to drive only for himself, not for the team (which, of course, conveniently ignored his often controversial blocking tactics and rough driving in order to aid a teammate).

The toadies in the Italian press were perfectly attuned to the nuances of the Ferrari position and quickly took Reutemann to their bosoms while awaiting a decision on Lauda's fate. It is probable that Ferrari would have signed up either Hunt or Fittipaldi in his place had they been available, but that not being the case, he was left with a driver he clearly believed to be traumatized to the point of paralysis behind the wheel of a racing car. Again, he was operating on the basis of second-hand information and had not the vaguest idea of the reservoirs of icy determination that lay within this Austrian star.

The 1977 season was doomed from the start in terms of team acrimony and dissent. Ferrari initially offered Lauda the job of team manager, thinking that he could utilize him as an occasional substitute driver and still keep him away from rival teams in the event he decided to get serious about racing again. What the old man did not realize was that Lauda had lost none of his resolve and, if anything, was determined to prove that he could win again with Ferrari, even if the cars (312T2s) were essentially unchanged (with the exception of large Fiat logos being emblazoned on the bodywork). Reutemann was labeled as the Number One driver on the team, but Lauda was openly disdainful of the Argentinian from the start. When asked by a journalist if he considered Reutemann a teammate or a rival, Lauda sniffed, "Neither."

There were changes in the management as well. Audetto was gone as team manager. Some, including Lauda, would claim he had been fired; others explained that he had merely been transferred within the Fiat organization to handle their rally team. Whatever the reason, his replacement was another longtime Ferrari crony. Roberto Nosetto had been a Scuderia employee for twenty years and was supremely loyal, although his allegiance did not transcend his zany superstition about the color green. Nosetto was inclined to cover himself in that hue from head to foot—a fetish which not only clashed with the Ferrari livery but made him the butt of endless jokes on the team. He was also obsessed by the number 7, being so fearful of the digit that he refused to rent an automobile if it appeared on the license plate. As the season progressed, Lauda came to despise him to a point where he became, in his words, "nothing but a speck of green."

As usual, the politics within the team remained superheated. Forghieri was under enormous pressure to improve the aging flat-12 engine, which was flagging against the ever-improving Cosworth. Moreover, Colin Chapman was in the vanguard of creating a whole new technology called ground effects. Several years earlier General Motors engineers had developed a Chaparral Can-Am sports car with a set of motor-driven fans that literally sucked the car onto the track surface. Chapman quickly recognized this boon to road holding and began putting reverse airfoils in the side panels of his Lotus 78 which would passively create aerodynamic downforce. His new car was to encounter teething problems but would lead a year later to the revolutionary Lotus 79, which would carry Mario Andretti to the World Championship and alter forever the basic technology of racing machines. This trend was initially ignored by Ferrari, who remained glued to the old notion that superior horsepower always beat superior road holding, and Forghieri and his staff—still without a wind tunnel—were left to tweak up the old flat-12 and hope for the best.

What they did not expect was a totally renewed Lauda, who won the second race of the season in South Africa (after Reutemann had been victorious in Brazil) and went on to record six second places and two more outright wins to claim his second World Championship. By the Dutch Grand Prix, about two-thirds through the long season, Lauda had convinced the doubting Thomases in Italy that he was back at his full powers. But now it was his turn for revenge. He secretly signed a deal with Bernie Ecclestone, the crafty, ambitious former London East End car salesman who was team manager at Brabham. Lauda would drive for that operation in 1978, although Ferrari was unaware of his plans.

But it was not long before the finely tuned Ferrari intelligence network transmitted news to Maranello that something was up with

Lauda and Ecclestone. Ferrari became noisily defensive, shouting insults at Lauda whenever the two met at test sessions and threatening anyone who departed the fold without his blessing. Lauda's friend and personal mechanic, Ermanno Cuoghi, who headed the six-man crew that had tended Lauda's cars since his arrival on the team, was also thinking about a transfer to Brabham by the time the Grand Prix circus moved to Watkins Glen. There Cuoghi received a late-night phone call from Ferrari. Would he stay with the team? Cuoghi was asked. The mechanic said that he hoped to confer with his wife when he returned to Italy. Ferrari would tolerate no temporizing and fired him on the spot. When Lauda arrived at the track the next morning, he found his old friend in tears.

At Watkins Glen, Lauda drove cautiously to fourth place, thereby clinching his second World Championship. But there was no joy. Nosetto, in green as usual, barely acknowledged the new champion and the post-race celebrations were muted and strained. Lauda was furious about the treatment of Cuoghi and drove off to Toronto, where he sulked in a hotel room for two days. He then sent a telegram to Ferrari curtly announcing that for reasons of health he would not drive in the final two races of the year at Mosport, Ontario, or in Japan.

The Austrian had done the unthinkable. He had defied the single most powerful man in motorsports, walking away from his red automobiles in a moment of triumph. The humiliation in Maranello was unbearable. Ferrari was apoplectic over the cheek of this cool professional. He raged to the press about the Austrian's lack of manners and his blatant treachery, but it did little good. After decades of treating his drivers like pawns, the tables had been turned on the great man, and it would be years before he would begin to forgive or forget the slap in the face.

But as always, legions of emotionally inflamed young men were standing in line to take Lauda's place. Even before he had made his shocking departure, a third driver had been signed as a temporary replacement for the Canadian Grand Prix. Several years earlier James Hunt had encountered a tiny, ebullient French Canadian in a minor

Formula Atlantic race in Quebec. He had returned to Europe stunned by the quickness of the twenty-five-year-old youth and arranged for a test with the McLaren team. The fresh-faced kid, who looked like a teenager, was named Gilles Villeneuve, and while he was not retained by McLaren, Ferrari offered him a tryout at Mosport in the absence of Lauda. He drove like a fiend, then went off the track, as he would many times again. But here was a tiger, a classic *garabaldino*, and Enzo Ferrari quickly understood that this man from faraway Quebec had the potential not only to erase the memory of Niki Lauda but to conjure up comparisons with past giants like Nuvolari and Ascari. For Ferrari there could be no wish stronger than for such a reincarnation.

18

Nineteen seventy-eight was a year of death and rebirth for Enzo Ferrari. Counterbalancing the arrival of the wildly enthusiastic Gilles Villeneuve on the team was Laura's death, on January 27 — they had been married fifty-five years. In the end he was saddened and depleted by her loss, having long since accepted a marriage that was more a détente than a union of love. To be sure, she had been more deeply involved in the business than anyone outside the household might have imagined. They had been lovers, partners, antagonists, co-conspirators, pals, enemies, rivals and embattled teammates through five decades of struggle that had seen them both rise out of the ruck of Italian lower-class life to major national prominence. Ferrari had done his best to keep her in the "private realm," and save for her odd incursion into the racing scene in the early 1960s, she had remained in the background, content to operate within the tight circle of the factory and its Modenese environs.

Laura Domenica Garello Ferrari died as she had lived, an enigma in terms of her health and her true relationship with her husband. Surely her behavior had been erratic in the later years, and she had become a more private person who steadily slipped into the shadows as the years passed. When able, she had spent more time at the Adriatic summer place, where Ferrari had often found refuge on the weekends. The cause of her death, at age seventy-eight—a few weeks before Enzo Ferrari somberly celebrated his eightieth birthday—remains a mystery. It is known that she experienced great difficulty in walking during her later years, as did Ferrari himself (although his lack of mobility is probably attributable to simple old age and the fact that for the last forty years of his life he did almost no exercising beyond heaving his large frame in and out of automobile seats). Laura was interred in the vast family mausoleum at San Cataldo, in a crypt on the opposite side of the circular, marble-walled room from Enzo's mother, father and brother and separated by a single blank enclosure from her son, Dino.

For a time Ferrari remained alone in the enormous house on Largo Garibaldi. Plans were being made to admit Piero, his wife, Floriana, and daughter, Antonella, as well as Lina Lardi following the legitimization process, but several years would pass before that would become appropriate. Until then his time was spent in seclusion, broken only by quiet dinners with friends, business meetings and the distractions of the racing operation. Enzo Ferrari refused to lose touch. He and his new driver, Dino Tagliazucchi (who later became the gatekeeper at the Fiorano test track), spent many weekends at Viserba, where constant communication would be maintained with the racing team via telephone.

The new team manager (*direttore sportivo*), Marco Piccinini, a facile, rather oily loyalist who was to become the Talleyrand of the operation (and because of his piety was to be nicknamed "the Monsignor"), spent most of his weekends on the telephone at the major Grand Prix tracks. Observers recall that during the early street races at Long Beach, California, the Ferrari pits were always located near the old, crumbling Breakers Hotel, where Piccinini was able to take up a station in a public phone booth in the lobby and transmit constant updates to his boss. These conversations took place only during practice and qualifying, with the race itself an anticlimax. Ferrari was able to watch the contests thanks to a worldwide television linkup that aired most of the Grand Prix schedule. For the most part he watched the contests alone, although one associate was occasionally invited to sit in. He recalls that Ferrari never showed emotion, no matter what the fortunes of his automobiles. "Ferrari simply sat there, silent and inert, never responding to either victory or defeat," he says.

His view of racing remained constant: the event itself was essentially meaningless. For him the stimulation came in the planning and preparation, in the creation of the machines, in the organization of the human beings who would man the team and in the endless wrangling with the press, promoters and sponsors. For Enzo Ferrari the race was over when the engines were fired and the cars leapt away from the starting grid. From then on it was in the hands of his drivers and there was nothing for him to do but to ease back in his chair— either in the lavishly converted farmhouse adjacent to the Fiorano test track (used specifically for entertaining important guests and for company conferences) or at the Viserba beach house—and let the race unfold. He was powerless from that point onward, which was his elemental reason for never attending the races.

But this made sense only up to a point. As the leader of the Scuderia, there is no question that he would have benefited from attendance at the competitions. He then could have seen for himself how his drivers and crew behaved, rather than depending on the eyes and ears of men like Forghieri and Piccinini. It was only Montezemolo who had dared to be totally candid, and the operation had been the better for it. Now Piccinini was the link between the Old Man and the track, and once again the messages became distorted and self-serving. Results—or the lack of them—were often modified simply to keep peace in the family. A devout Catholic, Piccinini was well connected with the Vatican and with the financial community. Big chunks of Ferrari's personal fortune were alleged to repose in his father's bank in Monaco, leaving the young man devoid of the leverage enjoyed by the Fiat-backed Montezemolo. It was said that Piccinini, a frustrated driver and racing-car builder, had gotten his job with the Scuderia thanks only to his family's financial links to Ferrari. Never openly religious himself, Ferrari and some of his old cronies enjoyed mocking Piccinini about his religion, mouthing ribald jokes about the Church and generally denigrating his faith. Piccinini, the consummate diplomat, did his best to take this offensive behavior in stride and carried on manfully. In the great schism that was to sweep through the Formula One community in the early 1980s, Marco Piccinini was to become invaluable as Ferrari's nuncio.

With Fiat's money pouring into Maranello, expansion was going on everywhere. In 1978 a massive addition to the manufacturing facilities was completed on the north side of the factory, bordered by a new road called the Via Musso. Plans were also in the works for more development in the future, with the entire sporting department (*gestione sportiva*) to be moved into vast new quarters to the west of the Cavallino restaurant—which would also undergo radical renovation and updating. Since the Fiat acquisition in 1970, the Ferrari works had more than doubled in size and by the mid-1980s would have about six times the square footage of the little factory created in the middle of World War II.

Fiat was aggressively trying to expand its presence in the United States—an undertaking that would end in dismal failure and cause Ferrari sales to stagnate in that critical market. The new, relatively low-priced 308 coupes and Spiders piled up on the New Jersey docks and worldwide reaction to the machines was tepid at best.

By then the Chinetti family had been completely shorn of its importer status and relegated to operating one of the many Ferrari dealerships in the United States. Lawsuits had been threatened and angry words exchanged, but the intervention of Fiat made changes in the Chinetti-Ferrari entanglement inevitable and unstoppable. The relationship between the two men had grown like Topsy and by the mid-1970s had become a rat's nest of verbal agreements, hazy contracts, wishful thinking and sentimental half-promises. Fiat's attempt to place the operation on a more formal basis caused the Manhattan showrooms to be closed and what remained of Chinetti's American operations to be transferred to more modest quarters in suburban Greenwich, Connecticut.

Fiat pressed ahead to make Ferrari its international "image" machine, and the corporate influence steadily increased within the operation. But Ferrari himself still held sway within the racing department. Always careful with his lire, he resisted the urgings of Forghieri and others to build a wind tunnel at Maranello. Until such a facility was completed, all aerodynamic research—which was becoming more essential by the hour—had to be undertaken at the Pininfarina facility in faraway Turin. The engineering staff had also determined that the Fiorano test track was too short and lacked a high-speed bend where the new science of ground effects could best be studied. To expand the track meant purchasing adjacent property and Ferrari was reluctant to spend the money. Therefore the team was forced to labor at an obsolete facility or drive to the larger track at Imola for test sessions as they had done in the old days. Worse yet, Ferrari's reluctance to buy additional land opened the way for the incursion of several ceramics factories along the Abetone road. The dust from their kilns constantly drifted onto the Fiorano surface, requiring that the track be swept prior to any high-speed running.

Although Ferrari's intellect was as sharp as ever, his body was showing definite signs of age. In addition to his creaking legs, he was losing weight and the once puffy cheeks were now beginning to hollow out, leaving his face dominated by his massive Roman nose. The eyes, sinking deeper into the skull, still glistened in sharp contrast to the mound of ivory hair that swept away from his forehead, but they evidenced a weariness that had been unknown when he entered the decade. He was now a national icon, a larger-than-life proconsul in the ancient welter of Italian culture, a national treasure whose automobiles were major show business on five continents. His power in the sport was enormous, and no decision was made within the international councils regarding rules or long-range policy without his blessing. A decision by the Scuderia Ferrari to boycott a Grand Prix race could send ticket sales plunging by 50 percent.

The collecting of his automobiles was becoming a quasi-religion. A prime example was Frenchman Pierre Bardinot, whose 375-acre estate at Aubusson, two hundred miles south of Paris, was being converted into a Ferrari shrine, complete with a three-kilometer racetrack, a museum housing a priceless collection of Ferrari cars and a restoration and repair shop manned by expert mechanics. Others would turn the veneration of his automobiles into virtually a fetish, although Bardinot was uniquely favored by being offered the privilege of purchasing a number of rare examples from the factory. But for all the homage paid him, Enzo Ferrari never evidenced the slightest interest in visiting the Bardinot collection or even beginning a formal collection of his own until a few years before his death. Even at age eighty, the glories of the past were but a foundation upon which dreams of the future could be mounted. In 1978 those dreams would surely be based on the monumental skills of the ebullient Gilles Villeneuve. If any man was to resemble a reincarnation of Tazio Nuvolari, it was this former snowmobile racer from Quebec. Ferrari took an instant liking to the young man. He was a commoner, one of the few drivers in Formula One who had made it on the basis of sheer talent and without the boost of a family fortune. He was open and uncomplaining, eager and willing to drive any car on the edge of control, never faltering. He seemed constitutionally unable to drive anything less than flat out, yet he was an unselfish team member, willing to subordinate his raw speed to guarantee a high finish for a teammate—as he would do on several occasions. To be sure, he crashed a great deal, especially in his first season. His overeagerness destroyed a number of expensive machines and sometimes infuriated Ferrari to a point where he briefly considered getting rid of him. But his exuberance, his loyalty and most of all his blindingly fast driving kept him in place. Surely Gilles Villeneuve was to be the last of Enzo Ferrari's great favorites, joining the likes of Nuvolari, Guy Moll, Peter Collins, Stirling Moss et al. in his personal pantheon of immortal drivers.

The 1978 season was something of a disaster. Reutemann, never a totally happy man in a racing car, seemed uncomfortable with Villeneuve nipping at his heels. Piccinini, in his first season, was too much the political novice to take command of the situation. Villeneuve salvaged much of the flagging team's honor at his hometown of Montreal when he won the Canadian Grand Prix on the tricky Île Notre-Dame circuit, thereby assuring himself a place on the team for the following season and gaining instant hero status with his fellow Quebeckers. Reutemann, on the other hand, was heading toward the door. He had complained loudly about Forghieri's new 312T3 chassis and its lack of road-holding qualities against the newer ground-effects cars from England. But the wide, flat-12 engine, now in its ninth season of competition, was unsuited to the new ground-effects technology, and new engine designs were being considered. But a driver whining about his equipment was verboten at Maranello and it was obvious that the Argentinian was on the way out.

A prime candidate for the team was a dour South African named Jody Scheckter. He had several excellent seasons with privateer Walter Wolf after overcoming a reputation as a fast but crash-prone novice. Scheckter was known as a loner, a totally committed driver who had little time for the press or the niceties of public relations or sponsor coddling. Following his retirement he would explain that he considered these to be distractions to his primary mission of winning races and did not mean to be impolite. But during his prime years Scheckter's reputation as a brutally frank, reclusive, tough-minded professional was universally accepted.

He was, as usual, approached in secret by the Ferrari operatives. Like Lauda, he sneaked into Maranello, having been met at the same autostrada exit and been escorted into the presence of Ferrari through a back gate. Neither he nor his new boss stood on ceremony and the interview was brief and to the point. "How much do you want?" Ferrari asked. "I am too young to talk about money," Scheckter shot back at Piccinini, who was acting as interpreter. The brash response caught Ferrari off guard and immediately let him know that this was a driver who would not be buffaloed by the Ferrari mystique. In fact, Scheckter was perfectly prepared to discuss money, and the two men engaged in a lively session, haggling over the pay and perks. The day had long since passed when Ferrari could entice a driver with the "honor" of joining the Scuderia. If top drivers were to be recruited, it would be purely on the basis of hard cash.

Scheckter's demands were high. Ferrari's offers were low, as was to be expected. The South African demanded and received a six-figure retainer, which he describes as "very competitive" with that being paid by other top teams, but the negotiations bogged down on the matter of prize money. Jody wanted 20 percent of the purse; Ferrari offered 10 percent. They argued for a while until Scheckter realized that the Old Man was fully aware of the wage scales in Formula One and knew that no driver got 20 percent. He capitulated and accepted 10 percent. Then Ferrari tried one final ploy. Scheckter demanded payment in dollars. Ferrari agreed, but offered Canadian dollars, which were about 20 percent less valuable than their American counterparts. Scheckter balked, Ferrari gave in and the deal was finalized.

At first Jody Scheckter, a veteran of the Formula One wars, was skeptical about having the boyish Villeneuve as his teammate. The older, more mature Reutemann seemed a better choice, and the two men arranged a meeting in the south of France to discuss the matter. Scheckter found Reutemann a nervous wreck after the pressures he had faced at Maranello. The Argentinian insisted on meeting secretly in a parking lot and refused to leave his car, lest the two be seen. Scheckter decided then that this scarred combatant in the Ferrari political wars was emotionally unprepared to carry on. Therefore Villeneuve became his teammate by default. It was to become a surprisingly happy relationship for both men.

Conventional wisdom in the motoring press held that the combination of the tough-minded Scheckter, the puppylike Villeneuve and the temperamental, Machiavellian Ferrari was a fatally poisonous brew. Nothing could have been further from the truth. The two drivers, perhaps because their approach to the sport was so divergent, got along well. Moreover, Enzo Ferrari seemed to respond to the general tranquility of the situation and settled back to watch the pair dominate the 1979 season. Thanks to a much-improved 312T4 from Forghieri that offered better ground effects and even more of the vaunted Ferrari top-end horsepower and low-end torque, the two drivers won repeatedly. Scheckter won the World Championship with a number of high finishes and three outright victories. Villeneuve also took first in three Grand Prix (as well as a nonchampionship race in Great Britain) but did not have as many high placings as his teammate. While Scheckter was a deserving champion, much of his success was attributable to the anvil-like reliability of his Ferrari. He failed to finish only twice in the seventeen-race season and on both occasions the car was not at fault (one failure was caused by a collision with another car, the second by a tire puncture).

Scheckter dominated the Italian Grand Prix, where he and Villeneuve finished one-two and sent the *tifosi* into rapture by clinching his first World Championship and the last in Enzo Ferrari's lifetime. In an act of supreme sportsmanship, Villeneuve obediently tailed his partner home in second place, never challenging for the lead, although many in the pits were certain the youngster could have offered up a mighty duel had he chosen. This was a display of teamwork and sportsmanship that stood in stark contrast to the 1956 Castellotti-Musso duel on the same circuit, where both drivers tossed discipline to the winds and tore their cars to pieces in the ensuing battle. For this act alone the French Canadian gained Ferrari's everlasting affection and gratitude.

Legend has it that Villeneuve was a devoted member of the Ferrari team and thus an Italian at heart. This was not the case, says Scheckter. "Gilles actually didn't like Italy much," he recalls, "which is why he lived in Monaco and flew in for tests and meetings by helicopter. On the other hand, I loved the Italians and their way of life. My wife and I had an apartment in Modena, and although I didn't speak Italian I felt very much at home. As a champion driver you are considered part of their family, and believe me, when an entire crowd of diners rises to applaud when you enter a restaurant, you appreciate how much they care about their sport.

"Gilles and I tried to stay out of the factory politics, although that was almost impossible. Forghieri could be a madman, but he was brilliant and we had numerous arguments. As for the Old Man, you couldn't help but be ill at ease around him. I tried to be very professional. I was there to win races. Nothing else. That was the difference between me and Gilles, I suppose. He was trying to win laps. I was trying to win races. As for Ferrari, he didn't want to hear about the failings of his cars, especially the engines. I remember one day I was briefing him on a race and I remarked that the Cosworths had more power than we did on certain parts of the track. Piccinini, who was translating, said, 'You can't say that. You can't tell him that the Cosworths have more power.'

"But Enzo Ferrari was a very special man. You've simply got to respect him for what he did. He was all business with his drivers. I remember after I won the World Championship he said nothing. No letters, no phone calls. Nothing. Then one day a few weeks later I saw him at Fiorano. He walked right past me. He gave me a little salute and said, 'Hey, champion.' That's the only word I ever heard from him about the title. At the end of the season they had their annual banquet and a bunch of us—mechanics, test drivers, engineers, the lot—received trophies. Mine was a Prancing Horse on a little wooden base. One of the legs had been broken off and brazed back on."

Scheckter and Villeneuve remained on the team for 1980, but by then the world of Grand Prix racing was becoming embroiled in a maelstrom of political intrigue that made the madness at Maranello seem trivial by comparison. The trouble had been brewing for nearly a decade, or since the sport had begun to attract major sponsorship and lucrative television contracts. At the root of the problem was, not surprisingly, money. The young lions of the sport—for the most part the British independent teams like McLaren, Brabham, Lotus and Williams, which had risen to prominence without major automobile-factory backing—were hungry for influence in the hallowed, rather musty halls of the French-dominated Fédération Internationale de l'Automobile. The FIA, through its subcommittee which governed international motorsports (FISA, or Fédération Internationale du Sport Automobile), had grown hidebound and isolated from the new realities of television and commercial sponsorship. FISA was aligned with the old grandees of the sport, the classic European car manufacturers like Porsche, Mercedes-Benz, Peugeot, Renault, Alfa Romeo and of course Ferrari, which had traditionally built automobiles for FIA-sanctioned races and rallies. It had been business as usual for the old establishment until the British upstarts formed FOCA (an acronym for Formula One Constructors' Association). The leader was the brash

Bernie Ecclestone, who had risen through the ranks of motor racing to gain ownership of the Brabham team and was leading the charge for more practical rules, lowered costs, higher purses and stronger team representation in the FISA councils.

The French and their handmaidens on the FIA countered by appointing a pompous Frenchman named Jean-Marie Balestre to defend the honor of the international body. It was to be a battle of the grandees versus the *garagistas*, and it would rage for five years until peace was finally reached. Balestre was an absurd man, given to overblown pronouncements and the levying of fines for supposed insults. He was a poltroon of comic dimensions and the target of endless lampoons in the motoring press. The Italians called him "the crazy Pope." Things got serious when an Italian magazine published photos of him in what appeared to be a German uniform. The story accused him of collaboration with the Germans in World War II as an official of the Vichy government. It was then revealed that Balestre had served time in prison, which he claimed was attributable to his being fingered by the Nazis as a Resistance spy. His critics countered that he had been jailed by the Germans after he had been found stealing from them. The argument boiled on for years, with no conclusive evidence ever produced to dislodge Balestre from his august position.

The two sides warred over all manner of issues, many of them arcane technical questions dealing with turbochargers, water-cooled brakes, adjustable suspensions, ground effects tunnels, and so on. But these blurred the larger issue of who was to control the future of Formula One racing—the teams or the aging establishment in Paris. This was the kind of struggle in which Ferrari was a master. He and his emissary, Piccinini, held high cards. The respected English motoring journalist Nigel Roebuck revealed in his column in the widely read British magazine *Autosport* that a survey in the early 1980s indicated that 30 percent of any crowd at a Grand Prix race came exclusively to watch the Ferraris. With this kind of box-office clout, Enzo Ferrari was courted by both sides in the endless, mind-numbing arguments over rules and policies. For the most part he sided with the

FISA establishment, for reasons of tradition, commonality of the French and Italian cultures and a general dislike of the upstart English who formed a majority of the opposition. Yet he and Piccinini artfully remained above the fray, using the power and influence of the Ferrari name to prod both sides in directions that best suited the Scuderia. He had lived through hundreds of such petty disputes, always involving the immense egos that were attracted to this most challenging of sporting endeavors, and he was not about to let his beloved Grand Prix racing be swept away in a flood of acrimony. On the one hand he could act as the impartial arbiter in the FISA-FOCA dispute, the aloof patriarch observing a scrap among lesser members of his flock, while using every error or sidestep by either side to his advantage.

It was clear to everyone that modern Formula One cars were becoming too fast. The new science of ground effects was escalating cornering speeds to absurd numbers. Traditional curves where hard braking was once necessary were now being taken flat out. Moreover, some teams were developing turbocharged 1.5-liter engines that were producing prodigious, if still unreliable horsepower. So-called skirts had been added to the lower bodywork of the cars, creating a tighter seal between the inverted wings that now formed the undercarriages. These were banned, which caused an enormous uproar. A minimum ground clearance was mandated, which prompted adjustable suspensions that could be lowered for racetrack operation, then raised for inspection in the pits. More acrimony. Water-cooled brakes, water injection, movable airfoils, limitations on fuel-tank capacity, turbocharging—all were the subject of endless wrangling between the two power structures. Races were boycotted and tempers flared in the expensive motor coaches that served as mobile offices for the various teams. Piccinini was forever in the middle of these disputes, generally siding with the FISA establishment, but always prepared to parlay with Ecclestone and his cadre of dissidents. While the minutiae of the rules formed the basis for the endless skirmishing, the real struggle involved control of Formula One competition and its millions in sponsorship and television fees. Enzo Ferrari was perfectly aware of this and positioned himself to be on the winning side, no matter who triumphed. While he continued to insist that his cars would remain "pure" in terms of non-automotive sponsorship, the logos of Marlboro cigarettes, Olivetti office machines and Longines timepieces began to appear on the bodywork. This was explained away as sponsorship that belonged to the drivers, not the team, although, of course, the Scuderia was taking a generous commission on the six-figure fees being paid by the brands. Stacks of dollars, francs, pounds and lire were pouring into the sport. For one who had struggled for decades with modest sums from the likes of Shell and Pirelli, this new infusion of wealth had to overwhelm even a man so commercially oriented as Ferrari. By the mid-1980s sponsorship fees would push past $5 million per car for major teams and drivers' salaries would edge past $1 million per year. Even Ferrari, who still lagged somewhat behind the leaders in terms of compensation, was forced to shell out over half a million dollars in retainer fees to assure the services of men like Scheckter and Villeneuve.

Although the Scheckter-Villeneuve duo produced a period of relative tranquility within the Scuderia, Ferrari was shaken by an event that took place outside the violent precincts of the racing world. In October 1979 a group of uniquely depraved individuals broke into the San Cataldo crypt and attempted to steal the body of Dino Ferrari. The criminals almost hacked their way through the metal casket, only to be scared off before their gruesome mission could be accomplished. They left behind several plastic bags, suggesting that they planned to steal the remains and hold them for ransom from a distraught father. The criminals were never apprehended and Ferrari said nothing to the press. His only action was to fit the tomb entrance with heavy decorative iron gates. In the final update of his memoirs published three years later, he wrote, "I never did imagine that the price of notoriety, which I have always paid at every point in my life, would include the destruction of the tomb in which twenty-six years ago I buried my son Dino. After many events I feel alone and almost guilty for having survived. At times I think that the pain is but an exasperating attachment to life when faced with the hallucinating fragility of existence."

At that point Ferrari was still feeling the shock not only of the grave desecration but also of the mysterious disappearance in the summer of 1978 of Carlo Bussi. He had been a close associate of Forghieri and Rocchi in the engine design department and was a loyal and respected employee. A member of a prominent Italian family, Bussi had vanished while vacationing in Sardinia. At first it was believed to be a terrorist kidnapping, but no messages were received and no clue as to the young man's whereabouts ever surfaced. Bussi's body has never been found.

Once again the big house at 11 Largo Garibaldi was amply occupied, although dozens of rooms above the now-defunct Tucano restaurant were vacant. In fact, the old neighborhood that Enzo Ferrari knew so well was rapidly changing. Modena was now a city of 150,000 and growing by the day. In 1975 the Fini family had moved their hotel to an elegant new building a kilometer east on the Via Emilia and the old Real-Fini had been turned into a bank. The Grand, scene of so many debauches, had also been converted to a bank, after Alejandro de Tomaso had opened the new Canal Grande in the medieval walled city that formed the nucleus of Modena. The Orsis' mansion was gone, and negotiations had begun to transform the original Scuderia on the Viale Trento e Trieste into a concrete ramp garage—surely a historical desecration of the first magnitude. Piero and his wife and daughter had moved into a large apartment in Ferrari's house, as had Lina, who now occupied her own separate quarters. Piero had been legally recognized as Ferrari's son. He had first begun to acknowledge the boy in 1975, when a press release referred to him as "a young man intimately related to me." Faithful to his promise to his mother, Enzo Ferrari initiated formal adoption proceedings following Laura's death and two years later Piero assumed the name Piero Lardi Ferrari. He took on more duties on the racing team as a kind of senior assistant to Piccinini, responsible for driver relations, travel arrangements, racetrack negotiations, and the like.

Piero was also a member of the senior council that included Forghieri, Piccinini, Ermanno Della Casa, Franco Gozzi and Ferrari's personal secretary, Valerio Stradi, plus a handful of technical specialists.

Despite the presence of the reigning World Champion and a driver universally acknowledged to be the most talented in recent decades, the 1980 Formula One season was an unmitigated disaster for Ferrari. The British-developed ground-effects technology had far outdistanced any horsepower advantages the team might have enjoyed, and the best Scheckter and Villeneuve could do during the fourteen-race schedule were three distant fifth-place finishes—mixed with a myriad of mechanical failures.

Worse yet, at Long Beach old Ferrari loyalist Clay Regazzoni crashed his Ensign Formula One into a concrete barrier at over 150 mph and was permanently paralyzed from the waist down. The plucky Swiss refused to lose his good humor, and after years of therapy he eventually learned to drive his hand-controlled Mercedes-Benz road car almost as well as before his accident. It was rumored that when he first visited Maranello in his wheelchair, Ferrari's opening query was: "Can you still make love?"

Scheckter, having spent a decade in the Grand Prix trenches, was sufficiently depleted to announce at mid-season that he would quit following the final race at Watkins Glen. His parting with Ferrari was to be amicable—a rarity among the Scuderia's World Champions. (Ascari, Fangio, Hill, Surtees and Lauda had all left in the midst of acrimony; only Hawthorn and Scheckter, both of whom had retired, departed peacefully.)

With Scheckter gone, Villeneuve was joined in 1981 by Frenchman Didier Pironi, a round-faced young man from a wealthy family with a reputation for arrogance. (He denied this charge, explaining that he was only shy, but this did little to improve his reputation among the motoring press or his racing colleagues.) Forghieri and the design team had produced an excellent 1.5-liter turbo motor, and newcomer Harvey Postlethwaite—who arrived in Maranello during the summer of 1981—came up with a chassis that would lead Ferrari out of the dark ages of all-metal monocoques and into the new land of plastic composites. A spare, crisp, formally trained engineer, Postle-thwaite had served with the March, Wolf, Fittipaldi and Hesketh teams as a designer of suspensions, aerodynamics and carbon-fiber chassis. His presence was intended to free up Forghieri to perfect the new 126C turbo motor while making the team competitive in the new science of advanced road holding and braking. Postlethwaite would be the first of a number of "Anglos" to enter the otherwise all-Italian precincts of the design department. He was quickly to discover that business at Ferrari was conducted in a unique fashion. He found a large, competent staff equipped with excellent computers. More than two hundred full-time specialists were assigned to the racing section, all veterans of the Formula One wars and—on the surface at least—devoted to the leader they simply called the Old Man. But Postlethwaite was disappointed to learn that there were no plans to build a wind tunnel or to lengthen the Fiorano test track.

"I quickly found out that business was done through the back door, through leaks from informants and from the masses of cronies that surrounded the Old Man. There was almost no debate or criticism aired during formal meetings," he recalls. "Ferrari had no interest in, much less control over, the passenger-car side of the business. In fact, he had total disdain for the people who bought the road cars. He called them *fools*, although they were in a sense supporting the entire racing operation. I was paid by check from Fiat, not Ferrari."

The Englishman found himself treading on untrammeled turf within the engineering department. "There was no interest on Ferrari's part in chassis, aerodynamics or brakes. He lived in the past, totally distracted by big horsepower to the exclusion of all else," he says. After BMW had early success with a turbocharged version of its 4-cylinder Formula One engine, Ferrari insisted that Forghieri begin development of such a model. "He heard that the BMW engineers were getting 1,100 horsepower on the dynamometer, compared with our 800 or so horsepower from the V6, and insisted that we begin work. He was told that it was not the way to go, that BMW's success was based on an advanced fuel-injection metering system, but he'd have none of it."

This immediately put Postlethwaite and Forghieri at loggerheads. Word was filtering through the sport that the FISA was considering reducing the capacity for turbo engines from 1,500 cc to 1,200 cc, and in this smaller configuration the 4-cylinder seemed to make sense. Therefore work was commenced on the 154C—a project that was to blight the excellent record of Mauro Forghieri and ultimately lead to his departure from the firm he had served with such loyalty and enthusiasm. Forghieri believed that the 4-cylinder would work, especially if the turbo formula was put into effect. Postlethwaite and a small group of friends and associates, which included Piero Lardi, were convinced the 4-cylinder was a technical dry hole. To open the rift further, Forghieri had little regard for Lardi, whom he described privately as a "nice guy, but an idiot." Ferrari was openly critical of Forghieri, complaining that his prominence was beginning to rival his own. Forghieri charged that Ferrari had been "dead for ten years" in a technical sense and that he had reached a point in his life where he was living in the past, with perfect recall of incidents from 1925, but little from 1975. Still, Ferrari's memory was prodigious. On numerous occasions a high-level conference of the technical staff would be interrupted by the boss recalling that a certain suggested design had already been tried years—even decades—before. An assistant would then be sent into the vast factory archives, to appear later with detailed engineering drawings from the hand of Lampredi, Colombo or Rocchi to verify the boss's memory.

While the new 126C turbo V6 appeared to hold great promise, Forghieri was rapidly losing steam. He had labored with sometimes maniacal zeal for the team since 1960 and now, at forty-five years of age, the endless intrigues and power struggles within the operation were beginning to grind him down. He would struggle onward, but his urge to labor endlessly for the glory of the Prancing Horse of Maranello was rapidly disappearing. In private he was heard to grumpily describe his boss of twenty years as "a good businessman, but as a man, a zero."

The new team of Villeneuve and Pironi seemed to mesh its gears with reasonable ease, although the two men operated at opposite ends of the social spectrum. Villeneuve was strictly a hamburger-and-french-fries type, open, uncomplicated and unaffected by his new celebrity. Pironi, on the other hand, was aloof, distant and faintly mystical in his imperious approach to driving. There was no question that Villeneuve was the faster of the two, although Pironi was never far behind in qualifying. Despite the failure of the new engine to withstand the relentless Villeneuve driving style, he did manage victories at Monaco and the Spanish Grand Prix at Jamara, while crashing in England, Austria and Holland. His impetuosity remained a source of criticism. Many complained that his lack of discipline would prevent him from ever being a consistent winner. But Ferrari defended him with increasing ardor, repeatedly comparing him with one of his all-time favorites, Guy Moll. He called them both *uno spudorato* (a shamelessly bold individual).

The 1982 season opened with the brash Villeneuve and the calculating Pironi mounted in new Postlethwaite-designed 126C2s built from bonded honeycomb fiber—manufactured in Belgium because Ferrari lacked the modern resin-bonding capability needed to make such sophisticated chassis tubs. The Englishman had wanted to use an even more advanced carbon-fiber composite, but decided that the Ferrari facilities (and mind-set) were so outdated that such a leap would be technologically unfeasible. He therefore opted for a more conservative approach with the idea of moving into purely contemporary realms once the rest of the design team got up to speed. In the meantime Forghieri had done extensive work on improving the throttle response on the turbo V6 power plant and it was expected that the engines would equal, if not outperform, the powerful but fickle Renaults that were leading the turbocharged revolution.

Ferrari had returned to Goodyear after three years of middling results with Michelin and its radical new radial racing tires. Leo Mehl, the giant Akron firm's racing director, had known Ferrari since Goodyear's earlier involvement with the team in 1974–79. He recalls the relationship with great warmth. "Ferrari was always very loyal. He was our most dependable member of the Grand Prix community and, at least in my experience, never fit his image. He was, quite unlike several other teams using our tires, never particularly difficult to please. I recall one race when our new radial tires were a total disaster. The Ferraris started on the front row, then, because of our tires, fell back and finished out of contention. I was expecting trouble when I ran into him at Maranello a few days later. But he smiled and said, 'I understand you had a difficult day.' I agreed, rather sheepishly. Then he said, 'Look at it this way: So far this season your tires have been equal to the competition in two races, superior in six and inferior in two. At the same time my cars have been equal in four, superior in two and inferior in six. So you are better than us.'

"To be sure, Ferrari cost Goodyear more than the other teams, but there was vastly more value."

The team encountered mechanical problems at the South African Grand Prix. Then Pironi had a serious accident during a test session at Paul Ricard that sent his car hurtling into a spectator area—which fortunately was vacant. He escaped with a bruised knee. The Brazilian Grand Prix saw Villeneuve hold the lead until he overcooked it in a corner while dueling with Brazilian Nelson Piquet and crashed. Pironi, still in pain, straggled home at the back of the pack. The Frenchman crashed again at Long Beach, while Villeneuve soldiered home to third. It was surely irritating for Enzo Ferrari, watching the race back in Italy, that the winner was none other than Niki Lauda, who had returned to competition with the British McLaren team and seemed very much back in his old form.

The FISA-FOCA battle continued to escalate. The British dissidents, without the new turbochargers, were trying all manner of tricks to stay competitive with their now-underpowered Cosworths. Brakes were being cooled with water, but it was a ruse to race below the minimum weight. The cars—Williamses, Brabhams, et al.—were weighed in at the start and with their water reservoirs full, they were legal. But as soon as the green flag dropped, the water was jettisoned and the cars became lighter and faster. The Brazilian Grand Prix winner, Nelson Piquet, and the second-place finisher, Keke Rosberg, were disqualified by FISA following a series of protests over the bogus brakes. This led to a ten-team FOCA boycott of the upcoming San

Marino Grand Prix set for Imola. The battle lines between what Ferrari called the *grandi costruttori* (grand constructors) and the *assemblatori* (the British "kit car" makers, or the old *garagista* gang) were sharply drawn as the race weekend began.

Enzo Ferrari had long made a habit of driving up the autostrada for test sessions and race qualifying, both at Monza and at the Imola circuit, now named in memory of his long-departed son, Dino. It is believed that this was the last year that he made the journey to the Imola circuit and thereafter seldom, if ever, traveled beyond the immediate environs of Modena.

With only the FISA loyalists on hand, and the pesky British sulking on their tight little island, the *tifosi*, well-oiled with the local *frizzante* wine, could savor the first chance for a Ferrari victory that season. Gilles Villeneuve was already a local hero, having survived an awesome crash two years earlier at the Tosa hairpin—an impact so severe that it tore his Ferrari apart and left the wreckage, containing its uninjured driver, smoking in the middle of the track.

The two turbo Renaults—fast but frail as always—started on the front row, but the flashy yellow-and-black machines of René Arnoux and Alain Prost failed in the race and left the Ferraris running one-two. While the fans rioted with joy, Villeneuve led Pironi by a few car lengths. In situations such as this, it was an unwritten law that teammates must hold position, the logic being that a duel would endanger both cars. As the race entered the closing laps, Pironi made several feints into the lead, but Villeneuve responded and retook the position, thinking that the Frenchman was merely playing to the crowd in what had otherwise been a dull parade. But as the two entered the final lap, Pironi once more drew up alongside Villeneuve. As they sailed into the Tosa, Pironi outbraked him and muscled his way into first. This time he refused to give way and charged through the twisty back section of the track, where passing was all but impossible, to take the checkered flag.

Villeneuve was furious. He stood in sullen, shocked silence on the victory podium and refused to speak to his teammate. He complained to Forghieri and others that he had simply let Pironi through, thinking that he would return to second place, as was the accepted practice. Piccinini, ever the smarmy diplomat, told the press that the team had no specific orders in situations such as that, and that Pironi was perfectly justified in passing Villeneuve. This infuriated Forghieri, who knew otherwise, and led him to hold the Roman *direttore sportivo* indirectly responsible for the tragedy that was to unfold thirteen days later.

Pironi was defensive, explaining to anyone willing to listen that he believed it was anyone's race and that he merely took advantage of the situation to pass his teammate. This was arrant nonsense, and numerous members of the Scuderia, including Piero Lardi and Forghieri, insisted that standing orders existed by which teammates were to hold position in the closing laps. They cited both Villeneuve's and Peter Collins's behavior in similar situations which assured wins for Scheckter and Juan Manuel Fangio. Be that as it may, the relationship between the two drivers was shattered, and the breakdown led to a series of events that, symbolically at least, triggered the unraveling of the entire Formula One operation for the rest of Enzo Ferrari's days.

During the two weeks between Imola and the Belgian Grand Prix at Zolder, the Villeneuve-Pironi feud percolated within the Ferrari operation. The Canadian simmered with anger, while the Frenchman was unrepentant. During practice around the tricky, undulating circuit laid out in the sandy, pine-covered hills north of Liège, each man was grimly committed to be quicker than the other. As the final qualifying session to determine grid positions wound down, Pironi recorded a lap one-tenth of a second quicker than Villeneuve's best. Vowing not to let his rival outqualify him, Villeneuve ripped out of the pits for one final kamikaze run. On the back part of the circuit he overtook the German Jochen Mass as the two hurtled into a corner. Impetuous as always, this "shamelessly bold" young man tried to nip by the slower driver. The Ferrari's left front wheel rode over the right rear of Mass's March 821, sending the red machine cartwheeling off the track. The Ferrari landed nose first in the sandy verge, killing Villeneuve instantly. A series of violent secondary impacts then tore the body of the car apart and flung the dead driver to the ground. Shock waves radiated through the world of motorsports. Surely the death of the most talented, naturally charismatic and popular driver of his era, Gilles Villeneuve, rocked even the ironclad Old Man in Maranello. While romanticists have described Ferrari as loving Gilles Ville- neuve, that is no doubt an exaggeration. He had seen too many eager young men die at the wheel of racing cars to let himself be seduced by their charm. He understood all too well the "hallucinating fragility of existence," as he had put it, to become emotionally entwined with anyone who risked his life at the wheel of one of his automobiles. He did state publicly on several occasions that he "loved" Villeneuve, but this was classic hyperbole, and several associates, including Forghieri, maintain that he evidenced little private remorse over the loss of his star, other than to fret whether or not Pironi could fill his shoes and win the World Championship.

After public wailing had died down, and the criticism of Postle- thwaite's chassis—which had shattered in the crash—had been dis- credited, Pironi did in fact seem headed for the title. His car was withdrawn from the Zolder race as a memorial to his teammate, but he returned to finish a strong second at Monaco two weeks later. Pironi then took third at Detroit before moving on to Montreal, where the Île Notre-Dame track had been renamed in honor of the local hero. Ferrari was now seeking a second driver to fill Villeneuve's place. But until a proper candidate could be found, Pironi was left to uphold the team's honor alone. This led to a series of disasters that would also end the Frenchman's career. At Montreal his engine stalled on the grid as the green flag fell and his Ferrari was rammed from behind by eager newcomer Riccardo Paletti. Though the colli- sion killed the luckless Italian, Pironi was unhurt and he reentered the race with a backup machine. He drove the car—untuned and untested—to a ninth-place finish. Pironi was involved in yet another crash a week later during a test session at the Paul Ricard circuit in the south of France. Again he escaped unhurt after a suspension wishbone fractured and the ensuing impact destroyed the car.

At this point the expert French driver Patrick Tambay was signed to take Villeneuve's place. A gentleman with impeccable manners both on and off the track, Tambay was believed to be the perfect team player, unlikely to threaten the self-centered Pironi, who now had a clear shot at the World Championship. Moreover, Tambay was a capable test driver who was willing to spend endless hours circling vacant tracks like Fiorano or Paul Ricard while experimenting with new engine and chassis setups. His selection caused an uproar among the jingoists in the Italian press. Why, they yelped, had not an Italian like Michele Alboreto, Riccardo Patrese or Elio De Angelis been selected, rather than yet another Frenchman? Ferrari, always easily nettled by the press, countered that Tambay was the best man available and he would brook no more criticism on the issue. His underlying point was clear and the Italian journalists understood it all too well: The cars— the crimson bolides of the Scuderia—were the transporters of the national honor, not the transient men who sat in their cockpits.

The new combination seemed to work well. Unthreatened by his colleague, Pironi won the Dutch Grand Prix while Tambay labored home with a balky engine to take eighth. The pair ran second and third at the British Grand Prix, then dropped back to third and fourth at Paul Ricard after a thorough trouncing by the Renault turbos driven by the brilliant Alain Prost and René Arnoux. By now Pironi was in a dominant position to win the title, but the sniping of the Italian press increased. His popularity had plummeted after Villeneuve's death and now he was being flayed for not driving hard enough, for simply seeking high finishes to pad his point total without charging for the checkered flag. A prideful man who tended to internalize such assaults on his honor, Pironi arrived at Hockenheim for the German Grand Prix determined to prove that he was as quick and courageous as anyone on the track.

The day before the race all of southern Germany was pelted with rain. The big, ominous, forest-shrouded track was puddled with water and visibility was limited. Still Pironi insisted on driving flat out, even though he had secured the pole position the day before. As he rocketed down one of Hockenheim's long straights, Prost's Renault loomed up in the gloom, and before Pironi could veer out of the way, he slammed into its tail. The accident was strangely reminiscent of the Villeneuve wreck, although this time Pironi's car tumbled to earth tail-first, which perhaps saved his life. The Ferrari finally lurched to a stop with its nose torn away and its driver suffering from two horribly broken legs.

Pironi was hauled off to the hospital, where he would begin a long, arduous recuperation. He would never race Formula One cars again, although he was to try his hand at offshore powerboats. (He would die off the Isle of Wight in one of the monster craft in 1987.) When Ferrari was notified of the Hockenheim wreck, he reportedly said nothing except *"Adieu, mondiale"* ("Goodbye, championship").

He hardly realized the dimension of his remark, for the loss of Pironi marked the last moment in his lifetime that a Ferrari driver would seriously contest for the World Championship. He would live to see his cars compete in seventy-eight more Grand Prix races, but only five victories would be recorded, complemented by a baker's dozen second places. Other teams, specifically one of the hated *garagistas* from England—McLaren—would dominate the proceedings like no one since the blitzkrieg days of Mercedes-Benz. Patrick Tambay carried the Ferrari colors for the remainder of 1982, joined for the final two races by American hero Mario Andretti. The 1983 team was composed of Tambay and yet another Frenchman, feisty René Arnoux. This pair won four races during what was a hopeful season for the team. Tambay's emotional win at Imola's San Marino Grand Prix—where a year earlier the Villeneuve-Pironi drama had begun—was a particularly emotional moment for the *tifosi*. Arnoux, whose career would be tainted with inconsistency, enjoyed a hot streak by winning in Canada, Germany and Holland. This burst of success doomed the courtly Tambay, who was replaced in 1984 by the popular Italian star Michele Alboreto—a selection that was cheered by the nation's press and the team's millions of followers. Unfortunately, the dismissal of Tambay left Ferrari with two untalented test drivers at a time when literally thousands of miles of development running were necessary for each lap of formal competition. Tambay's forte was development work, whereas Arnoux and Alboreto were unskilled at and diffident about such drudgery.

Many point to the Tambay firing as an indication of Ferrari's failing judgment. He became increasingly isolated from the day-to-day workings of the operation and depended more and more on the eyes and ears of his agent, Piccinini. Still he remained the centerpiece of the team, the final arbiter of every detail of policy and a mystical figurehead to the outside world. Each day his entrance into the factory would be witnessed by the faithful who had traveled to Maranello to catch a glimpse of the great man. Doctors, politicians and self-assured tycoons stood on the curbs of the Abetone road like Canterbury pilgrims, palms sweating, eyes bulging with anticipation, in the hope of sighting a man who—to the naïve and the gullible—had reached the level of a quasi-deity. He would pass by, seated beside his driver, and reward them with a regal wave, his face unsmiling and inert behind his opaque sunglasses. For the supplicants, this was a gift from the heavens.

Inside the humming factory, the political infighting maintained its manic pace. Piccinini was the envoy to the FOCA-FISA wars, which had reached such a destructive state that both sides were seeking a truce. The Monsignor performed to perfection, oozing into meetings across the continent dressed in his long, double-breasted blue woolen topcoat even in the heat of summer. He artfully managed to be on both sides of the issue, keeping Ferrari simultaneously aligned with the Ecclestone and Balestre sides. "He was extraordinary," recalls one who was deeply involved in those struggles. "Piccinini was everyone's friend at the same time. Everybody was taking sides, but only Piccinini and Ferrari were taking *both* sides at the same time."

As usual, the *gestione sportiva* swirled with acrimony. Forghieri openly despised Piccinini and was uncomfortable with Postlethwaite, who clearly threatened his position as the overall leader of the technical department. Piero Lardi Ferrari hovered in the background, surely the man who would have enormous influence should his father suddenly die or give way to what was now clearly declining health. Still, the Ingegnere, the former Commendatore, remained totally in command. He passed his eighty-fifth birthday with more titles, honors, gifts, presentations and simpering homages crossing his desk. His position as the international patriarch of motorsports had long been assured, but such accolades meant nothing in comparison with the declining fortunes of the race cars. There he focused the laserlike intensity of his entire consciousness and there he was still able to exert enormous pressure on his minions to *forza!* in the face of the mounting challenges from England.

The team won but a single race in 1984, Alboreto at Belgium, while Arnoux drove erratically at best. The new star on the Formula One circuit was the TAG McLaren, an English chassis carrying a Porsche-designed engine financed by a mega-rich Saudi. The engine was an example of a sophisticated new technology that produced both power and fuel economy. A new compact between the warring FISA-FOCA factions had mandated that a limit of 220 liters of fuel be consumed during a race, which created the need for exotic fuel-monitoring systems that offered maximum performance with minimum fuel consumption. The fact that a Porsche-designed engine was best able to accomplish this feat was particularly nettlesome to Ferrari. The brash Stuttgart firm had taken the Prancing Horse's place in international endurance competition, and many experts claimed that its high-performance 911 turbos and 928s were superior to the Ferrari road cars in engineering and quality, if not outright performance. Worse yet, tension between Forghieri and Piero Lardi Ferrari had long since reached intolerable levels. The younger man had been elevated to general manager of the team and he had repeatedly butted heads with the emotional, totally committed Forghieri. The engineer resented a man he considered a dilettante. This was exacerbated by Piero's repeated absences while visiting his elegant home in Cortina. At the end of 1984 the engineer was "elevated" to the directorship of a new advanced research office (*ufficio ricerche studi avanzati*), where he was assigned to noodle over exotic future designs while Postle-thwaite—a close friend of Piero's—was left to oversee Formula One engineering. The proud Forghieri considered this an unforgivable slap in the face.

The situation worsened in 1985. Not only were the TAG Porsches faster, but Honda arrived on the scene with the British Williams team. The Japanese company, which had experienced little success during its first foray into Formula One in the mid-1960s, had returned with a vengeance. In a campaign that was said to have cost over $300 million, Honda introduced a V12 that was enormously powerful and significantly more reliable than anything from Maranello. Arnoux lasted one race before being canned in favor of a journeyman Swede named Stefan Johansson. Alboreto managed to win the Canadian and German Grand Prix (run on a shortened, sanitized version of the Nürburgring immediately nicknamed "the Green Party Ring"), then entered a slump that encompassed thirty-four Grand Prix races without a victory.

Alboreto and Johansson labored fruitlessly through all of 1986. Frustrated, Ferrari reached out for help, shocking the sporting world by hiring the much-honored British designer John Barnard. This rather taciturn mechanical engineer had first gained prominence in the racing world when he designed the Chaparral that won the Indianapolis 500 in 1980 (for which he received no public credit), then went on to create the magnificent TAG Porsche-McLarens. There was no question that Barnard was one of the most creative of all the racing engineers and that he could help the struggling Ferrari operation. But Barnard rejected the initial invitations by Piccinini and others, flatly refusing to leave England for Maranello. In an extraordinary move that must have been sanctioned by Enzo Ferrari, Barnard was given carte blanche to set up a Ferrari design studio near his home southwest of London. In a clever play on words the new establishment would be called GTO (Guildford Technical Office). It was widely reported that Barnard would receive $500,000 per year for his labors plus generous operating expenses.

The move was but one of a number of outrages that finally ended the long relationship between Mauro Forghieri and Enzo Ferrari. The man who had created more brilliant designs—and more winners—

than any other head of the engineering section finally left the place he had called his home for over a quarter century. Not yet fifty and still full of energy, Forghieri was quickly retained by Lamborghini Engineering to develop a new engine for that firm, which had been taken over by the Chrysler Corporation and was now eager to enter the Formula One lists. He was to join another old Ferrari hand, Daniele Audetto, in an elaborate operation set up in a former brass-bed factory located in an industrial park on the eastern edge of Modena.

Barnard's selection troubled Postlethwaite as well. He had, in effect, been demoted by the hiring of the Englishman, and once again rival power structures were set up within the operation. Piero Lardi Ferrari had established a solid relationship with Postlethwaite and opposed the hiring of Barnard, as did driver Michele Alboreto, who said that having the chief designer working in faraway England was like having "a doctor doing brain surgery by telephone."

One of Barnard's first acts was to advocate the cancellation of a nascent project to develop an Indianapolis car. Once more Ferrari had turned his eyes toward that great 500-mile race, and plans were on the drawing board for a new chassis and engine designed specifically for that event. But Barnard could see that the Formula One program was in such disarray that any diversion of effort would be a disaster. He prevailed, thereby ending Ferrari's last chance to triumph in the one major race he'd never won. But there was little that John Barnard or anyone else could do against his former teammates. The McLaren operation had been blessed with not only the magnificent Honda engines but the two best drivers in the business, canny Alain Prost and his young, volatile, moody, strangely complex Brazilian cohort, Ayrton Senna. They were simply unbeatable, and try as he might, Barnard's cars could not keep up during the early part of the 1987 season. Alboreto managed a couple of distant third places at San Marino and Monte Carlo, after which the team plunged into a funk over repeated engine, transmission and suspension failures. This led to an attempted palace coup by Piero Lardi Ferrari and Postlethwaite. In June of that year they retained French aerodynamics expert Jean-Claude Migeot to aid in the construction of a new Ferrari Formula

One car that would incorporate Postlethwaite's advanced notions about a composite chassis and a new suspension. Barnard seemed to be hopelessly entangled in plans to perfect an electronically controlled gearbox, and the team was floundering. Led by Piero, the insurrection showed promise until Ferrari himself discovered the plot. An enormous row between father and son ensued. Rumors of blows being struck filtered through the racing shop. Piero was fired from his position as team boss and demoted to a hazy position with little authority. Postlethwaite was tossed out, and Migeot's brief involvement was ended. The rift was to cause a strain between father and son that was never completely healed.

Ferrari was left alone. His love affair with Lina Lardi had long since faded, and he had never been particularly close to Piero's wife or daughter. When the *ferragosto* holidays arrived in August, Ferrari found himself isolated in the big house at 11 Largo Garibaldi. The city of Modena was deserted. His favorite haunts were empty. His few remaining friends had departed for the mountains or the beach. He was heard to plead with his driver, Dino Tagliazucchi, while sobbing openly, "Please don't leave me alone!" The great man who had always opted for respect over love was now paying for that cynical choice.

By 1987 the fight between the FISA and FOCA forces had been settled by a "Grande Concorde." Jean-Marie Balestre was the titular head, but the diminutive Englishman Bernie Ecclestone was the power behind the throne, handling the pompous Frenchman like a rather awkward puppet. Ferrari and Piccinini had played their cards masterfully and remained above the fray, ostensibly friends to all involved but spiritually attached to the old Paris establishment and its new leadership. The turbocharged engines of the technical wizards at Honda and Porsche were now producing over 1,000 horsepower in short bursts, and a movement was afoot to replace them with lower-powered, less expensive, more reliable unblown units. FISA had proposed moving to 3.5-liter engines, un-supercharged, with an interim year to phase out the turbos. Such a transition would be costly, and a number of the small FOCA teams, having heavily invested in turbo technology, were fearful of having to make such a radical change. Again a schism appeared to be in the making. Again the Ferrari forces sat on the fence, smiling in both directions. A meeting was held in Paris to discuss the issue, primarily from the point of view of the smaller constructors. Ferrari representatives were present to support the case. On the same day Balestre and Ecclestone were at the old Fiorano farmhouse meeting with Enzo Ferrari. They reached an agreement that the 3.5-liter formula would be adopted. The patriarch had given his blessing and all was well—and the proceedings in Paris were no more than useless rhetoric.

At least the 1987 season ended on a hopeful note. Johansson had been replaced on the team by a feisty, good-natured German named Gerhard Berger, and he responded with a pair of autumn victories at Suzuka in Japan and at Adelaide in Australia (where Alboreto finished second). The furor over the Barnard-Postlethwaite contretemps seemed to have been favorably resolved, and the Italian press once again rhapsodized over the possibility that the hated Honda-McLarens could be defeated in 1988.

While Enzo Ferrari remained detached from the passenger-car operation on a daily basis, he did not fail to respond to the accolades being heaped on Porsche's amazing 959—a twin-turbocharged, four-wheel-drive, 200-mph road car that would easily outperform the vaunted top-of-the-line Testarossa. This prompted him to counter with the development of the twin-turbo F40 (celebrating the fortieth anniversary of the Scuderia), a winged 200-mph coupe with a gutted interior similar to that of a racing car. The automobile delighted Ferrari, and when prototypes began to run at Fiorano, he exclaimed to a friend, "This car is so fast it'll make you shit your pants."

There is no question that his contribution to the F40 project was significant. This semi-racing car was the basis for Ferrari's reputation as the manufacturer of the fastest road cars on earth (an issue that could be debated by Porsche and Lamborghini loyalists, but the F40 would remain a significant contender in that exclusive league by any measurement). He made one other significant contribution, albeit a negative one, to the passenger-car side of the business. In 1987, the same year the F40 was formally introduced, the Fiat management was on the verge of designing a four-door Ferrari on the theme of the luxurious, automatic-transmission, V12 400i coupe. Ferrari was furious over this plan to emasculate his machinery into a tamer, four-door boulevardier. He protested so vigorously that the entire project was canceled.

Enzo Ferrari celebrated his ninetieth birthday on February 18, 1988. A gigantic party, catered by the Cavallino, had been organized in a decorated section of the newest assembly hall in the Maranello factory. A total of 1,770 guests had been invited. Ferrari was seated at a corner table, joined by Piero, Marco Piccinini, Franco Gozzi, Ermanno Della Casa and the overall general manager of the Ferrari factory, a Fiat man, Gianni Razelli. The sumptuous affair featured local dishes and wine, as was to be expected. The room was carpeted in red, with white-and-red streamers covering the factory girders. The tablecloths were yellow, the color of Ferrari's native city. Twelve enormous cakes were served for dessert, each decorated with the Ferrari emblem. The Old Man was presented with a special cake adorned with the first triangular version of the *Cavallino Rampante* that dated from the early 1930s. A single candle was placed in the center.

This was a special occasion, for Enzo Ferrari was becoming more reclusive and less visible in public. His legs were failing him, and travel had become almost impossible. The big house on 11 Largo Garibaldi was now like a tomb, with Piero, his wife, Floriana, his teenage daughter, Antonella, and his mother, Lina, often absent. While they appeared close in public, the rift between father and son remained and the two were seldom seen together outside business situations.

Still the fortunes of the racing team obsessed him. Berger and Alboreto were kept on for 1988, but the McLaren-Honda MP 4/4 machines continued to operate in a class by themselves. Not only were the engines and chassis still superior to the competition, but the team, lavishly funded by Marlboro cigarettes, still enjoyed the services of Prost and Senna, although the pair were feuding openly. This did not prevent them from dominating the 1988 season as they had done the year before. Ten straight races fell to them between the opener in Brazil in April and the Hungarian Grand Prix on August 7. Berger had managed second at Brazil and Monaco but only thanks to mechanical failures by the Hondas. Otherwise the Ferraris were relegated to docile thirds and fourths behind the red-and-white machines.

An air of futility descended over Maranello. The Old Man was now almost totally infirm. He could not walk without assistance, and it was known that his kidneys were failing. The defeats on the track seemed to deplete him more. Modena, now a boomtown reputed to be one of the richest cities in all of Italy, was less and less focused on the mystique of Ferrari. Other businesses—ceramics, machine tools, foodstuffs and clothing—were rising up to rival the prosperity of the famed Maranello operation. The old Abetone road, where the red cars once ripped toward Formigine at 180 mph on misty mornings, was now lined with new factories and clogged with truck traffic. The town had changed from a dusty corner of Italy into a bustling center of commerce. New high-rise apartment developments had gone up around the old autodrome, which was now being converted to a park. Only the control tower remained among the bulldozed rubble. The revered birthplace of it all, the original Scuderia Ferrari garage on the Viale Trento e Trieste, was being torn down, to be replaced by an atrocious concrete ramp garage that also housed the new downtown offices of the racing team. Historians and car enthusiasts alike denounced the destruction of the old building, claiming that it was a landmark of the Italian automobile industry, but the demolition proceeded. Many blamed it on an insensitive Piero, but it was Enzo himself who had sanctioned it.

In Maranello, construction of a Ferrari museum, a project long delayed by indecision, was beginning to show progress. A plot had been cleared on the new Via Dino Ferrari, but no real construction had started and only a few cars had been collected. A replica of the first Tipo 125 had been built, and Valerio Stradi was contacting the Ferrari clubs around the world (478 of them!) in hopes of obtaining more early examples. But the market in used Ferraris was zooming like a runaway gold rush. GTOs were being bought for as much as $5 million apiece, with speculation rife that the price would triple following the patriarch's death. Other excellent examples from the 1950s and 1960s were commonly trading for $1 million or more, which made the acquisition of museum material now staggeringly expensive. Worse yet, there were almost no old race cars; dozens of them had been tossed into sheds behind the factory to rust or to be cut up into pieces. Ferrari had contracted with a Modenese construction firm to build the museum, but had demanded the return of his $250,000 deposit when months passed and no building was erected. The contractor said that he was prefabricating the building at another location, hence the apparent lack of progress at the site. The dispute further delayed the development of a museum, which most enthusiasts agreed ought to have been created thirty years earlier.

Ferrari ceased to see any visitors, other than close business associates who had now taken over most of the day-to-day operation of the team and the factory: his ever-loyal *consigliere*, Franco Gozzi; Fiat men Cesare Romiti, Piero Fusaro and Gianni Razelli; Piero, who was now being groomed to take over the operation of the passenger-car factory; the new team manager, Cesare Fiorio, who had handled Fiat's rally operations; Luca di Montezemolo, who had remained close to the Ferrari team; and the omnipresent Marco Piccinini. The Old Man now spent more and more time in bed as spring gave way to the steamy, oppressive days of summer.

He was forced to miss the visit of the one man in Italy who was better known than he. Pope John Paul II toured the Po Valley in June 1988 and stopped at the factory. The visit had been arranged by Don Galasso Andreoli, a Catholic priest from the nearby village of Baggiovara, who was not only a close friend of Ferrari's but pastor of the metalworkers' union of Modena. Ferrari, who in later years was sarcastically called "the Pope of the North," had numerous friends among the clergy, including "the Flying Priest," Don Sergio Mantovani, who raced a Ferrari himself in the late 1950s and early 1960s (with a rosary clenched in his teeth!). Also much in evidence during the Pope's visit was the pastor of the Maranello parish, Don Erio Belloi, who had since 1983 rung the bells of his church to celebrate every Ferrari victory.

On the bright, clear day John Paul arrived at the factory, Ferrari was bedridden and unable to greet his famed guest. Piero acted as his substitute, giving the Pope a tour of the facilities and taking him on a lap of the Fiorano track in one of the company's glittering new sports coupes. The Pope responded by blessing the race cars before they left for the Canadian Grand Prix (both broke down in the race).

Clearly it was a disappointment that Ferrari himself could not be present, and his absence only underlined the severity of his illness. John Paul did, however, speak to him by telephone. The content of that conversation remains a secret, but it is assumed that it involved Ferrari being welcomed back into the arms of the Catholic Church. It is also assumed that the Pope heard Ferrari's confession. The old Modenese, who had long before lamented that he lacked "the gift of faith," was now repenting in the electronic presence of the father of his church.

Then Luigi Chinetti appeared. He was a regular visitor in Modena, driving down from Paris to see his son, who was building several custom-bodied Ferraris in a small fabricator's shop on the edge of town. He asked to see his old friend one afternoon in early August while he was visiting the Maranello factory on other business. Chinetti was told that Ferrari was too ill to see anyone.

He was about to leave and drive back to the Fini Hotel when word came: he must stay. Ferrari wanted to see him.

Chinetti waited in silence until a small retinue of men spilled out of an office door. In the middle was Ferrari, being supported at the elbows by two strong young men. He clearly could not walk by himself. He displayed the same gray visage Chinetti had recalled from his visit in 1946. His face looked drawn. There was no smile. The eyes remained shielded by the sunglasses that had been part of his persona for twenty years. Luigi Chinetti, himself a man of flinty strength, watched his old friend and adversary approach, defeated as he was by the uselessness of his limbs. Suddenly the urge to forget all that had gone before overcame him and he walked toward him. "May I embrace you?" he asked impulsively. In silence, Enzo Ferrari opened his arms and the two men hugged, groping perhaps for the life that was slipping away from them. There was a brief display of tears, quickly daubed, and then business began. They discussed an old Ferrari Formula Two car.

Chinetti agreed to lend the automobile to the proposed museum. He was well aware of the feeding frenzy of Ferrari collectors around the world that had made the car in question worth perhaps $1 million or more. But for his old friend, he proposed a deal: $40,000 would have to be donated to four charities, including the poor children of Maranello—a town, like Modena, dominated by the Communists and not known for private charity. Ferrari balked. He would pay no more than $30,000. Not a penny more. Chinetti could not help but be amused. Here was this man, ninety years old, mortally wounded, as it were, by the passage of time, yet still able—yes, eager—to haggle over a few dollars. The old rascal, worth at least $40 million, Chinetti reckoned, intended to go to his reward with his last lira. The deal. Always the deal. It was classic Ferrari, and Chinetti for one last time joined in the shopkeeper's bargaining that so enlivened Italian life. Finally an agreement was reached and the two men parted, perhaps closer than they had been in years.

The end came quietly at 11 Largo Garibaldi in the early morning of Sunday, August 14, 1988. At his bedside, said the unconfirmed reports, were Piero and Floriana. Not noted was the fact that he had received the last rites of the Catholic Church.

Don Galasso Andreoli gave him extreme unction and presided over the small private funeral attended only by immediate members of the family. Though the world had anticipated a massive outpouring of grief and a gigantic funeral to mark his passage, there was nothing of the sort. Whether it was according to his own wishes, or deemed appropriate by his son, Enzo Ferrari's death was not announced until he had already been interred beside his father in the San Cataldo crypt.

(Thirty days following his death, a funeral mass was celebrated in

Modena's immense thirteenth-century Duomo. The huge turnout included many dignitaries, including Fiat scion Gianni Agnelli.)

It was over, a strangely disappointing anticlimax to a life that had spanned most of the history of motorsports. His health had been in severe decline for months and the end had been expected. While his mind remained quick and facile to the last, the powerful body had long since crumbled under the sheer weight of years.

Editorial writers around the world groped for words to express what Enzo Ferrari had meant. Many tried to describe him as an automotive pioneer, which he was not; others called him a great racing driver and engineer, which he was not. He was, however, exactly what he had repeatedly said he was: an agitator of men. And he remained true to his credo to the day he died.

If there was one essential quality about the man it was his ironbound tenacity, his fierce devotion to the single cause of winning automobile races with cars bearing his name. From 1930 onward, for nearly sixty years, hardly a day passed when this thought was not foremost in his mind. Win or lose, he unfailingly answered the bell. In that sense his devotion to his own self-described mission was without precedent, at least within the world of motorsports. For that alone he towered over his peers.

A mere two days after his passing, Fiat S.p.A. officially announced that it would exercise its option to obtain the remaining 40 percent of the outstanding shares of Ferrari. Piero Lardi Ferrari would retain the other 10 percent. Production would be increased, business would be stabilized and the last vestiges of the old handcrafted ways would disappear. Even the vaunted engines, the heart and soul of the old Scuderia, would be bastardized. Hereafter some of the passenger-car power plants would be assembled in Bologna by Ducati, the motorcycle manufacturer, and trucked to Maranello. Ferrari was now just another automobile factory.

Three weeks after the death of Enzo Ferrari the Formula One teams returned to Monza for the Italian Grand Prix. The McLaren-Hondas were still unbeatable, and the Ferraris of Berger and Alboreto were coming off a Belgian Grand Prix at the redesigned Spa circuit, where both of their cars had once again broken down in the race. Even the fanatic legions of Ferrari rooters at Monza were resigned to defeat when the race began. But Prost fell out with mechanical troubles, letting Senna take over a wide lead. Berger and Alboreto, always the bridesmaids, dragged along, doomed to second and third. Then the impulsive Brazilian tried a foolish pass on French driver Jean-Louis Schlesser and touched wheels. He spun off the track. An act of providence! Berger took the lead and rushed on to victory, with Alboreto second! A stirring, inspired, miraculous Ferrari win on the hallowed home ground. Berger was mobbed at the finish line by a tumbling sea of red, yellow and black flags and bellowing *tifosi*. A banner was unfurled which read: "Ferrari, we followed you in life and now in death." Marco Piccinini emotionally hailed a new era for Ferrari. For that brief, heady moment at Monza, it appeared perfectly reasonable to the bubbling mobs that they had in fact witnessed a resurrection. And for the Honda team, which had been unthreatened and undefeated so far that season, the Ferrari victory no doubt was worrisome — although their total domination of the sport was to continue for two more years. Piccinini's prediction of a new era might have been correct, but it would be a hollow representation of the old.

Enzo Ferrari, the last of the great automotive titans, was gone, never to be replaced.

When Enzo Ferrari died on August 14, 1988, the automotive world bowed its head and held its breath. What would become of the company? Would there be growth or stagnation? Enzo's strength and personality were legendary and his loss created a vacuum. Who would take over and lead? At the time of his death, the passenger car division was on the cusp of bankruptcy, and the racing team hadn't won a Constructors' Championship (formerly the Manufacturers Championship) since 1983.

Immediately following Enzo's death, as the company faced hemorrhaging losses that threatened to cripple all aspects of Ferrari, Piero Ferrari was installed as vice chairman. Initially, continuity was the key as Piero, the heir apparent, stepped into the breach, bringing familiarity and skill to a position not many men could, or would, embrace. Fiat also stepped into the void, using its optioning power to expand its control of the company, taking on a 90 percent stake while leaving Piero the remaining 10 percent, which effectively assuaged any fears by adding the financial backing Ferrari needed to move forward.

While this collaboration between Piero and Fiat worked in the short term, the company was floundering, and unless major changes took place, the passenger car division would most likely fail, leading to the demise of the racing division as well.

In 1991 Gianni Agnelli, realizing the future of Ferrari was at stake, brought back Luca Cordero di Montezemolo, the protégé of both Enzo and Agnelli, to lead Ferrari. Montezemolo had begun his career as Enzo's assistant in 1973 and became head of the Formula One division in 1974. His canny ability to spot talent and enact change within the division created a renaissance period and breathed new life and excitement into a racing team that had been struggling under Enzo's strict authority and myopic unwillingness to innovate.

Agnelli's move was genius. Montezemolo was essentially Enzo 2.0. While stylistically different and certainly more charismatic, Luca had learned at the feet of Ferrari and had the same iron will and stern business philosophy for the company: Ferrari would flourish and win, no matter what. His promotion set into motion an amazing turn-around, based on Enzo's core values, that would leave Ferrari debt free, dominant once again in Formula One, and poised for future growth on a level Enzo had always hoped for but failed to attain in the last few decades of his life.

Leading Ferrari into its next chapter was not a quick or easy proposition. It had taken decades for the company to reach the breaking point, and it would take years for institutionalized changes to pull it back from the brink. Montezemolo's mission was to rebuild and re-imagine the business model from the ground up.

In 1991 the automotive world was still digesting the impact of Enzo's death on Ferrari as a whole. As Enzo's successor, Montezemolo's strong management style and familiar strength of character added continuity, but he also brought a strong financial plan and business model which ultimately moved the company into an explosive period of growth and earnings never before seen in its history. Enzo's legacy was thus reborn under the tenacious and brilliant mind of Montezemolo. This was never more apparent than in 1999.

After almost a decade Montezemolo had enhanced the brand of Ferrari. He understood that the company had to walk the knife's edge of keeping its exclusivity while allowing growth in the passenger car division. Without increased production, profits would stagnate, but increased manufacturing ran the risk of devaluing the brand. In the past, Enzo's relative apathy and disdain for his road cars, and the fact that American buyers were anathema to Ferrari, had created a toxic environment within the company that disavowed anything other than the racing team. This short-sighted philosophy failed to recognize that without the capital from the road cars, the racing team could not sustain itself financially. Montezemolo saw this right away. He understood the value of the brand, reveled in the interest in the road cars and realized the need to expand production while keeping exclusivity. This being the case, Ferrari had a new tenet for its business model: the passenger car division and the racing division were symbiotic and as such would hold equal importance within the company. While this made infinite sense theoretically, the reality was very different.

THE PROTÉGÉ'S RACING PHILOSOPHY

When Montezemolo took over the company, he had two mandates: expand production of the passenger cars, thus ensuring profitability, and create a winning racing team from manufacturing to paddock to podium. In the grand scheme of things, the first mandate was relatively simple: increase production and keep the exclusivity of the cars within the confines of the economically elite while growing the luxury brand of Ferrari to a larger portion of the global market. Essentially, if you could not afford the car, you could still embrace the brand, immersing yourself in the ethos, luxury and mystique of Ferrari. Consumers could purchase a Ferrari keychain, watch or jacket, ensuring that the prancing horse emblem became more recognizable and more coveted, thus guaranteeing a larger piece of the economic pie for the company, especially in a growing global market. Under Montezemolo's shrewd guidance, profits soared. While the first mandate was implemented and achieved, the second mandate proved a much trickier proposition.

Racing teams encompass a diverse mix of skill sets and personalities combined to achieve one goal: to win. It is a predominantly male-dominated industry where testosterone, talent, genius and ego merge and chemistry is key.

Many enthusiasts either do not know or fail to remember that there is a reason for the use of the word "formula" in Formula One racing. Unlike NASCAR, with its "run what you brung" origins in bootlegging, street car racing and stock cars, Formula One has always been about extreme engineering and "open-wheel" racing at its finest based on strict parameters each team adheres to. This formula to run Grand Prix races is governed by the FIA (Fédération Internationale de l'Automobile) and includes both sporting and technical regulations. The technical regulations deal with the car's engine, transmission and suspension, outlining the rules for size, weight, cubic inches, horsepower, aerodynamics, ground effects, fuel mixture and a multitude of other aspects pertaining to the form and function of each vehicle. The sporting regulations, on the other hand, which also include rules issued under the Concorde Agreement, define how each race is run—from start to finish, the race itself, the media and the money. This being the case, a team able to win consistently has essentially captured lightning in a bottle. They have checked every regulatory box and not only mastered and engineered the finest vehicle but also have the best driver, crew and management possible, creating a cohesive unit which dominates the podium.

Understanding the balancing act needed to achieve wins, Montezemolo's second mandate in running Ferrari was therefore based on engineering a great car while at the same time finding the appropriate individuals with the skills, determination and ability to check their egos and create a chemistry strong enough to make them winners. It took eight years, but Montezemolo accomplished his goal, launching Ferrari into the annals as the winningest team in history.

1999 was a watershed year. Ferrari was in the black thanks to increased automobile production and the hefty fees it was earning for licensing Ferrari's brand to other manufacturers of luxury goods, but the racing team was still failing on every level and was unable to reach the podium. Losses notwithstanding, Montezemolo's institutional and managerial changes were beginning to bear fruit.

When Jean Todt was hired as team director in 1993, it set the stage for positive changes within the racing team. Through perseverance and the astute ability to surround himself with extremely talented individuals, Todt wound up assembling a dream team of engineers, including Ross Brawn, and Ferrari was finally able to hire Michael Schumacher away from Benetton. Under Montezemolo's leadership, Todt, Brawn and Schumacher convinced Rory Byrne to join their ranks as chief designer, filling the final position necessary to ensure victory. This diverse yet cohesive group of men had a win-at-all-costs philosophy, controlling every aspect of Ferrari's racing team, which led to an unprecedented six consecutive Constructors' Championships and five Drivers' Championships. Ferrari was seemingly unbeatable from 1999 to 2004.

Between Ferrari's expanding profitability and its racing dominance, Montezemolo's vision and direction seemed unimpeachable, but as most insiders know, you are only as good as your last win or season, and unfortunately 2005 would mark the beginning of the end of Ferrari's supremacy in Formula One.

FERRARI STAGNATES

In 2005 Formula One instituted a new rule that, in part, though obviously never stated, was meant to curtail Ferrari's domination of the sport. After almost six years of consecutive wins for Ferrari, races had lost their excitement and spectators were growing restless, because it was a foregone conclusion that Ferrari would achieve a podium finish. Under the guise of a simple technical change, Formula One banned tire changes during races, creating a disaster for the partnership between Bridgestone and Ferrari. Since Bridgestone tires were not full radial construction, their sidewalls were rigid, creating more movement in the tread. This caused treads to run hotter during races, incurring more wear and tear, diminishing their efficacy and forcing multiple tire changes during a single race. Since Ferrari relied heavily on Bridgestone tires, this new rule gave a serious advantage to those teams using flexi-sidewall full radial Michelins, which were better able to sustain viability throughout an entire race. In light of this rule change, Ferrari, with Schumacher at the wheel, won only the U.S. Grand Prix that year (and, to be honest, that win happened because many of the top teams pulled out of the race due to a safety issue with Michelin that made running the race more dangerous than usual). In light of this situation, FIA ratified its stance on tires in 2007, forcing all teams to use the same manufacturer, thus ensuring continuity on the track.

While 2005 marked the end of Ferrari's consistent wins, 2006 marked the end of an era with the disbanding of Ferrari's dream team. Jean Todt was promoted to CEO of Scuderia Ferrari, Brawn left on a supposed sabbatical, Schumacher retired as a driver and Byrne shifted into an advisory capacity, vacating his role as an active member of Ferrari's team. This dissolution changed the team dynamic, and no matter how seamlessly it tried to ease the transition, Ferrari has not been the same since. Unbelievably, the company was unable or unwilling to recognize the spiraling trajectory of its once-unstoppable team, and no one stepped in to initiate the changes needed to stop Ferrari's continued fall from grace.

Although Montezemolo was always an engaged leader, during the height of Ferrari's dominance he remained focused on growing the brand of Ferrari and expanding the profits of the passenger car division. He trusted in the continued strength of the racing team and failed to recognize how the loss of his key leaders would ultimately affect its future success.

Montezemolo's error was twofold: he had become accustomed to a hands-off approach to the racing team and he was so focused on road cars that he was slow to recognize disturbing trends in other areas of his company, a miscalculation that would cost the Scuderia dearly. Years earlier, Enzo's inability to recognize the value of the passenger car division had almost bankrupted the company, and now

Montezemolo's inability to realize that the racing division was beginning to crumble under the weight of too much spending, too many employees and not enough talent and innovation led to one of the most disastrous periods in Ferrari's racing history, one it has still not recovered from even more than a decade later.

Those who are unwilling to learn from their mistakes are destined to repeat them, and this was certainly true for Scuderia Ferrari and Montezemolo himself. The problem with an autocracy is that no one is willing to speak truth to power for fear of repercussions. This fear creates an environment where voices are muted, innovation is stunted and it is safer and easier to tout merits and achievements rather than to dissect and analyze poor choices, ineffective leadership or errors in manufacturing.

In this same vein, Montezemolo's inability to improve research and development, as Enzo himself had done, created a legacy of stagnation in which the company's ability to improve cars and win races was determined by its inability or unwillingness to change. Oddly enough, the one change Montezemolo did embrace during this time was a bad one. He jumped on the ecological bandwagon and made ill-advised (though, some would argue, ethically correct) choices by embracing green technology and touting the benefits of hybrid innovation—this in an industry in which ethics rarely held pride of place if they impacted speed, unless it was pivotal to a driver's safety. His insistence on researching and implementing hybrid technology showcased Ferrari's inadequacies in ground effects and innovative aerodynamics, compounding its already tenuous position in Formula One. Ferrari was simply no longer fast enough or good enough to reach the podium.

CHANGING OF THE GUARD

By 2014 the Ferrari racing team had become merely a footnote on race day. The once-unstoppable team was no longer considered a threat in highly contested races, and its history of wins had become a vague memory. Outsiders recognized that Ferrari had reached another tipping point in its history, and the company was once again off-balance.

When Ferrari failed to attain a podium finish in the 2014 Italian Grand Prix, Sergio Marchionne, as head of Fiat, had had enough. Despite Montezemolo's successes in the passenger car division, Marchionne was disgusted with the prospect of Ferrari being defined as an also-ran for future races. Montezemolo was out.

The ink was barely dry on Montezemolo's recently signed three-year contract extension. Marchionne made it clear that no single person was more important than the company itself; since Montezemolo was no longer able to secure the results necessary for a successful racing team, Ferrari needed to be taken in a new direction. That direction led directly to Sergio Marchionne.

When Marchionne had been hired as CEO of Fiat in 2004, many in the tight-knit, incestuous world of the automotive industry had questioned the choice. While his business acumen as the former CEO of SGS, a Swiss company testing toasters and baby toys, was beyond reproach, Marchionne had no prior experience with cars, and insiders wondered if a manager with no working knowledge of the intricacies of the auto industry could truly be effective. While Marchionne was considered a bean counter, naysayers quickly learned that he was a new kind of CEO for a new millennium, with a hands-on management style that shook Fiat to its core and brought it back from the brink of bankruptcy through brilliant financial restructuring. Marchionne returned the company to profitability within two years, easily silencing any doubts which may have lingered.

Marchionne continued his successful strategy at Fiat, and in 2009 Fiat acquired 20 percent of the failing U.S. automaker Chrysler, which had filed for bankruptcy and entered Chapter 11. Marchionne's guidance proved pivotal; within two years of the alliance, Chrysler was profitable once again. Soon after, Marchionne's intention to fully acquire Chrysler became apparent; by 2014, the companies had become Fiat Chrysler, creating the seventh-largest automotive company in the world.

The acquisition of Chrysler heralded a changing philosophy within the companies, and 2014 marked an extraordinary year in Fiat's and Ferrari's remarkable history. Marchionne now had a proven track record of success, and Fiat Chrysler and Ferrari were finally not only profitable but also poised for explosive growth of the global market. Surprisingly, this was also the year Marchionne announced Ferrari's independence from Fiat.

On the surface, the Ferrari IPO was a simple, if financially complex, business deal. Fiat, under Marchionne's leadership, would spin Ferrari off into an autonomous, publicly held company that would succeed or fail on its own merits. To those outsiders who did not understand the complicated relationship between Marchionne, Fiat, the Agnellis, and Ferrari, the move was questionable. Why would Fiat, which was finally profitable, allow Ferrari, the jewel in its crown, to go public as an independent company? In order to understand the complicated dynamic of Ferrari's autonomy, it is necessary to look back a decade to the days following Gianni Agnelli's death in 2003.

At the time, Fiat had been losing two million dollars a day, with no end in sight. Gianni's death was quickly followed by that of his brother Umberto in 2004. John Elkann, Gianni's eldest grandson, was now the undisputed head of the vast Agnelli family, but at twenty-eight years old, his business acumen notwithstanding, he simply did not have the experience to run a company the size of Fiat. Giuseppe Morchio, the CEO of Fiat, was pushing for the dual title of CEO and chairman, but certain factions within the company were questioning his leadership skills and instead wanted Luca di Montezemolo to become chairman of Fiat as well as Ferrari.

A risk assessment was conducted examining the ramifications of a power struggle between Morchio and Montezemolo, and in light of this situation, it was decided to quietly send a representative of Fiat to Switzerland to speak with Sergio Marchionne about becoming CEO of Fiat in the likely event that the Morchio quit. John Elkann was the obvious choice to approach Marchionne, ensuring a seamless transition if necessary.

As a sitting board member of Fiat and the CEO of Exor, an investment company located in Turin, Elkann had an unimpeachable résumé. Essentially, Elkann was the Agnelli family, and by extension, the Agnelli family was Fiat and Exor, with no decision being made without Elkann's express approval.

Elkann's choice of Marchionne was brilliant. As a primary investor in SGS, Elkann understood Marchionne's business philosophy as well as his loyalty and leadership skills. The two men met in secrecy and discussed all possible permutations of Morchio quitting, Montezemolo being hired and Marchionne taking the lead. The meeting was extremely successful, since Morchio did indeed quit, leaving an opening for Marchionne. Elkann, a true tactician, played his hand well and set in motion a changing dynamic for Fiat and, ultimately, for Ferrari as well. The nuance and subtlety of his strategy were incredible, and his ability to see the long game and to strategize so far out was stunning in its complexity and continues to bear fruit to this day.

Flash forward to 2014. Montezemolo "quits" after years of a contentious relationship with Marchionne, unwilling to take ownership of the race team's failings and disgusted with the Americanization of his beloved Ferrari. Marchionne becomes the last man standing in Maranello, heralding the beginnings of a new era of modernity and technological advances finally unshackled from the old guard, as he confidently announces Ferrari's autonomy, outlining the rationale for its IPO to an industry and a global market that are stunned by the rapid changes they are witnessing.

A NEW BUSINESS MODEL

When Marchionne took over Ferrari from Montezemolo, he had a three-part plan of action: (1) expand current production of cars from 7,255 in 2014 to more than 9,000 per year; (2) get the racing team back on the podium; and (3) bolster retail sales of branded goods into the exploding luxury market. It was a period of considerable volatility. Many observers predicted a Ferrari SUV and station wagon, envisioning a future in which Marchionne created an assembly line of affordable cars, delegitimizing and devaluing the exclusivity of its passenger cars for the sake of sales. The internet was abuzz with supposed insider information and gossip, but industry insiders understood that change was necessary. Marchionne's vision would continue to move Ferrari forward without risking its place in the highest echelons.

During the CEO shuffle at Ferrari, Marchionne vehemently insisted that Ferrari could not continue to grow as a manufacturer without the Scuderia on the podium. Losses led to diminished sales, because buyers wanted the cachet of a winning vehicle, but Marchionne was acutely aware that this was an antiquated philosophy. During Enzo's lifetime he was a staunch believer in the theory that Ferrari wins push Ferrari sales. His belief was that the racing team's stamina, history of wins and endurance on tracks were essential to promoting sales and garnering customers, but in today's brand-centric and fickle marketplace, car companies can no longer count on their racing teams to generate interest in their street-legal namesakes. Formula One is certainly the stomping ground for innovation and a showroom for the excellence of Ferrari craftsmanship, but reliance on the European racing circuit as the primary promotional channel for cars is a business model that has failed countless times. The Formula One team may be reliant on the profits derived from the passenger car division, but passenger cars are no longer reliant on the competitiveness of Ferraris on the track. Until this reality is truly confronted the two divisions will continue to be at odds.

THE AUTONOMY OF THE BRAND FERRARI

As a savvy strategist, Marchionne was fully aware that the market had changed. For a while he had two speeches emphasizing completely different points depending on the audience. To the racing community he touted the merits and importance of Ferrari's team to the continued profitability of the company, but to potential investors in Ferrari's IPO, his sales pitch changed. He stopped referencing Ferrari's current wins and losses and focused on the company's brand.

Marchionne understood the modern marketplace, recognizing that brand-centric millennials, with little interest in history, base their purchasing decisions on the opinions of others, especially those on social media, and he was fully aware of the expanding Asian, South American and Middle Eastern markets, where anything associated with Ferrari is desirable. In his approach to the IPO, Marchionne compared Ferrari to Gucci, Hermès and Tiffany, selling investors on the notion of growth through branding and merchandising, and on no longer pigeonholing Ferrari simply as a car manufacturer or a racing team. Ferrari's potential as a luxury brand was, in theory, limitless. Add to the equation the creation of Ferrari theme parks in Abu Dhabi, northeastern Spain and (someday) China and the Ferrari brand has more weight and relevance in today's economy than its racing competition.

Having sold investors on the branding of Ferrari, the IPO, announced in 2014, hit the market in October 2015 with a share price of $52 on the New York Stock Exchange under the symbol RACE. At this price point, Ferrari's market capitalization was approximately $9.8 billion, with the company floating about 17 million shares to the public, which represented nearly 9 percent of the company. This reduced Fiat Chrysler's stake in the company from about 90 to 80 percent and raised nearly a billion dollars in public capital. Marchionne's strategy worked, and while shareholders have certainly reaped the benefits, the Agnellis, Exor and John Elkann made millions while still retaining control of the company.

There have been bumps in the road since Ferrari went public. The stock had a fairly significant slump in the first eighteen months after the public offering based primarily on a 30 percent drop in sales in the Asian market, causing the price to drop to a historical low of $37 a share. (It has since rebounded and risen to over $300 a share.) While no company's market domination and global viability are ever assured,

Ferrari's desirability and exclusivity continue to seduce the buying public, creating interest on a level Enzo could never have foreseen.

HELL FREEZES OVER

Marchionne continued to build on this interest in Ferrari, and although he once assured the public that hell would freeze over before a Ferrari SUV went into production, it looks like the forecast is, at a minimum, calling for jackets and mittens. In the summer of 2017, to the horror of purists, news broke claiming Ferrari would release an "FUV" (Ferrari Utility Vehicle) in 2022. The Ferrari Purosangue was released in September 2022 for just under $400,000, and sold out immediately. Semantics aside, historians can pontificate that the stage had been set for an SUV when Ferrari introduced the four-door concept Pinin in 1980 and that the Ferrari GTC4Lusso sealed its fate, but the writing was on the wall when Porsche introduced the Cayenne in 2003.

In this new reality, in which top-flight automakers began embracing soccer moms, the appearance of the Cayenne helped make it safe for SUVs to enter the once-elite world of high-performance vehicles. Ferrari and Porsche are far from alone in embracing the SUV market: Maserati, Lamborghini, Bentley and Jaguar have all moved in this direction. In the reality of today's marketplace, conceptual or hierarchical purity has no place. Ultimately, profitability pushes production, and SUVs are profitable.

THE OLD IS NEW AGAIN

One of the interesting aspects of Ferrari's continued growth is that, in a world of planned obsolescence, where newer is always better and *vintage* is often a synonym for "worn-out," the classic car market is exploding. In 2013, during the Pebble Beach Concours d'Elegance in Monterey, California, RM Auctions sold a 1967 Ferrari 275 GTB/4*S NART Spyder for $27 million. The energy in that room was unlike anything I had ever encountered. Outside the venue, crowds massed, desperate to view the Ferrari that was about to make history; many jockeyed to purchase the coveted passes that permitted entry onto the auction floor. Inside the packed room, a quiet, almost reverent solemnity marked the passing seconds. The auction lasted mere minutes, but the monetary implications would prove far-reaching for the vintage market. This sale price has proven not to be an anomaly, and auction prices have continued to increase: a Ferrari 250 GTO sold for over $38 million in 2014, and in August 2018 a 1962 Ferrari 250 GTO was sold at auction for $48.4 million, with rumors circulating that another GTO sold in a private sale in June 2018 for upward of $70 million. Whatever financial ceiling had once been in play for vintage Ferraris no longer exists. Today, buyers are willing to spend and spend big, based on Ferrari's history, exclusivity and craftsmanship. Most insiders believe these escalating sales are here to stay and the value for these purchases will hold and grow over time.

FERRARI'S FUTURE

While Ferrari S.p.A. and Scuderia Ferrari have had a meteoric rise in the past three decades, it is paramount to look at the totality of the company's history to get an accurate view of and appreciation for Enzo Ferrari's creation. Under the strict and deft leadership of Enzo Ferrari, Luca di Montezemolo and Sergio Marchionne, with the financial backing of Gianni Agnelli, John Elkann, Fiat and Exor, the company has navigated two world wars, countless political intrigues, dire financial struggles, catastrophic losses on the track and at home, changing market trends, myopic planning and extraordinary innovation—and it is still standing when so many others have failed. Ferrari has proven time and again that it is stronger and more resilient that any one man—as it must be, given the reality it is currently facing.

On July 25, 2018, Sergio Marchionne died of a cardiac arrest due to complications from shoulder surgery. There is speculation that he may have had an underlying medical condition that exacerbated his trauma, but whatever the case may be, Fiat and Ferrari now face a future in flux. Marchionne had already decided to step down from Fiat in April 2019, grooming Michael Manley to assume the role of CEO after his retirement, but many in the industry assumed he would continue to be a guiding, if silent, force within the company. Ferrari, unfortunately, was blindsided by Marchionne's death.

Marchionne's intention was to concentrate on Ferrari, continuing to move the company forward as a powerhouse in the industry, and while his loss was devastating, the company has a track record of sustainability. John Elkann, no longer an inexperienced twenty-eight-year-old but rather a highly respected and canny businessman, fully able to tackle all aspects of Fiat and Ferrari, installed Louis Camilleri as the new CEO of Ferrari. Camilleri's background as the head of Philip Morris International raised questions for industry insiders who felt his lack of working knowledge in the automobile industry would create problems in the future, but these same fears and concerns were raised when Marchionne, with his background in toasters and toys, took over fourteen years earlier. Ferrari faced another hurdle when the entire world shut down in March 2020 due to the COVID-19 pandemic. For seven weeks, the longest period of closure in Ferrari's history (including two world wars), the factories in Maranello and Modena ceased production, turned off their lights, and sent everyone home. Luckily, Ferrari produces so few automobiles each year that the forced shutdown barely registered. They were able to restart production on May 8, with very little impact to their bottom line. The pandemic did create issues when Ferrari's CEO, Louis Camilleri, contracted COVID-19 and soon thereafter quit for personal reasons, with many positing that continued health concerns may have been behind his resignation. John Elkann became interim CEO until 2021, when Benedetto Vigna was hired as the new head of Ferrari. Frédéric Vasseur was then named new team principal of Scuderia Ferrari in

December 2022. The years since Marchionne's death have been challenging, but if history is any indication, Ferrari will persevere, proving once again that this is a company built to withstand the vagaries of time.

ELECTRIC VEHICLES COME TO MARANELLO

While slow to embrace the technology, Ferrari announced in 2022 that a massive 800,000-square-foot factory would be built in Maranello to produce electric vehicles and next-generation power trains. The aptly named E-building, set to begin production in the summer of 2024, will be able to produce both electric and combustion vehicles, allowing the company to finally compete on the world stage with many other luxury brands already marketing electric and hybrid automobiles. This expansion is perfectly timed for the release of Ferrari's first EV, which is expected by the end of 2025. Between the creation of the Purosangue SUV, a new electric vehicle, and a plan to be carbon neutral by 2030, Ferrari stands at the precipice, ready to capture a new generation of customers. Although, given the price point, limited inventory, and strict set of guidelines needed to purchase a Ferrari, it may take a generation until individual consumers can actually own any of these new vehicles.

THE HOLLYWOOD CONNECTION

With so much attention focused on Ferrari, it is unsurprising that Hollywood has taken notice of the epic rise of Ferrari and embraced the importance of the man who created his empire in Maranello, Italy. In 2017 a documentary entitled *Ferrari: Race to Immortality* was released, and two major motion pictures were created to highlight and examine both Enzo Ferrari and the company he founded. The first film, *Ford v Ferrari*, was released in 2019 and starred Christian Bale and Matt Damon. The second, set to release on December 25,

2023, is directed by Michael Mann and stars Adam Driver, Penelope Cruz, and Shailene Woodley. Mann's version is a biopic, based on this book—a fitting tribute, both to Enzo's life and to Brock's biography.

Ferrari has come light years since Enzo's death in 1988. It is a trajectory that Enzo might not have foreseen but one he would have definitely embraced. Brock's book is without question a biography, but the symbiotic nature of Enzo and his eponymously named company cannot be divided out on the basis of the man and his creation and thus, while not strictly biographical, the additional information adds context and weight to an already fascinating legacy. Ferrari will always be a reflection of Enzo, defining and redefining his strength of will and determination, cementing in perpetuity the legendary status of this complicated, gruff, often-feared yet incredibly respected bastion of industry who created a company that has stood the test of time and given rise to a legion of acolytes, while also becoming the pride of a nation.

ACKNOWLEDGMENTS

Without the kind assistance of the following, this book would not have been possible. The author wishes to extend his sincere thanks to them all.

In particular to five men for their extraordinary efforts: Bernard Cahier, Luigi Chinetti, Jr., Mauro Forghieri, Ray Hutton and Franco Lini.

And to these friends and associates as well: Jesse Alexander, Richard Anderson, Frank Arciero, Daniele Audetto, Roger Bailey, Bill Baker, Derek Bell, Giberto Bertoni, Bob Bondurant, Tony Brooks, Dr. Stephen Brown, Aleardo Buzzi, Leopoldo Canetoli, Luigi Chinetti, Sr., Gabriella Coltrin, David E. Davis, Jr., René Dreyfus, Dott. Giorgio Fini, John Fitch, Gaetano Florini, Lewis Franck, John Frankenheimer, Professor Donald Frey, Olivier Gendebien, the late Richie Ginther, Franco Gozzi, Larry Griffin, Augusto Guardaldi, Dan Gurney, Phil Hill, William Jeanes, Denis Jenkinson, James Kimberly, Louise Collins King, Michael Kranefuss, Frank Lubke, Karl Ludvigsen, Count Gianni Lurani, Peter Lyons, John Mecom, Jr., Leo Mehl, Kimberly Meredith, Dr. Michael Miller, Eduardo Moglia, Don Sergio Montavani, Craig Morningstar, Stirling Moss, Stan Nowak, Alfredo Pedretti, Earl Perry, Jr., Fillipo Pola, Chris Pook, Harvey

Postlethwaite, Chuck Queener, Brian Redman, Franco Rocchi, Gianni Rogliatti, Peter Sachs, Jody Scheckter, Steve Shelton, Dr. Nick Stowe, John Surtees, Romolo Tavoni, Nello Ugolini, Chick Vandagriff, Cris Vandagriff, Jacques Vaucher, Brenda Vernor, Dott. Corneli Verweij, Luigi Villoresi, Jean-Pierre Weber, John Weitz, H. A. "Humpy" Wheeler, Kirk F. White and Eoin Young.

CHAPTER TWO

The Ferrari Papers

The primary sources for Enzo Ferrari's early life are, as luck would have it, only his own. These works, often confusing and truncated, were begun in 1962 with *Le Mie Gioie Terribili (My Terrible Joys)*, published by Casa Editrice Licinio Capelli of Bologna. Two years later Ferrari published an updated version, *Due Anni Dopo*. A large-format edition, *Le Briglie del Successo*, lavishly illustrated, was published in 1970 and 1974. *Ferrari 80* was the final edition.

These were the recollections of an openly egocentric man with a legitimate gift for writing in the flowery prose that was in fashion during his boyhood. The efforts were artfully crafted, but self-serving and often deceptive, designed to create the impression of a modesty that did not exist, of a compassion that was rare, and to conceal a sense of humor that was ribald and often ironic. These books by Ferrari have been read as much for what they do not contain as much as for what they do.

It has been widely assumed that these works were written exclusively by Ferrari himself, but in fact much of the material was ghostwritten. *My Terrible Joys* was written by Gianni Roghi, a Milanese journalist who spent nearly a year with Ferrari before composing the book, for which he was paid with a new 250GT coupe. Sadly, Roghi was killed by a rogue elephant in Kenya two years later while doing a story on African wildlife. As for the remaining works, there is no question that Ferrari's deft hand is apparent, although insiders say much of the actual writing was done by his loyal aide Franco Gozzi.

A Swiss automotive journalist, the late Hans Tanner, in 1959, published a valuable history, titled simply *Ferrari*, based on notes taken during interviews with the subject. It is now in its sixth edition, with updates by the fine English historian Doug Nye. *Ferrari* remains perhaps the most broad-based history of the marque, although the early personal history for the most part contains only the sparse information that Ferrari himself chose to offer. Tanner is also said to have used as a source a small volume, *Appunti di Storia (Footnotes to History)*, by Carlo Mariani, which dealt with the development of Shell racing gasoline but also treated Enzo Ferrari's youth in some detail.

The Family Automobiles

In the first edition of *My Terrible Joys*, Ferrari chooses to omit the fact that the family had three automobiles. These possessions are hardly consistent with his rather strident efforts to describe his family as dirt-poor working folk. Only in the later editions of his autobiographies did he reveal that his father owned the cars, thereby contradicting the notion that his first contact with exotic machinery had come on the 1908 trip to the race in Bologna.

Father Ferrari's Business

For some mysterious reason Ferrari radically altered the size of his father's work force in a 1985 interview with the Swiss magazine *Hors Ligne*. He totally contravened his earlier claim that the business employed "15 to 30" workmen by saying, "I was the son of a mechanic who had a little garage with five or six workers." In view of the affluence implied by the ownership of the aforementioned automobiles, as well as his own references to his comfortable childhood, the earlier estimate of employment makes more sense. But then, who is to dispute the source, no matter what number he chose to select? It is possible that the employment level varied depending on the work at hand. Consider, too, that when he granted the 1985 interview he was eighty-seven years old and perhaps, understandably, suffering from a failing memory.

The Young Journalist

Ferrari's contributions to *Gazzetta dello Sport* are discussed in an informal biography of Ferrari published in 1977 by the prominent Italian television journalist Gino Rancati. Titled *Ferrari Lui (Ferrari the Man)*, the book is a loose collection of anecdotes and selected incidents intended to flesh out a man who had by then reached legendary status among the Italian public. Rancati is adroit in his treatment, clearly trying to establish his personal friendship with a man whom others claim had no friends. It is obvious that Rancati was handicapped by the understanding that a truly candid appraisal of his subject would cost him dearly, not only in Maranello but perhaps in all of Italy. Still, the book is a valuable source in that Rancati recounts numerous private meetings with Ferrari during two decades in which he was alternately in and out of favor.

Ferrari's Mother

In the same 1985 interview with *Hors Ligne*, Ferrari makes this seemingly pre-posterous statement: "Being an orphan at an early age . . ." What prompted him to make such a claim can only be explained by Italian usage. Ferrari was refer-ring specifically to the loss of his father, which made him an "orphan." His mother, Adalgisa, to whom he was devoted, lived into her nineties in his Modena household. She is recalled by senior management employees as occasionally barging into morning meetings at the Scuderia to ask, "How's my little boy?" He makes no mention of her in his writings. This might seem curious, considering the traditional Italian male's preoccupation with his mother—a fixation so pow-erful, notes Luigi Barzini, that the Roman Catholic Church found it sufficiently important to create the cult of the Virgin Mary and to open a doctrinal rift be-tween it and Protestantism so wide that it may never be bridged. Enzo Ferrari never found it necessary to publicly mention his mother simply because she was part of his "private" life and was therefore separate from his business and public existence. It has nothing to do with his known affection for her.

Early Contact with the Sport

In Valerio Moretti's excellent examination of Ferrari's actual race-driving career, 1919 to 1931, *Enzo Ferrari, Pilota* (Edizioni di Autocritica, 1987), there is a photograph of Milanese carmaker Giuseppe de Vecchi seated in one of his tour-ing cars prior to the start of a 1911 regularity run in Modena. At his side is Ugo Sivocci, but there is no indication that Enzo Ferrari, then aged thirteen, was present or for that matter was even aware of the event. If we accept Ferrari's his-tory, his involvement with motorcars came only through the two earlier races he attended and whatever coverage he was able to glean from the press.

CHAPTER THREE

The Monza Autodrome

The Autodromo Nazionale di Monza was a masterpiece of engineering, laid down by architect Alfredo Rosselli. It included an immense 4.5-kilometer (2.79-mile) banked oval and a 5.5-kilometer (3.41-mile) road circuit featuring a series of very fast, unbanked bends. The two layouts could be combined to create a massive super-speedway of 10 kilometers (6.214 miles) for major Grand Prix events. Over the years a number of permutations would be fashioned by cutting off various sections of the circuit, either to reduce or to increase lap speeds, de-pending on the type of automobiles being raced. To this day Monza remains one of the two or three most challenging and prestigious racetracks in the world. The track was officially opened on September 3, 1922, with the initial victory going to Fiat ace Pietro Bordino. Enzo Ferrari tested Alfa Romeos there, but never ac-tually raced on the track.

The 2-Liter Formula

Enzo Ferrari's first up-close exposure to Grand Prix racing came during the so-called Golden Age of motor racing (one of many, depending on one's orientation and enthusiasm) from 1922 to 1925. The formula was instituted by the Automobile Club de France, which in those days controlled international motorsports policy, and was simple in concept: No engine would exceed 2 liters (122 cubic inches) in displacement and no car would weigh less than 650 kilograms (1,433 pounds). From this formula major manufacturers from England, France, Italy, Germany, Spain and the United States created the first truly efficient lightweight high-performance automobiles. Using features developed by Swiss engineer Ernest Henry for the 1912 Coupe de l'Auto Peugeot—twin overhead camshafts and four valves per cylinder (still considered the optimum to this day)—the cars from Mercedes, Fiat, Miller, Duesenberg, Sunbeam, Ballot, Bugatti, to mention but a few, came to battle with a new device, the supercharger, which had been perfected during World War I for aircraft engines. Superchargers appeared in America and Europe at roughly the same time, and increased power in the 2-liter engines to up to 140 hp and permitted speeds approaching 140 mph. Moreover, the Duesenberg brothers, Fred and August, introduced four-wheel hydraulic brakes in 1921 and that further enhanced performance. Efforts were made to streamline bodies with rounded noses, smooth contours and long tails. Only the suspensions remained antediluvian, with wagon-style leaf springs and solid axles unchanged from the turn of the century. In many ways this concentration on horsepower, with little attention given to handling or road holding, would influence Enzo Ferrari's philosophy about racing cars for the rest of his life.

Unlikely Competitors

Because Ferrari drove in a number of lesser events, he competed against many gentlemen enthusiasts and rank amateurs who viewed motorsports as a diversion, not a profession. Two men with whom he raced would go on to become major influences in the fortunes of the Ferrari manufacturing operation. Edoardo Weber raced against Ferrari on several occasions before retiring to Bologna, his hometown, and beginning the manufacture of the well-known dual-throat carburetors that were to become a trademark of Ferraris until they were replaced by fuel injection in the 1960s. Weber was shot by Partisans in 1945 as a Fascist sympathizer. Years later the great designer Pininfarina (born Battista Farina, who later legally changed his name, literally, to "little Farina" to differentiate himself from his father, also a coach builder, and from his nephew, the champion race driver Dr. Giuseppe "Nino" Farina) would muse with Ferrari how he had beaten him in the 1922 Aosta–Gran San Bernardino hill climb—a race that Pininfarina won outright. The two men would team up in 1952 to create perhaps the most beautiful of all the Ferrari cars.

Racing Safety

When one considers the safety precautions taken by today's modern racing driver, it seems a miracle that anyone survived during the woolly days when Enzo Ferrari competed. Crash helmets, while used for motorcycle racing, were considered not only hot but unmanly and did not enjoy acceptance until the mid-1930s, when they came into use in the United States and England. However, major European Grand Prix drivers did not start wearing them until the early 1950s. Seat belts were unheard of, it being the current wisdom that being tossed from a crashing car was the best bet for survival. Considering that the big fuel tanks were unprotected and easily ruptured, and that the cockpits were without reinforcement or roll bars of any kind, the ejection theory may have made sense. Clothing was of *no* use in a crash, although some drivers wore leather jackets to protect their skin while skidding along the road surface. Others wore linen coveralls, mainly to be fashionable and to keep mud and grease off their underclothes. Add to the lack of any body armor or automotive crashworthiness the fact that the tracks were without barriers and were lined with trees, ditches, stone walls, kilometer markers, ravines, houses and meandering farm animals, and the relatively low fatality rate borders on the miraculous.

The Ferrari Racing Record

The historian Griffith Borgeson accurately notes that Enzo Ferrari tended to inflate the results of his racing career and says that the official record of the ACI (Automobile Club of Italy) lists 21 events in which he is supposed to have competed, winning nine times.

This fine historian wrote, "Most of the events he took part in were bush-league affairs or hill climbs." But there is more to the story. Valerio Moretti's excellent study of Ferrari's early racing career (*Enzo Ferrari, Pilota*) credits Ferrari with 38 race entries from 1919 to 1931, including the French Grand Prix at Lyon, which he never started. Moretti gives him ten victories and three second-place finishes, all in relatively minor races and hill climbs within the borders of his homeland. However, Italian journalist Giulio Schmidt, in his 1988 book on the subject (*Le Corse Ruggenti: La Storia di Enzo Ferrari Pilota*), credits Ferrari with running in, or at least entering, 41 events. Schmidt's study is the most extensive and his data must be respected. But among the 41 events listed Schmidt notes three races—Monza, 1922; Mugello, 1923; Lyon, 1924—in which he was entered but did not compete. Ferrari's record also includes six hill climbs and five trials, straightaway runs, rallies, etc., none of which can be defined as classic, wheel-to-wheel motor races. Neither Schmidt nor Moretti lists Ferrari's entry in the 1930 Monte Carlo Grand Prix, although during my own interview with René Dreyfus he firmly recalled that Ferrari was listed as a contestant but again did not appear. A number of historians have claimed that Ferrari finished ninth with Giulio Foresti in the 1930 Mille Miglia. Neither Schmidt nor Moretti credits him with such an accomplishment. The confusing element is the presence of at least four other Ferraris, all unrelated, who raced as amateurs during the 1920s and 1930s. They included Giuseppe, Girolamo, Bartolo and Valerio Ferrari, none of whom made an appreciable mark on the sport. It was no doubt one of this foursome who finished ninth in the 1930 Mille Miglia. Whatever the actual number of automobile races to be credited to Ferrari, the central point remains: Enzo Ferrari drove for the most part in minor Italian races. His major accomplishments were his second-place finish in the 1920 Targa Florio and his victory in the 1924 Coppa Acerbo. Aside from those, his efforts were restricted to regional hill climbs and local road races. Ferrari's only attempt to drive in a major Grand Prix, at Lyon, ended in a withdrawal that was to be the subject of controversy for years to come. At best, his record marks him as a semiprofessional sports-car specialist with no more than second-rank credentials.

Alfa Romeo Production

Lest anyone think that Alfa Romeo was a giant automotive concern during the time that Enzo Ferrari was associated with it either as a racing driver, sales representative or team manager, Borgeson notes that during the two decades, 1920–39, leading to World War II, annual production averaged only 473 cars—many of them bare chassis sent off to custom coach builders for special bodywork. Only once, he says, did the company build more than 1,000 units in a year, that being 1925, when 1,115 sports cars were manufactured. Only six of the famous P2 racing cars were built, and most of those ended up in the hands of private sportsmen in the 1930s. By 1936, when production had shifted to aircraft engines and military vehicles, a mere ten automobiles were completed. Alfa Romeo is now part of the immense quasi-governmental consortium headed by Fiat and produces hundreds of thousands of small and medium-sized sedans annually.

CHAPTER FOUR

The Racing Formula

While sports-car racing of the 1920s and 1930s was essentially a Formula Libre (or what American Southern stock-car racers would later refer to as "run what you brung"), the Grand Prix races were then organized under strict rules set by the AIACR (Association Internationale des Automobile Clubs Reconnus), a French-dominated collection of national automobile clubs which had risen from the first formal Grand Prix, the French Grand Prix of 1906, organized by the Automobile Club de France. To this day, the ruling body for international motorsports remains based in Paris. As the speeds of the 2-liter cars of the early 1920s increased (cars like the Alfa Romeo P2, for example), the AIACR cut the engine size to 1.5 liters (91 cubic inches) for the 1926 and 1927 seasons. This proved unpopular with manufacturers like Delage, Fiat, Bugatti and Alfa Romeo, and entries fell off. The doyens of the sport responded by creating a free formula (Formula Libre) for 1928, permitting any size engine, but requiring car weights between 1,200 and 1,650 pounds. Two-man bodies were still mandatory. Ironically, single-seaters were permitted in America from 1923 to 1930, when the so-called production-based "Junk Formula" was introduced to reduce costs and a riding mechanic was again required. This added nothing except the chance of killing or injuring a helpless passenger. At the same time that the riding mechanic was being reinstated at Indianapolis, the AIACR removed him. But until 1932 the two-seater body style was mandatory, although no actual passenger was required. Racing regulations, especially those created by the French, have traditionally been complex, often ludicrously contradictory, and have sometimes evolved more from the demands of the competitors than from the arcane rulings emanating from Paris. In 1931 the AIACR insanely decreed that all Grand Prix races would run for at least ten hours! This in a depression! (In justice to the French, the original suggestion for such an absurd race length came from Vincenzo Florio.) Confusion reigned across Europe until 1934, when finally a formula was agreed upon that seemed to make sense. It required that all cars weigh no more than a maximum of 750 kilograms (1,650 pounds) less driver, tires, fuel and other liquids. The idea was to slow cars down, the current wisdom being that no engine of significant size could be fitted to automobiles that light. But the Germans—and in some cases Alfa Romeo and the Scuderia Ferrari—managed to use the rules to design and build some of the fastest racing cars of all time.

Racing Colors

Until major sponsorship arrangements brought a myriad of colors to Grand Prix racing cars in the 1970s, it was accepted that cars would race in the livery of their home country. The idea is believed to have originated with early Polish motorist and sportsman Count Zbrowski, who first suggested the idea around 1903. Initially the United States was awarded the color red, but it was soon assigned to Italy, with the United States receiving white with a blue stripe. Britain's cars were green, France's blue, Germany's white (which, de facto, became silver in the 1930s); Belgium's yellow, etc. These national colors remained a part of Grand Prix racing until the late 1960s, when the livery of various sponsors obliterated them. Today only Ferrari adheres to the old tradition.

The 1930 Mille Miglia

The contest between Tazio Nuvolari and Achille Varzi has been immortalized in numerous novels, historical references and the 1955 Kirk Douglas motion picture *The Racers*. Legend has it that Nuvolari stalked Varzi in the predawn run to the finish at Brescia with his headlights doused, thereby lulling his rival into thinking he was holding an easy lead. Then, with the finish in sight, it is said, Nuvolari sped up and passed a shocked Varzi for the victory. This was clearly not the case.

Count Giovanni Lurani's superb history of the Mille Miglia clearly states that Nuvolari was in an advantageous position from the beginning of the race. He started ten minutes behind Varzi and therefore could determine his rival's progress at each control or fuel stop. Worse yet, Varzi lost time with two punctured tires south of Bologna and as he headed for the finish he surely knew he was behind. (The finish was determined on who completed the course in the least amount of time, regardless of when he started. Therefore, because Nuvolari started ten minutes later, any less of a difference would have placed him ahead of Varzi.) After the race Varzi said that he and his riding mechanic, Canavesi, sighted Nuvolari's distinctive triple headlight 1750 Alfa Romeo at least 120 miles from the finish and knew then they had lost. Journalist Gino Rancati recounts an interview with Nuvolari's riding mechanic, Giovanbattista Guidotti, in which this witness to the incident said that Nuvolari did indeed turn off his lights, but only for a minute or so, as they moved up on Varzi. Guidotti noted that both drivers were aware of the other's presence, and that Nuvolari switched off his headlights hoping to deceive Varzi into thinking he had stopped. But why? According to the format of the race, Nuvolari had only to tail along behind Varzi in order to win. Surely it was Nuvolari's fiery, prideful urge to shame his rival by passing him before the finish and to reach the screaming throngs at Brescia first that prompted his decision. Causing Varzi to slow, even for a few moments, made the task of passing him somewhat easier, even though his overall victory was already assured. Guidotti later gained more sedentary employment as a test driver with Alfa Romeo. Varzi, by the way, avenged his defeat by decisively whipping Nuvolari a few weeks later in the Targa Florio. He also abandoned Alfa shortly thereafter to race first for Maserati, then Bugatti, and did not return to the fold until 1934.

The Teams

While the Scuderia Ferrari was hardly the first private team to be formed in Europe, its quasi-official relationship with Alfa Romeo was rather unique. More traditional was the arrangement in 1927 and 1928 of Nuvolari and Varzi, who simply bought cars from Ettore Bugatti and raced them on a purely independent basis. At the other end of the scale were the full-blown factory teams that employed professional drivers. All manner of financial arrangements were used. Some drivers were members of the engineering or sales staff; others were independent contractors, racing for either a fixed salary or a percentage of the winnings, or both. Rich sportsmen often bought cars from a manufacturer at favorable prices and paid to have them maintained, much as Ferrari did for his customers in the early years of the Scuderia. The riding mechanics of the day, such as Pepino Verdelli and Giulio Ramponi, were generally employees of the factory or members of independent teams who volunteered for the extremely dangerous job of riding along in the races. They served no function; by the late 1920s racing cars were sufficiently reliable so that emergency tire and spark-plug changes or repairs needing a mechanic were not required. These men were simply extra bodies to be crushed in the event of a crash and participated only for the glory and a small percentage of the winnings. The last riding mechanics in major open-wheel motorsports were seen in America at the Indianapolis 500 in 1937. By then literally hundreds of them had been killed and maimed for no rational reason, other than as mindless slaves to tradition.

Maserati

In the late 1920s and 1930s Officine Alfieri Maserati S.p.A. was a considerably more serious automotive presence than anything operated by Enzo Ferrari. The brothers Carlo, Bindo, Alfieri, Ettore and Ernesto were natives of Bologna and had links with the industry dating to the turn of the century. The eldest, Carlo, who died in 1911, raced for Fiat and Bianchi, while Bindo and Alfieri worked for Isotta Fraschini. During World War I the two brothers manufactured spark plugs under the family name. In the mid-1920s Alfieri designed and built several racing cars for Diatto, including a straight-8 Grand Prix machine. When that firm dropped out of competition in 1926, the brothers took over the racing cars and began to manufacture updated and improved models in their small family garage near the Ponte Vecchio in Bologna. The logo of their company, which is still in use, was Neptune's trident, the symbol of their home city.

The acknowledged leader among the brothers was Alfieri, a pleasant, universally liked man with a broad understanding of engineering. In early 1932 Alfieri crashed at Messina and died on the operating table at age forty-four. His loss was a devastating blow to the firm, which never produced more than a handful of single-seaters and sports cars annually. The surviving brothers were hardworking and talented, but lacked the business acumen to expand the operation much beyond a small machine shop capable of producing a few superb automobiles and continuing their modest spark-plug business.

It has been estimated that no more than 130 "street" Maseratis were built before 1957, when the first production 3500GT was offered for sale. By then the brothers were long gone. In 1936 production had lapsed to a mere nine cars, and a year later the three remaining brothers, Bindo, Ernesto and Ettore, sold a majority interest in the operation to the Orsi family of Modena. The senior Orsi, Adolfo, and his son, Omer, signed the brothers to a ten-year management contract, after which the Maseratis formed OSCA (Officine Specializzate Costruzione Automobili) back in their hometown of Bologna. These high-performance cars enjoyed substantial success in the smaller-displacement classes until the aging brothers sold out to MV-Agusta in 1967. In the meantime the Orsis continued to operate as rivals of Ferrari but finally were forced out of Grand Prix competition by sagging business ventures in Argentina at the end of 1957.

Under the stewardship of the Orsis, the Maserati operation retained a reputation as a pleasant, more subdued place to work and race than the intrigue-infested Ferrari factory in nearby Maranello. By the early 1960s Maserati began building high-volume grand touring cars and a few sports-racing cars. The original flavor of the company was hopelessly diluted. In 1966 the Orsis entered into a partnership with the French Citroën company and the end came in 1975 when Argentinian Alejandro de Tomaso took control. He had become adept at swallowing up such bankrupt operations as motorcycle manufacturers Benelli and Moto Guzzi, as well as mini-car builder Innocenti. This brought forth such disasters as the Maserati BiTurbo and an ill-fated working arrangement with Chrysler before the last vestiges of the proud firm were absorbed into the Fiat empire in 1989. Like so many honored makes of the early years, Maserati is destined to become a mere nameplate on a series of homogenized passenger cars, or, ironically, may one day share certain components with its archrival of yore, Ferrari.

For those intrigued with the early, glorious years of Maserati, required reading is Luigi Orsini's superb *Maserati: A Complete History from 1926 to the Present* (Libreria dell' Automobile, 1980), with pictures from the vast historical collection of Franco Zagari.

CHAPTER FIVE

The Tripoli Scandal

Many accounts of Italian motorsports in the 1930s conveniently ignore the scandals of 1933 and 1934, but there is no question that they occurred in all their luridly zany glory. Even historians who take note of the incidents often decline to point the finger at such luminaries as Nuvolari, Varzi or Giovanni Canestrini, but little doubt remains about their involvement. Hans Tanner, who spent years immersed in the inner sanctums of Italian motorsports, did extensive research on the subject and provided detailed background material for a story published in the May 1965 issue of *Car and Driver*. The incidents are also treated in some detail in Chris Nixon's *The Silver Arrows* (Osprey Publishing, 1986). The involvement of Enzo Ferrari, as noted, is entirely circumstantial, but it borders on the impossible to believe that a man who was so deeply involved in (and so totally informed about) the goings-on in Italian motor racing on a daily basis, and whose best drivers were co-conspirators, would not have been privy to the plan. In 1985 Chris Nixon calculated the winnings of some of the major Italian drivers based on the Bank of England's estimate that the Italian lira, which was extremely unstable in the 1930s, had an exchange rate of 93.75 to one English pound (which in turn was worth five American dollars). Therefore Varzi's 975,000 lire winnings in 1934 (the best of the Italian stars) was estimated to be worth about $890,000 in 1985. Not bad earnings, even by today's bloated standards. This total, published in *Gazzetta dello Sport* in 1935, did not include Varzi's under-the-table winnings from the Tripoli Grand Prix, so that the dour Italian no doubt won over $1 million in 1985 value for the season.

The Tipo B *Monoposto*

This excellent design by Vittorio Jano was actually a derivative of the older 8C-2300. It was Europe's first true single-seat Grand Prix car. It employed a similar straight-8 double-overhead-camshaft engine of 2.6 liters. The block was actually a pair of 4-cylinder alloy units joined at the middle by a central gear train which drove the camshafts, the oil pump and the twin Roots-type superchargers. The power was claimed to be 215 hp at 5,600 rpm, which some historians question as overly optimistic. About 180 hp seems a more realistic figure.

The uniqueness of the car came in its twin drive shafts that formed a

30-degree "Y" from the back of the engine and were connected to a bevel drive at each rear wheel. Jano chose this layout to reduce unsprung weight—not to lower the center of gravity, as some have speculated. This twin-shaft arrangement sapped power through frictional loss and was never copied, but it was apparently beneficial in terms of improved handling. The P3 was known to be a very light and forgiving car. "It's like a bicycle," exclaimed Louis Chiron following his first drive after a long career in cruder Bugattis. After the rise of the German teams in 1934, the Tipo B P3 was fitted with an engine bored out to 2.9 liters, but by then the car was hopelessly outclassed. This increased power and torque put added strain on the transmission, and by the end of 1935 the car was no longer seen in any major races. It is, however, considered to be one of the best-designed cars of its era and a tribute to Jano's creativity.

The Italian Air Force

Italo Balbo was an aviation pioneer of sorts and in 1933, just prior to the Tripoli race, had led a formation of flying boats across the Atlantic. He was an avid proponent of strategic bombing, but Benito Mussolini believed that his Air Force ought to be primarily a tactical wing in support of his highly mobile, mechanized (and lightly armored) Army. By 1933 the Air Force was equipped with Fiat CR-32 biplanes, and they served well in the Spanish Civil War, but they were hopelessly outclassed by the German and British aircraft that were developed almost concurrently. When Mussolini invaded Ethiopia in 1935, his troops were supported by 400 bombers and fighters. That they succeeded against the enemy's 12 antique war birds offered little indication of their true potency. Still, Mussolini could boast of real strength in sheer numbers of combat aircraft, and several of his later fighters—including some of the first jets—were excellent machines. But by then a totally bankrupt war effort blunted their effectiveness.

The Trossi Duesenberg

Because of its name, the implication is that Didi Trossi obtained one of the front-line Duesenbergs that had been so successful during the 1920s and early 1930s at Indianapolis and in other American board and dirt-track races. That is not the case. The car was actually one of a pair of specials built by a former Harry Miller mechanic named Skinny Clemons in association with August Duesenberg, who had left the family firm (owned by E. L. Cord since 1926) several years earlier. His brother Fred, the true genius behind the immortal cars, had been killed in August 1932 while descending Ligonier Mountain, near Johnstown, Pennsylvania, in one of his powerful SJ roadsters. The Trossi car was originally built to the Indianapolis "Junk Formula" of 1932–36, in which large-displacement, production-based engines were permitted. The engine in the automobile was based on the old 1920–27 Duesenberg Model A power plant and displaced 269 cubic inches (by comparison, the 8C Monzas and Tipo B P3s from Alfa Romeo displaced about 158 cubic inches). The Clemons-Duesenberg special was delivered to Italy with the help of the Champion Spark Plug Company and fell into

Trossi's hands at a time when a number of Miller and Duesenberg Indianapolis cars were being imported for speedway-type competition on the continent. Big, banked ovals were located at Brooklands (at Weybridge, outside London, near the present Heathrow Airport), at Montlhéry (near Paris) and at Monza. The car began life as a two-seater, but before being rather tardily delivered to Trossi, it was converted to a *monoposto*. As with its American brethren, its lack of four-speed transmission and large brakes, both of which were superfluous on oval tracks, handicapped it for road racing competition. Trossi sold the car in 1934 to American expatriate Whitney Straight, who drove it to several class records on the Brooklands bowl, averaging 138 mph. Still in red Italian livery, the car was later campaigned by a series of British gentlemen drivers, including gentleman steeplechase rider Jack Duller, "Buddy" Featherstonhaugh and Dick Seaman— who was to become England's most successful prewar Grand Prix driver. After the war the car was restored and was owned by the well-known dean of British motorsports journalists, Denis Jenkinson.

CHAPTER SIX

Tazio Nuvolari

Although much has been written about Tazio Nuvolari that implies his career was little more than inspired victories punctuated by a series of lurid crashes, this is oversimplified nonsense. He was a complicated man, mercurial and of course egocentric (no great racing driver has been known for his modesty), but surprisingly disciplined. His driving technique, which involved controlled power slides, now called four-wheel drifts, was original and highly unorthodox. To the uninitiated it appeared as if he was abusing the car, but Nuvolari's finishing record was very good, considering the underdog equipment he was often saddled with, not to mention his spirited style. Had it been otherwise, he would not have won the Le Mans 24 Hour race (with Raymond Sommer) in 1933 and taken the Mille Miglia in 1930 and 1933. Those were classic long-distance endurance contests that required great restraint and patience in order to preserve the delicate mechanical components of the machinery at their disposal in the 1930s. Nuvolari was something of a clothes horse, wearing on occasion either a red or a blue leather flying cap, a leather vest, a turtleneck sweater and sometimes, as in the case of his *bimotore* record run, a conventional pair of white linen coveralls. It was only after he joined the Auto Union team in 1938 that he assumed a costume sporting the earlier-mentioned D'Annunzio turtle brooch and his initials embroidered on the right breast of his shirt.

The Ferrari–Alfa Romeo *Bimotore*

Owing to their weight and gruesome appetite for tires, it was apparent to Ferrari and Bazzi that the *bimotores* were doomed to be novelty items in the Scuderia's automotive armory. Tires were an incurable Achilles' heel, and the two automobiles were quickly sold after Nuvolari's 200-mph run. The car Chiron drove was scrapped; historian Doug Nye says the Nuvolari car was sold to British gentleman driver Arthur Dobson after Ferrari installed a pair of smaller, less powerful 2.9-liter Alfa engines. Sharing the car with Fairey Aviation's chief test pilot, Chris Staniland, Dobson drove the car in the British Racing Drivers Club 500-kilometer race and later ran as fast as 132 mph on the high-banked Brooklands speedway. The car then passed through a series of owners, finally being cut in half, with the rear engine being hauled away and installed in another English special. The remainder of the emasculated car ended up in New Zealand, fitted with a GMC truck engine. A restoration of the grand old machine has been reportedly completed by a New Zealand enthusiast. It should be noted that while the *bimotore*'s 200 mph was noteworthy, it took two Alfa engines totaling 6.3 liters to achieve the speed. By comparison, the brilliant Californian Frank Lockhart seven years earlier had constructed a tiny rear-drive streamliner powered by a 3-liter V16 (two Miller 8s on a common crankcase) and had run over 225 mph on the sands of Ormond Beach. This amounted to 25 mph more speed from an engine less than half as large as the *bimotore*'s. Sadly, Lockhart cut a tire on a clam shell and the Stutz Black Hawk carried him to his death before any official record could be recorded. It was estimated that the car could have achieved 300 mph under the proper conditions—or almost 100 mph faster than the existing land speed record.

German Motorsports

No one is truly sure how much the Nazis spent to support the Mercedes-Benz and Auto Union teams during the 1930s. In 1947 a unit of British intelligence issued a report on the subject ("Investigation into the Development of German Grand Prix Racing Cars between 1934 and 1939, Including a Description of the Mercedes World's Land Speed Record Contender," by Cameron C. Earl, Technical Information and Document Unit, British Intelligence Objectives Sub-Committee, April 22, 1947), which stated: "The Nazi Party was fully aware of the part motor racing could play in building up an international reputation that Germany's engineers were supreme. Recognizing this potential value of motor racing to Germany, Hitler agreed to award a sum of 500,000 Rm [reichsmarks] [£41,600] per annum to firms producing successful Grand Prix machines. This grant by no means covered the expense of the production and maintenance of a Grand Prix team, but the firms concerned were also favored with large armament contracts as an additional incentive." Chris Nixon notes that Cameron Earl interviewed William Werner, the former technical director of Auto Union, while collecting data and was informed that his company was spending 2.5 million reichsmarks per year, which was far more than the government subsidy. By Nixon's reckoning, the Nazi donation amounted to about 322,500 reichsmarks for both Auto Union and Mercedes-Benz. Karl Ludvigsen, the excellent historian and writer, who had intimate access to Mercedes-Benz prewar racing files when preparing his *The Mercedes-Benz Racing Cars* (Bond Parkhurst Books, 1971), states that the Transport Ministry was to pay 450,000 reichsmarks annually to the maker of a Grand Prix racing car, with bonus payments of 20,000, 10,000 and 5,000 reichsmarks, respectively, for first-, second- and third-place finishes. Whatever the true number—which will no doubt remain the subject of some debate—the fact remains that Auto Union and Mercedes-Benz each spent lavishly on their programs and were rewarded in kind with enormous armament contracts and insider status with the Nazi Party.

While the individual members of the racing teams behaved in public with the proper salutes and party pandering, they held a diversity of private opinions about the Nazis. Rudi Caracciola, for example, refused to join the party and from 1937 onward lived in Lugano, Switzerland. During the war the Nazis cut off his pay and pensions, and rumors were circulated that he was a Greek or Italian, rather than a legitimate German (although his family had lived in Remagen, on the Rhine, for four hundred years). On the other hand, brash, blond Bernd Rosemeyer was transformed into a Teutonic idol and at the time of his death in early 1938 had been appointed an honorary SS-Obersturmführer in the dreaded Schutzstaffel. Other drivers, including Hans Stuck and Hermann Lang, held essentially honorary officerships in the NSKK, but their function was meaningless. It is known that Rosemeyer and others, including Neubauer, would mock the oratorical posturing of Hitler at private parties and after the war the survivors claimed to be apolitical (as did most of their countrymen). The teams had no problems in hiring foreigners, despite their nationalistic zeal, and Italians like Fagioli, Varzi and Nuvolari were employed, as was the brilliant Englishman Dick Seaman, but the hard core of the team was composed of Germans. Ferdinand Porsche, the creator of the Auto Union racing cars (in addition to the Volkswagen, the Porsche sports car and a number of Wehrmacht military vehicles), was imprisoned by the French for two years after the war as a war criminal, although some believe the incarceration was in retaliation for the jailing of car mogul Louis Renault as a German collaborator.

Ironically, the quintessential German of the lot, Manfred von Brauchitsch, ended up in East Germany, where he lived in baronial, decidedly non-Communist splendor for many years. René Dreyfus, on the other hand, was more than qualified to be a driver for the German teams, but was shunned because his father was a Jew and his family was connected with the celebrated military martyr Colonel Alfred Dreyfus. Stuck also had problems because his wife, Paula, had a Jewish grandfather. Hitler himself had to intervene to save her.

Italo Balbo

The racing community rather narrowly recalled Italo Balbo as a gracious host in Tripoli and the man who built one of the finest motorsports facilities in the world. But he was also a hard-bitten military expert who had long been associated with the Fascist cause. Balbo first appeared as a youthful opponent of Mussolini in his hometown of Ferrara, but by 1921 he was an avid recruit, leading a strong and ruthless band of Fascist militia through central Italy, displacing socialist local governments and trade union leaders and killing his more vocal opponents.

If any single man besides Mussolini convinced the Italian peasantry that Fascism was the wave of the future, it was Italo Balbo. After Mussolini came to power, he was rewarded with the position of air marshal of the Air Force, then became the governor general of the Tripoli Protectorate (Italy had enormous territorial possessions in Africa prior to the war, all of which were ultimately lost). The death of Balbo over Tobruk in 1940 remains something of a mystery. It is generally accepted that his aircraft was mistakenly shot down by Italian gunners on the ground, but a suspicion lingers that he was the victim of an on-board bomb. The truth will probably never be ascertained. Still, Italo Balbo remains, with Ciano, one of the most vivid characters in the bizarre Mussolini experience.

CHAPTER SEVEN

Wifredo Ricart

After Enzo Ferrari's departure from Alfa Romeo, Ricart carried on with a series of ambitious projects, including a daring 16-cylinder 3-liter Grand Prix car called the Tipo 162, which might very well have equaled the German best in outright performance had the war not intervened. Its twin-supercharged engine developed nearly 500 hp in tests, which was competitive with the figures being generated by Mercedes-Benz and Auto Union. Ricart also is credited with the creation of the ill-fated 512, a mid-engine 1.5-liter *voiturette* that was never sufficiently sorted out to be raced. It is likely that Colombo was deeply involved in the design of the 512 and it was probably this car—or one strikingly similar to it—that Ferrari rejected during his first meeting with Colombo to discuss the Alfetta. As was mentioned, the Alfa Romeo engineering staff continued to fiddle, Nero-like, with racing cars as the war raged around them. The 512 was first tested three months after Italy entered the fighting and Attilio Marinoni was killed on the Milan-Varese autostrada testing a modified 158 only a few days following the declaration of war. (Contrary to Ferrari's claim—surely made to discredit Ricart—the longtime mechanic-driver did not die at the wheel of a 512. His 158 may have been fitted with a 512 suspension, speculated Griffith Borgeson, but no matter, the accident occurred when Marinoni collided with a truck at high speed.)

Much of Alfa's wartime racing activity must be attributed to Mussolini's belief that the fighting would be short-lived and that Europe would quickly return to peace. Obviously he was wrong, and by 1941 Ricart and his staff were deeply immersed in the design and construction of what Borgeson called his "magnum opus," the 28-cylinder turbocharged Tipo 1101 aircraft engine intended to develop 2,500 hp. Had this project survived the repeated Allied bombings of the Alfa works that began in October 1942 and continued through much of 1943 and 1944, this engine would be remembered as one of the most powerful reciprocating aircraft power plants in history. Ricart also labored over some automotive projects during the war, including a mid-engine grand touring car and a sports car called the Gazzella. He remained with the company until the war ended, then returned to Spain to design the technically advanced Pegaso—a government-backed project that failed through no fault of the machine, but rather due to internal mismanagement and corporate lack of interest. (This permitted Ferrari one final shot at Ricart: "But the new Pegaso, in fact, never managed to beat Ferrari in a race and has today been forgotten. The name is still to be found only on some motor trucks that curiously resemble better-known makes.") Ricart in fact left the company in 1958, and only about 125 automobiles were built before Pegaso concentrated on the trucking business. There is no question that Wifredo Ricart was a strong-willed engineer with controversial notions. But he was a talented and highly creative designer whose reputation has been tainted by the carping of Ferrari and his apologists over the years. He died in his native Barcelona on August 19, 1974, aged seventy-seven years.

Galeazzo Ciano, Conte di Cortelazzo

Ciano was by all accounts a rather sympathetic character who, after marrying Mussolini's daughter, Edda, rose to become foreign minister in 1936. He at first favored the Berlin-Rome axis, but later opposed Italy's entry into the war. He joined the Grand Fascist Council in July 1943 in voting to depose his father-in-law and paid for his decision with his life. He was executed in Verona by the German puppet government run by Mussolini, called the Republic of Salo, in January 1944. Ciano was a genuine motor-racing enthusiast. Luigi Orsini says that during the 1936 running of his own Coppa Ciano, he entered the Ferrari pits and suggested to Ugolini and Bazzi that either Brivio or Pintacuda be called in and their car given to Nuvolari, whose 12C had broken down at the start. They complied, and Tazio took over Pintacuda's older, ostensibly slower 8C. He drove one of his greatest races, pressing the Auto Unions of Rosemeyer and Varzi so relentlessly that both broke down. Nuvolari won, leading Brivio and Dreyfus to a stunning one-two-three finish for Alfa Romeo.

The Jano Firing

The fact that Vittorio Jano was dismissed from Alfa Romeo has no real bearing on his much-deserved reputation as one of the greatest automobile designers of all time. As Colombo wrote, the political atmosphere at Alfa was so chaotic and the budget constraints so strict that it borders on the unbelievable that he was able to create any new machinery at all. Moreover, all of the best designers, including Colin Chapman, John Barnard, Harry Miller, Ettore Bugatti, Mauro Forghieri, etc., produced notable failures during their illustrious careers. Jano's slump at Alfa was hardly unusual. He more than redeemed himself at Lancia.

The American Connection

Certainly one of the most humiliating Alfa defeats during the dark days of 1936–37 came at the July 1937 Vanderbilt Cup when the popular, abundantly talented Rex Mays beat both Nuvolari and Farina and their new Scuderia 12Cs with a much-raced year-old 8C. Mays was a brilliant driver by all accounts, and his effort in the race—his first on a road circuit—was extraordinary. But there was a reason that his old car was so quick. Following the 1936 Vanderbilt Cup, which Nuvolari won, a wealthy American Indianapolis car owner, "Hollywood" Bill White, arranged to purchase one of the Scuderia's 8Cs that had been brought over for the race. According to the late race official and former mechanic Frankie Del Roy, White also persuaded Attilio Marinoni, who accompanied the Ferrari team, to remain behind in America for several weeks. Del Roy told me that Marinoni was treated royally with wine and women and repaid the hospitality by spilling all of the tuning secrets of the Alfa and Scuderia mechanics. This permitted Mays not only to run a surprising third in the Vanderbilt (behind the Auto Union of Rosemeyer and the Mercedes of Dick Seaman) but earlier to drive the car in the 1937 Indianapolis 500. There it was fitted with a two-man body, per the regulations, but retired with overheating troubles after twenty-four laps. On the basis of his marvelous drive in the Vanderbilt, Mays was invited in late 1937 to visit Italy and drive for Alfa Romeo (whether the invitation came from Alfa Romeo directly or from the Scuderia is not known). On his way to Europe, the smiling Californian stopped over in New York, where he was informed that a tuxedo would be a necessary part of his wardrobe. Mays had no evening wear and declared that he intended no such purchase. He turned around and went home. Had Rex Mays gone on to Italy, he would have been the first American to join the Scuderia, thereby predating the arrival of another Californian, Phil Hill, by almost twenty years. The Mays Vanderbilt 8C Alfa was last seen in the late 1950s serving as an advertisement outside White's restaurant on Glendale Boulevard in Los Angeles.

The Mercedes-Benz 165s

So much has been written about the titanic struggles of the Mercedes-Benzes and Auto Unions during the 1930s that any attempts to add further information here would be futile and redundant. The development of the Mercedes-Benz 1.5-liter cars is explained in detail in Karl Ludvigsen's superb book *The Mercedes-Benz Racing Cars* (Bond Parkhurst Books, 1971). During the fighting most of the racing cars of Alfa, Mercedes and Auto Union were hidden away (as well as numerous other European exotic machines), and most survived the war. The Auto Unions, located as they were in what was to become the Russian sector, were hauled off to Soviet technical schools and dismantled. At least two are now in the West. Several Mercedes-Benz W154s were saved by enthusiasts as they too were being shipped behind the Iron Curtain. The 165s were smuggled out of Germany at the end of the war, and Rudi Caracciola attempted to enter one in the 1946 Indianapolis 500. Military red tape prevented shipment and the cars drifted through private hands until being returned to Daimler-Benz. They are now part of the company's splendid collection in their Stuttgart museum. The 158 Alfas were concealed behind the false wall of a dairy and returned to racing in 1946 as the nemesis of their progenitor, Enzo Ferrari.

CHAPTER EIGHT

Don Giovannino Lurani Cernuschi, Conte di Calvenzano Patrizio Milanese

One of the great names in international motorsports, "Johnny" Lurani had a career that spanned almost as great a time period as that of Enzo Ferrari. Born into the Italian nobility, he became an automobile-racing enthusiast at an early age and by the 1930s was well known in Italian motorsports circles as a respected journalist, fine amateur driver, team owner and race official. Over the years he authored a number of books, including a biography of Nuvolari and a history of the sport. Lurani was an effective administrator in the upper echelons of the Italian motorsports authority and was the creator of the Formula Junior racing class that served as an international training ground for the best road-racing drivers in the early 1960s. Lurani knew Ferrari well for nearly fifty years, and while his public remarks are phrased with gentlemanly circumspection, he is privately critical of Enzo Ferrari and the historical revisionism that has been undertaken by his sycophants.

The Company Name

Enzo Ferrari's Auto Avio Costruzione was not replaced by a firm bearing his own name as soon as the Alfa Romeo restrictions ran out. By 1943 "Scuderia Ferrari" had been added to the corporate sales brochures, although the company name was not formally changed until 1946, when the firm was renamed Auto Costruzione Ferrari. In 1960 the company was restructured as a public entity and became Società Esercizio Fabbriche Automobili e Corse Ferrari. The company logo remained the Prancing Horse, although in 1946 the currently employed rectangular badge replaced the shield that had been used since 1932.

Italian Military Hardware

It was as much a fault of logistics as design that Mussolini's mechanized equipment did not perform more effectively. Although the Italians practically stopped work after the Germans took over, their production output under Mussolini was modest at best. But many of their planes and tanks were well designed and well built. As early as 1933 the Italian Macchi MC72 was among the very fastest of all military aircraft, and a year later it captured the prestigious Schneider Trophy race at a speed of 441 mph. Caproni-Campini had an experimental jet aircraft—the CC2—flying as early as November 1941, but it was heavy and slow compared with the models being flown by Heinkel and Messerschmitt. The Macchi C202 was an effective piston-engine fighter plane, but in its final permutation carried a Daimler-Benz DB601 in-line engine. By 1942 any serious efforts to create new aircraft had collapsed and all Italian aircraft were considered obsolete. Most of the Italian military cars had been developed in the 1930s as "colonial" machines—suited for rugged off-road work in the African possessions. Fiat built a majority of them, although Alfa Romeo produced the four-wheel-drive 6C-

2500 that featured coil springs at the front and torsion bars at the rear and a variation of the Jano-designed twin-overhead-camshaft engine used in their 6C racing cars. From the standpoint of design, these exotic machines outclassed the agricultural-level jeeps used by the Allies, but they were much too complex to operate and maintain under combat conditions.

Fiat built most of the tanks and heavy gun carriers that were for the most part outdated when the war began. A majority of them were either captured or destroyed in the African campaign.

CHAPTER NINE

The Early Ferrari Cars—How Many?

Because the Ferrari archives (assuming there are any) are inaccessible even to the most devoted and responsible automobile historians, the exact number of cars produced by the firm in the early years is difficult to determine.

The excellent work *Tipo 166: The Original Sports Ferrari*, edited by Angelo Tito Anselmi with Lorenzo Boscarelli and Gianni Rogliatti (Libreria dell' Automobile, 1984), provides what is probably the most complete listing of Ferrari road cars (not racing machines) built between 1947 and 1950. The authors note that they, like all other historians, were denied access to the factory records and had to work through Italian provincial and Automobile Club registries. Because so many of the early cars were rebodied and sometimes rebodied a third and fourth time, an accurate count of how many were constructed cannot be determined. Some historians have blithely claimed that three cars were built in 1947, but Anselmi et al. say that the exact number is unknown—although it probably did not exceed five chassis. Enzo Ferrari was an inveterate hot-rodder who constantly fiddled with his products, moving an engine from one car to another, switching transmissions, modifying racing cars into road cars and vice versa. By the best estimates, it is probable that during the first four years in which the factory was making automobiles, 1947–50, *total* production of racing cars and road cars did not exceed ninety units, including single prototypes.

The Ferrari Finances

By war's end, Enzo Ferrari not only possessed the old Scuderia in central Modena but owned substantial property on both sides of the Abetone-Brennero road in Maranello, on which was located a well-equipped factory of about 40,000 square feet. He was certainly a man of means, and considerably better off than most of his countrymen. But what was the source of this prosperity? Who provided the funds to build the factory in the middle of the war? Some say his wife, Laura, was a donor, but that seems unlikely. Romolo Tavoni, a longtime team manager and associate of Ferrari's who now operates the Monza autodrome, says that Ferrari assigned the Maranello property (or part of it) to Laura for legal reasons, and this could be the source of confusion over her financial role. There is no question that Ferrari received a lucrative financial settlement when he was fired by Alfa Romeo, but it is doubtful it was sufficient to construct a new factory. Did Ferrari have wealthy silent partners? It is possible, because he was well connected with some of the richest men in northern Italy, many of whom were tied to Alfa Romeo and its Fascist management. Moreover, Ferrari probably saved a considerable sum during the Scuderia Ferrari days. He lived simply and that, plus loans, institutional or otherwise, may have provided him with sufficient financing. While the machine tool business gave him some income during 1947, the cost of creating the Tipo 125s and 159s from the ground up and without a customer base or any racing income had to have dug deeply into his resources. Yet, unlike the mid-1950s, when funds got tight, there is no indication during this early period that he was ever seriously cramped for money.

Ferrari vs. Maserati

Because the two firms were such intense rivals, and because both built new competition sports cars during 1947–48, it might be interesting to compare the two machines. It should be noted that the Maserati A6GCS was first campaigned in September 1947, whereas the 166SC was not actually introduced until the spring of 1948. Yet it was merely an improved version of the original 125s and 159s of the previous year and offers a valid counterpart to the Maserati:

	FERRARI 166SC	MASERATI A6GCS
ENGINE:	60 degree v12, Single Overhead Camshaft	In-Line 6-cylinder, Single Overhead Camshaft
DISPLACEMENT:	1,995.02 cc	1,978.7 cc
COMPRESSION RATIO:	8:1	11:1
CARBURETORS:	Three Weber 32DCF	Three Weber 36D04
MAXIMUM POWER:	125 hp at 7,000 rpms	130 hp at 6,000 rpms
WHEELBASE:	98.4 inches	92.5 inches
WEIGHT:	1,499 pounds	1,478 pounds
TRANSMISSION:	five-speed manual	four-speed manual
MAXIMUM SPEED:	112 mph	115 mph

Both cars carried two-seat, cycle-fendered aluminum bodies mounted on similar tubular frames fabricated from GILCO oval sections. The brakes on both cars were four-wheel drums, hydraulically actuated. The Maserati had an independent coil-spring front suspension; the Ferrari, also independent, had a single transverse leaf spring. Both cars used solid rear axles suspended by leaf springs. The Ferrari was probably the stronger of the two, while the Maserati may have been a bit tidier in the handling department. Both machines seemed plagued by transmission troubles, which may have been due in part to the shortage of high-quality alloys in Italy for the first few years after the war. Both the 166 and A6 racing cars formed the basis for lines of special-bodied touring automobiles— with Ferrari's body made by Carrozzeria Touring and the Maseratis by Pinin Farina. In a purely technical sense, the two machines were about equal, although the 166 would far outstrip the A6 in terms of success. The machine's reliability, which brought it major victories at Le Mans and in the Mille Miglia, was the critical difference between the cars. While Ferrari committed himself solely to the manufacture of automobiles, the Orsis at about the same time chose to ex-pand the machine tool and electric truck business, no doubt to the detriment of the continued development of the A6 series.

CHAPTER TEN

The First American Ferraris

The first Ferrari to reach America was the 166SC (Spyder Corsa), chassis num-ber 016-1, that Luigi Chinetti sold to Briggs Cunningham in the autumn of 1948 for $9,000. The car was raced a number of times by Cunningham and his friends during 1949–50 at amateur road race venues such as Watkins Glen and Bridge-hampton, Long Island. The car was hardly overwhelming. During the 1949 Wat-kins Glen races, Cunningham drove it in the four-lap Seneca Cup event over the bumpy, twisting 6.6-mile road circuit, finishing second behind George Weaver's prewar Maserati single-seater. Later in the day he again ran second in the 99-mile "Grand Prix," trailing Miles Collier's hot-rodded English Riley sports car pow-ered by a mundane flathead Ford V8 engine.

Sadly, the Spyder carried Miles's brother, Samuel Collier, one of the stal-warts of early American sports-car racing, to his death during the 1950 Watkins Glen race. The little car slid off course on a sweeping bend and flipped into a field, killing its popular, talented driver. It was the first death among the Ameri-can sports-car-racing set and a distraught Cunningham tried to sell the car. How-ever, he found no buyers and finally exhibited it in his superb California car museum until the collection was sold in 1987. The 166SC now resides in a place of honor at the excellent new museum in Naples, Florida, run by Samuel Col-lier's nephew, Miles Collier, Jr.

The Tommy Lee *lusso* (luxury or deluxe) Barchetta, chassis number 022-1, exhibited at the 1948 Turin show, arrived in America several months after the Cunningham car and spent most of its life in California. (For a time it carried whitewall tires!) Lee was a Los Angeles Cadillac dealer and automobile enthusi-ast. During the 1920s he hired a young artist, Harley J. Earl, to customize Cadillacs being sold to his Hollywood clientele. His results were so striking that he was hired by Cadillac to design the original LaSalle in 1927 and to spruce up the entire Cadillac lineup. This led to his appointment as General Motors' first head of the "art and color" section, which became the first styling staff in the industry. Earl rose to become General Motors' vice president of styling in 1940 and is credited with such mass-produced automotive benchmarks as concealed running boards and spare tires, curved rear windows and the first "hardtop convertible" bodies. Tommy Lee was also the sponsor of several Indianapolis cars in the years immediately following World War II. The Lee Ferrari is currently in a private collection.

The Farina Connection

Three distinct members of the Farina family played roles in the Ferrari saga. The most prominent in the early years was Dr. Giuseppe "Nino" Farina, the tough, temperamental Turin native who drove so diligently for the Scuderia and for Alfa Romeo before and after the war and became the first World Champion in 1950. He was the son of the older of the two brothers who operated the venerable (est. 1896) and respected Stabilimenti Farina coach-building firm in Turin. His uncle, Battista Farina, broke away from the family firm in 1930 and formed the renowned Carrozzeria Pininfarina, which produced so many superb body styles for Ferrari over the years. "Pinin" or "Little" Farina first called his new coach-building firm Pinin Farina, then finally had his name and the company's changed to a single word, Pininfarina, in 1958. His son, Sergio Pininfarina, at that point took over the immense design and body-building operation in Turin. Stabilimenti Farina built a handful of custom Ferraris on 166 and 212 chassis in the early years, but played an insignificant role compared with the other Farinas—Nino, Pinin and Sergio.

Early Passenger-Car Production

While Carrozzeria Touring produced about 100 of the earliest Ferrari bodies, including approximately 36 of the milestone Barchettas, numerous other superb Italian coach makers—the two Farinas, Vignale, Ghia, Bertone, Scaglietti, Zagato, to name but the most important—created some stunning bodies for the bare Ferrari chassis. This period of Italian design, 1950–60, generated perhaps the most sensuous, beautiful, riveting automotive styling in history. While Enzo Ferrari could lay no direct claim to this creativity (lovely bodies were created by the coach builders for dozens of other marques as well), his V12 road cars, because of their emotion-generating sound and performance, surely inspired the designers to lofty heights of expression. For the most part these road cars were equipped with detuned racing engines and drive trains adapted for road use with somewhat wider frames, softer suspensions, less radical brakes, etc., and given beautiful bodies in Turin and Milan. They were temperamental and difficult cars to drive on the road and stand today more as automotive sculpture than as examples of sensible, reliable, useful over-the-road transportation. As for the quantities actually built and sold by the company, historians have had great difficulty pinning down exact numbers. In many cases, chassis were rebodied several times, whether on the whim of the new owners or because of accidents. Moreover, Ferrari often converted sports-racing cars to road cars, which further confused the issue. Best estimates for the early years are as follows (customer cars only): 1948, 5 cars; 1949, 21 cars; 1950, 26 cars; 1951, 33 cars. While passenger-car production slowly increased during the 1950s, it was not until 1957 that over 100 cars were manufactured in a single year; in that year 113 machines rolled out of the Maranello works. It is important to understand that Ferrari built three distinct types of automobiles during the classic, pre-Fiat years: pure Formula One and Formula Two single-seat racing cars; lightweight open and closed two-place sports-racing cars; and grand touring customer cars, which were based on the engines and drive trains developed for the racing machines. Historian trying to determine the exact production of those early machines have encountered massive confusion because the cars often served one or two functions before being sold or destroyed or lost forever.

CHAPTER ELEVEN

The Ferrari Road Cars

As of 1990, about 55,000 Ferrari passenger automobiles have been built, with a majority constructed in the two decades since Fiat S.p.A. assumed control of the company. Prior to that time the road cars were little more than mildly detuned, limited-production, custom-bodied versions of the Ferrari racing machines. Until the late 1960s the principal power plants were variations of the original Colombo and Lampredi "long" and "short" V12s. The documentation of these automobiles has been the subject of lengthy study by noted automotive historians and fills a number of books. It being the intention here to deal with Enzo Ferrari personally, it would be inappropriate to try to cover in detail each and every customer model produced by the company—especially since Ferrari cared little for the machines other than for their ability to enhance his reputation and ultimately fill his coffers.

Insiders insist that after 1980 the company's road cars were simply rubber-stamped by a uninterested Ferrari and that he was openly disdainful of the men and women who bought them. No matter, the automobiles retain enormous value and mystique—especially those manufactured with the classic V12s before the Fiat acquisition. For those who wish to study the Ferrari road cars in more detail, the following books are recommended. (Other books deal with specific models in the Ferrari line and, while often excellent, are too numerous to be listed here.)

Hans Tanner and Doug Nye, *Ferrari* (Haynes Publishing Group, 1985)
Pete Lyons, *Ferrari: The Man and His Machines* (Publications International, 1989)

Angelo Tito Anselmi, *Tipo 166: The Original Sports Ferrari* (Haynes Publishing Group, 1984)

Gianni Rogliatti, *Ferrari and Pininfarina* (Ferrari Story, 1989)

Ferrari: Catalogue Raisonné 1946–1986, edited by Augusto Costantino in collaboration with Luigi Orsini (Automobilia, 1987)

Warren W. Fitzgerald and Richard F. Merritt, *Ferrari: The Sports and Gran Turismo Cars* (Bond Publishing Company, 1979)

Ferrari: The Man, the Machines, edited by Stan Grayson (Automobile Quarterly Publications, 1975)

The Colombo and Lampredi V12s

Because these two engines set the theme for Ferrari automobiles for decades and remain, symbolically at least, the heart and soul of the marque, a comparison of the two power plants may be of interest. Keep in mind that both engines were produced in a variety of permutations and in various displacements. The Colombo engine was made in both supercharged and normally aspirated form. The Lampredi engine was never offered with a supercharger. Both were 60-degree V12s, and many engineers claim that Lampredi's version was derivative of the earlier Colombo design, although extensive work was necessary to double the displacement, as was achieved with the first Lampredi 3.3-liter "long" engine in 1950. Both engines were built with single and double overhead camshafts and with twin-plug cylinder heads. The Colombo engine fell into disfavor during the early 1950s, then was resurrected in 1955 with the 250GT. At 3 liters, this engine was a superb unit, developing well over 300 hp in a variety of competition permutations and forming the basis for a number of superb road cars. Some versions were built at 4 liters. The Lampredi engine was finally increased in displacement to an immense 6.2 liters for Can-Am sports-car competition in 1969, but was most effective in the 4.1- to 4.9-liter range, where it generated up to 390 hp. Both engines were generally attached to five-speed, non-synchromesh transmissions and were fed both gasoline and methanol-based racing fuels through Weber four-throat carburetors. In racing applications, the engines were extremely reliable and had no intrinsic weaknesses. For road use, however, they were temperamental and inclined to foul spark plugs and overheat in heavy traffic. They were hard to start in cold weather, leaked oil around the valve guides (therefore producing smoky exhausts) and were extravagantly expensive to repair and maintain. Still, both engines are considered to be milestones in terms of their engineering and their aesthetics.

The "Thin Wall Special" Ferraris

These machines are unique in that they were among the very few Ferrari racing cars to be sold into private hands and to be successfully raced without factory help. There were actually three "Thin Wall Special" chassis owned by Guy Anthony Vandervell, the British industrialist and sportsman whose family owned the European rights to the American Clevite "thin wall" bearing that had been so important in making the early Colombo V12s reliable at high rpm. The first Vandervell car was an early 125GPC 1.5-liter supercharged Grand Prix model that he received prior to the 1949 British Grand Prix. It behaved so badly that it was promptly returned to the factory, where it was exchanged for a later model with two-stage supercharging and a slightly lengthened wheelbase. Ascari drove the second car once in the 1950 International Trophy race at Silverstone, but spun off the track. That machine was also found to be lacking in both quality of fabrication and design. Once again the abrupt, tough-talking Vandervell sent the car packing to Maranello. A year later he received the same car back, but this time carrying one of Lampredi's early 4.5-liter single-plug V12s and the latest Ferrari De Dion rear suspension. This car, in the hands of capable English veteran Reg Parnell, ran well in a number of races, achieving a fourth in the French Grand Prix. Vandervell, however, wanted still more, and he scrapped the Ferrari drum brakes and installed Goodyear-designed disc brakes (to be adopted by most front-line racing machines, although Ferrari would resist the change for years). The final "Thin Wall Special" was delivered in 1952. It had a long-wheelbase Indianapolis chassis, although its Italian body was replaced with a cleaner, more aerodynamic version fabricated in Vandervell's racing shops. Like the others, it was painted British racing green, and carried the name "Thin Wall Special" across the cowling—much to the displeasure of the priggish Royal Automobile Club, which considered such vile commercialism appropriate only in the barbaric American colonies. This last car was extremely successful in the hands of such eminences as Gonzales, Taruffi, Farina and Mike Hawthorn. As noted, Vandervell became irritated at Enzo Ferrari's increasing isolation and arrogance and in 1954 began producing his own racing cars under the Vanwall banner. In 1958 and 1959 his 2.5-liter Grand Prix machines, in the hands of Stirling Moss and Tony Brooks, were clearly faster than the "bloody red" Ferraris, as Tony Vandervell had come to call them. He died in 1967.

CHAPTER TWELVE

Lampredi and Colombo

There is no question that the early design work of Giaochino Colombo and Aurelio Lampredi created the engineering dynamics that established Ferrari's reputation. Yet both men left embittered, not only over their treatment by Ferrari but over the disputes concerning their contributions to the firm. For years afterward they sniped at each other in the sporting press and in interviews, although they were too discreet to openly criticize Ferrari himself. After designing the 250F and the powerful 450S sports car for Maserati, Colombo was retained by Bugatti in October 1953 to lay down the plans for the unique Type 251 Grand Prix car, which featured a transversely mounted straight-8 engine. That machine, plus a smaller sports car and a 4-liter military engine, were all unsuccessful. In 1956 he went to work in Turin for Carlo Abarth, where he designed a powerful 750-cc twin-cam version of the tiny Fiat 600 power plant. This led to a long association with the motorcycle firm of MV and subsequent work as an independent design consultant. In his later years he worked on electric vehicles, gearboxes and fuel-injection systems. He died in 1987. Lampredi remained with Fiat from 1955 to 1977, working as the director of power-plant engineering. Over the years he de-signed most of the significant passenger-car engines for both Fiat and Lancia—after that honored company was brought under the Fiat corporate aegis in 1969. From 1972 to 1982 Lampredi also worked for Abarth designing racing engines for the Fiat-based rally cars which won the World Rally Championship in 1977, 1978 and 1980. Retired from Fiat, Lampredi continued as a consultant and as a lecturer at the Turin Polytechnic Institute. He died in 1989, honored as one of the most respected of all the Italian engineering titans.

While it can be argued that Lampredi was more productive for Enzo Ferrari than Colombo (if one discounts the prewar Alfa Romeo 158), he gets short shrift in Ferrari's revisionist memoirs. In *My Terrible Joys,* which is surely the most candid of all his efforts, Ferrari dismisses Lampredi with but two references, the most elaborate being: "Aurelio Lampredi, who is now with Fiat, worked with me for seven years. He was undoubtedly the most prolific engineer-designer Ferrari has ever had. Starting from the 1.5-liter 12-cylinder, he first produced the 3-liter, then the 3,750-cc, then the 4-liter; these were followed by the 4,200-cc, the 4.5-liter and the 4,900-cc—all 12-cylinder engines. He next gave his attention to the 4-cylinder 2-liter, the 3-liter, the 3.5-liter; after which he turned out a straight-6 and finally a 2-cylinder 2.5-liter." Ferrari, of course, chose to ignore the fact that among that output came the firm's first Grand Prix winner and a machine, the Tipo 500, that dominated Grand Prix competition like no other automobile until the McLaren-Hondas of the late 1980s. He was kinder to Colombo, perhaps be-cause of his link to the old Jano fraternity and because, unlike Lampredi, he left the operation when it was not in a black slump. He referred to him as "my old friend" in whose ability he had "faith" and who was the "father" of the 12-cylinder. Still, considering the massive contribution both men made to the formation of the Ferrari myth, neither man enjoyed in Ferrari's writings anywhere near the recognition he deserved. Suffice it to say that, for all their differences, there would have been no Ferrari cars of any kind had it not been for Lampredi and Colombo.

The 4-Cylinder Grand Prix Cars

After he created the simple, purposeful Tipo 500s, the mystery remains as to how and why Lampredi got so far off the track with the 2.5-liter 625s, 553s and 555s. The 625 was essentially a Tipo 500 with a slightly enlarged, oversquare 2.5-liter update of the old 2-liter 4-cylinder. The engine developed about 245 hp and in the old upright chassis was a predictable, if somewhat underpowered, machine. Lampredi's Tipo 553 Squalo (Shark) was lower and wider, with its fuel tanks mounted on the side of the chassis to lower the center of gravity. The car was light (about 1,300 pounds) and quite responsive. But in the hands of old-timers like Farina, the car seemed twitchy and unforgiving. It was not as popular as the less sophisticated Tipo 625. Gonzales, on the other hand, was occasionally happy with the newer machine and gave it several excellent drives. But overall the Squalo was despised, and if given the chance, most team drivers opted for the 625s. During this period several 4-cylinder engine variations were tried in a number of chassis, which led to the Tipo 555 Supersqualo, a similar car with a modified, stiffer chassis made of smaller tubes. These machines were cursed with understeer and poor reliability, and while they were among the most impressive-appearing Grand Prix cars of the day, they were abject failures against the faster, better-handling and infinitely better-prepared Mercedes-Benzes that carried all before them in the 1955 season. Over the years the Ferrari works produced a number of automobiles, both for the road and for the track, that were wide of the mark, but none missed worse than the single-seaters that cost Aurelio Lampredi his job and the good graces of Enzo Ferrari. But was he entirely responsible? The boss was never hesitant in taking credit for the successful cars that rolled out of Maranello, but what was his involvement in the Squalo disaster? Surely he did not sit by as a benign spectator while Lampredi rushed ahead with the 4-cylinders. Ferrari had to have consented to the general course laid out by Lampredi and therefore must bear substantial responsibility. However, money was tight at the time and perhaps he had no choice. It is possible that the cost of developing a V8 of the type being used by Lancia or a straight-8 on the Mercedes-Benz theme was simply beyond the resources of his company. He may have been tied to the Lampredi 4s whether he liked them or not. In the end it may not have been the fault of the cars, but rather the enormous onslaught of Mercedes-Benz that doomed them. When driven by Gonzales and Hawthorn, the 4-cylinders were able to take the measure of the Maserati 250Fs on most days and, had they not been radically overstressed in order to stay in the wake of the Mercedes-Benz W196s, it is possible that they would have been remembered as decent automobiles. Also the fact that Vittorio Jano—a man he had known and respected for thirty-five years—could come out of nowhere and create a better car in the Lancia D50 than Lampredi no doubt only increased Ferrari's ire and frustration with the younger man.

CHAPTER THIRTEEN

Dino Ferrari and the V6

Exactly what role did young Dino Ferrari play in the creation of the V6 engine that ultimately brought his father the World Championship? Enzo Ferrari dutifully wrote in 1961: "I remember how carefully and with what competence Dino read and discussed all the notes and reports that were brought to him daily from Maranello. For reasons of mechanical efficiency, he finally came to the conclusion that the engine should be a V6 and we accepted his decision." Ferrari avoids mentioning that Jano, Massimino and Vittorio Bellentani were no doubt the source of the "notes and reports" Dino received. They may have already decided that a V6 was the answer. Historians, of course, offer various versions. The definitive Tanner-Nye book does not dwell on the issue, but states that "Massimino, Jano and, to an extent, Ferrari's son, Dino, developed a V6 concept." Alan Henry, in his excellent *Ferrari: The Grand Prix Cars*, tries to be charitable: "In his memoirs, Enzo Ferrari states that it was Dino's idea that a V6 engine configuration should be used for the Formula Two car, but one is bound to conclude that Jano's enormously experienced guiding hand lay behind the young Ferrari's advice to his father. Dino Ferrari quite obviously had a willing, inquisitive and active technical mind and he may well have scaled considerable heights had he developed his ability fully, but what he might have achieved had he survived must necessarily be left to the realms of conjecture. Enzo Ferrari has so publicly opened his heart on the subject of his only [*sic*] son that one feels bound, out of respect, to accept his assessment of Dino's contribution to the design concept of this V6 engine. However, one must always remember that Jano's enormous experience was not something that could be discounted on this particular score." Historian Mike Lawrence, who gives Ferrari no quarter in his *Directory of Grand Prix Cars 1945–65*, goes for the jugular: "Ferrari, that old myth-maker, tells how his dying son, Alfredino, took an active interest in the development of the new engines and it was his decision to use a V6 layout. In fact, before he left, Lampredi had already started designing a V6, while Jano had been responsible for some fine V6 engines for Lancia. The engine, however, was named 'Dino' in memory of Ferrari's son." Italian historian Piero Casucci, who has written extensively on Ferrari, has claimed on several occasions that the V6 was created by Lampredi. In his book *Enzo Ferrari: 50 Years of Greatness* (Arnoldo Mondadori Editore, 1982) he wrote: "It is said that the idea of the V6 engine was conceived by Dino Ferrari, but it was actually built by Aurelio Lampredi and then developed by Vittorio Jano." Other historians artfully describe Dino's role as "suggesting" the V6, but it appears that he was no more than a kindred soul in the project. The notion of three engineering giants—Jano, Lampredi and Massimino— implementing the theories of an inexperienced twenty-four-year-old, no matter how potentially gifted, is sentimental to the point of absurdity.

The engine, when it finally reached racing form, was a compact 65-degree V6 with two banks containing three cylinders each. The blocks were aluminum alloy and the cylinder heads carried double overhead camshafts and two spark plugs per cylinder. In 1.5-liter form for Formula Two, the Dino engine developed about 190 hp at 9,200 rpm. Increased to 2.5 liters for Formula One and running on aviation gas (mandated for the 1958 season), it developed a prodigious 270 hp at about 8,500 rpm. In terms of raw power, the Dinos were at the head of the Formula One class, but the chassis that held the engines were hopelessly outdated.

The Mystery of the Missing Cars

Enzo Ferrari had no consistent policy regarding his old racing cars, other than to crush, scavenge or dismember most of them. Some of his early examples, the Tipo 500s, 375s, etc., survived the wrecking ball, but after the Squalos, Super-squalos and Lancia D50s appeared on the scene, they were shut up inside the factory walls, to be left to rust, picked over for usable parts or sawed up and melted down. The wonderful D50 Lancias, which were later metamorphosed into the ill-handling 1957 Ferrari 801s, were all destroyed, save for one example, which now resides in the Turin Automobile Museum. The same fate befell the late Tipo 156s, lovely, shark-nosed 1.5-liter machines that carried Phil Hill to the 1961 World Championship. All were destroyed by the factory and not one example remains. In later years Ferrari was slightly more generous, occasionally giving an outdated car to a favored driver or selling one to a major collector like Frenchman Pierre Bardinot—who has his own staff of mechanics and a private road circuit on his large estate specifically to exercise his Ferraris. Over the years Ferrari often spoke of his love and respect for his automobiles, but he displayed that affection by setting teams of mechanics to work ripping up the chassis and melting down the aluminum-alloy bits. Why? "Old cars were useless to him. He thought only of the new," said one former company engineer. "Maybe he was ashamed of them," Phil Hill commented to me, only half in jest.

In one of his flourishes of prose on the subject, Ferrari wrote in the early 1970s about his retirement from competition and his alleged love for his cars: "I wasn't capable of making the car suffer." But if so, why did he allow his mechanics to shred the obsolete machines in the foundry like so many cadavers in a Nazi concentration camp—a sort of automotive Auschwitz for racing cars outside his office window?

CHAPTER FOURTEEN

Robert Daley, who left *The New York Times* to write such best-sellers as *Prince of the City* and *Year of the Dragon*, was one of several American journalists to discover Grand Prix racing in the mid-1950s. He met Count Alfonso de Portago at the 1956 Winter Olympics in Cortina, Italy, where the Spaniard had collected three of his friends, all novices at bobsledding, and entered the bobsledding competition. Piloted by Portago, the Spanish sled finished fourth, a wink of a second away from a bronze medal. Fascinated with the brash nobleman, Daley profiled him several times in magazine stories and in his tough-minded book on international racing, *The Cruel Sport*. He also served as a rough prototype for the main character in his later novel *The Fast One*. Another American writer who did much to popularize Grand Prix racing in America was Ken Purdy, whose rather lurid stories appeared regularly in *Playboy*. Purdy was a fine journalist who was overwhelmed by the glamour of the Grand Prix scene and the characters who inhabited it. Inevitably, the persona of Enzo Ferrari and his crimson racing cars became part of the story. In 1966 John Frankenheimer directed his excellent Cinerama motion picture *Grand Prix*, which contained a character openly based on Ferrari. During the decade 1956–66, the efforts of these men, plus the constant, adoring coverage of Ferrari in the enthusiastic motorsports press, were enormously influential in creating the man's mystique—much of which was embraced by a gullible American public.

Stirling Moss, who is often ranked with Fangio and Nuvolari as the absolute best ever produced in Grand Prix racing, ought to have won the World Championship on numerous occasions. He had finished second to Fangio in 1954 and 1955 and again in 1957. During the 1958 season Hawthorn won but a single race, while Moss won four. However, a series of second places accumulated sufficient points for Hawthorn to win the title. Moreover, Ferrari won only two races during the 1958 season, while Vanwall took a total of six and was clearly the superior machine. Beginning with the British Grand Prix of 1957, the Vanwalls had won nine of the fourteen races they entered, while Ferrari was victorious but twice. Still, Vandervell was depressed by the death of his friend Stuart Lewis-Evans and, faced with declining health, announced in January 1959 that the team was being disbanded. It may have been a propitious departure, for no rear-engine car was on the drawing boards, and as Enzo Ferrari was to learn the hard way, the old-fashioned design with the engine ahead of the driver was destined for the scrap heap.

Mike Hawthorn's highway accident was hardly unprecedented. Grand Prix stars of the day were notoriously fast drivers on the relatively vacant public roads of Europe and many of them were involved in serious crashes. Fangio barely escaped death in an accident with a farm wagon several years earlier, and Nino Farina—he of uncounted racetrack crashes—was to die in 1966 in the Alps when his Lotus-Cortina hit a telephone pole near Chambéry. Mike Parkes, a fine test driver and engineer for the Ferrari works in the late 1960s, was killed on the way to Turin in a high-speed crash in 1972. Numerous other drivers were involved in such accidents, but none was quite as shocking or dramatic as the one that took Hawthorn at the prime of his life.

The patriarch of the Mille Miglia, Count Aymo Maggi, was unrepentant after the government banned the race in the summer of 1957. He appeared to age rapidly following the news and seldom left his estate to visit his old friends at the Brescia Automobile Club, which had sponsored the event from the beginning. Suggestions that the race be modified to a rally format, with regulated speeds, infuriated him. "I will never consider anything different. In motor racing the one who runs fastest must win. He who gets rid of his inhibitions will always beat the one who lifts his foot. Motorsport is like that. The Mille Miglia has always been like that and will never change." (Quoted in Peter Miller's *Conte Maggi's Mille Miglia*, St. Martin's Press, 1988.) Ironically, it did change. Beginning in the mid-1980s, a revival of the race was begun, with historic cars dating back to 1957 and earlier. The old route was retraced at lower speeds, and the new version, while considerably tamer than the original, has become enormously popular in Italy once again. Piero Taruffi, the final winner, died in 1988, convinced that he had won outright, although there is little question that Trips followed Ferrari's orders and let the aging star and his crippled machine through to victory.

CHAPTER FIFTEEN

Not surprisingly, there was life after Ferrari for most of the men who left the hallowed precincts of Maranello. Tony Brooks retired from racing in 1961 after a short, but brilliant career and set up a successful car dealership in Weybridge, Surrey, adjacent to the fabled Brooklands track. Romolo Tavoni became the general manager of the Monza autodrome. Carlo Chiti went from the failed ATS operation to design a series of racing cars for Alfa Romeo before retiring to start his own automotive consultancy in Milan; he passed away in July 1994. Phil Hill raced until 1967, primarily in sports cars near the end, then retired to his native Santa Monica, where he opened a prestigious classic-car restoration firm. He remained active in the sport as a commentator and writer until his death in August 2008. Olivier Gendebien lived in elegant retirement in Belgium until he died at the age of seventy-four in October 1998, while Dan Gurney left driving in 1970 to concentrate on his thriving race-car-building operation, called All-American Racers. His Eagle cars won numerous Indianapolis car races, including the vaunted 500, before he teamed with Toyota to produce a series of sports-racing prototypes. Gurney sadly passed away from natural causes in January 2018. Richie Ginther stopped racing in the mid-1960s after winning the Mexican Grand Prix for Honda. He slipped into an obscure retirement, touring for a number of years in a motor home before settling in Baja California. Thanks to fellow journalist Pete Lyons, I was able to locate him for research on this project. Richie died in 1989 of a heart attack at age fifty-nine while on a European vacation. Stirling Moss never recovered sufficiently from his Goodwood crash to race seriously again. It may have been possible for him to compete, but Moss was such a perfectionist that he would not drive unless he believed himself to be at an absolute peak of health. He apparently knew that his eyesight and reflexes were somewhat dulled and therefore would not run again—although even with slightly diminished capacities Moss could probably have outdriven most men. He is now in retirement in London and makes numerous personal appearances and drives occasionally in vintage competitions. In partnership with Lister Cars, Stirling introduced the Stirling Moss Lister Knobbly in 2016, but after a serious illness he fully retired from public life in January 2018. Franco Rocchi remained active in Modena following his retirement from Ferrari. One of his final projects was the design of an exquisitely crafted, highly original W12—three 4-cylinder blocks mounted radially and powering a single crankshaft. After developing heart disease, he retired from engineering to become an artist, ultimately dying in February 1996.

The Championships

The Fédération Internationale de l'Automobile (FIA) over the years created a myriad of championships for sports cars, Formulas One, Two and Three, hill climbs, grand touring cars, rallies, etc. These titles were awarded to manufacturers who accumulated the most points over a season of competition. Before his death Ferrari cars won no fewer than fourteen titles in sports and grand touring car competition and another six in Formula One. From 1950 to 1989, the period during which formal World Championship points were kept in Grand Prix racing, Ferraris entered 442 races (out of a total of 469) and won 97, or about 2.5 per year over the 39-year span. His cars made 1,072 starts, with 77 drivers at the wheel. This is an exemplary record, not because of the frequency of wins, for the factory had eight terrible seasons (1957, 1962, 1965, 1967, 1969, 1973, 1980 and 1986) in which not a single victory was recorded, but because, with few exceptions, Ferraris were at the starting line in every Grand Prix race on the schedule. Critics might dismiss this as merely overwhelming the competition with sheer numbers, but no racing operation ever exhibited the same unflinching determination to compete, regardless of the odds, as Ferrari. For that alone great credit is due.

CHAPTER SIXTEEN

Ford Versus Ferrari

The struggle between these motoring giants was described at the time as a David-and-Goliath mismatch, but that is hardly accurate. While the Ford Motor Company surely brought vastly more money to the duel, Ferrari enjoyed infinitely more racing experience and a tighter, more flexible organization. Gearing up (or, more correctly, gearing *down*) a massive corporate entity like Ford to engage in a focused project like creating special Le Mans cars was more difficult than one might think. On the other hand, Enzo Ferrari had three decades of experience to call upon, plus a small race-hardened staff ready for instantaneous duty. This paid off for him with an easy victory in 1964, although the P3s and P4s from Maranello were extremely competitive machines. Had Ferrari paid as much attention to aerodynamic detail and handling nuances as Ford, he might have won. In fact, the Ford victories came thanks to veteran American racing organizations like Shelby American and Holman-Moody, as well as a bevy of British motorsports professionals, all of whom acted as subcontractors for the Ford Motor Company. A special tribute was given to Shelby engineer Phil Remington, who devised a scheme to change the red-hot disc brakes of the heavy Fords in less than a minute during routine pit stops. That, plus niceties like being able to charter a Boeing 707 to haul spare windshields from America on the eve of the 1966 race, meant the difference. However, make no mistake, with the financial support of Fiat, Shell, Dunlop, Firestone, etc., Ferrari was hardly the pauper his supporters claimed him to be. With better planning, a bit of luck and better human engineering—such as not permitting Surtees to walk off the job over an issue of petty politics—the Scuderia might very well have gone four for four. But in the end, the superb Mk IV of Dan Gurney and A. J. Foyt that carried those two excellent drivers to victory at Le Mans in 1967 marked the technological pinnacle of the last great corporate duel in endurance racing.

John Surtees and the British Masters

After leaving Ferrari, "Big John" Surtees raced briefly for Cooper-Maserati and Honda before starting his own Formula One team. This operation never functioned on all eight cylinders and he finally disbanded it in the mid-1970s. He retired in Surrey, England, and passed away in March 2017. Surtees was but one of a group of brilliant drivers who surged out of the British Commonwealth during the late 1950s and 1960s. Joining him were Jack Brabham, who won three World Championships; Jackie Stewart, who also won three; Jimmy Clark, who won two more before dying in a racing crash in 1968; Denis Hulme and Graham Hill, who each won titles. This extraordinary group of drivers, which also included New Zealanders Chris Amon and Bruce McLaren, dominated the sport from the retirement of Stirling Moss until the rise of Emerson Fittipaldi and Niki Lauda in the mid-1970s. While Surtees won the World Championship only once, his duplication of the feat on Grand Prix motorcycles is unprecedented in the modern annals of the sport. Between 1959 and 1971 Commonwealth drivers won the World Championship no fewer than eleven times, with only the American Phil Hill (1961) and the Austrian Jochen Rindt (1970) breaking the domination. After 1971 their influence waned, much as it had for the Italians and their Argentinian cousins who, thanks to Farina, Ascari, Fangio, Gonzales, Castellotti, Villoresi et al., dominated racing in the 1950s. There is no accounting for these surges in supremacy by one nation or another in sports, and currently the Brazilians and French seem to have massive pools of talent. The Italians still remain in the background. As much as they yearned for a countryman to win the title aboard a Ferrari, Alberto Ascari's crowning in 1953 remains the last such achievement. In recent decades no Italian driver has ever achieved much while driving for the Scuderia. Michele Alboreto worked valiantly for the team in the 1980s, but gained little success. As this is written, in 1990, Ferrari's drivers are Frenchmen Alain Prost and Jean Alesi, with no rising Italian stars on the horizon. Alesi was a mainstay for team Ferrari through much of the 1990s, until Michael Schumacher and Eddie Irvine joined the Scuderia in 1996. Ferrari had its shining star in Schumacher, much to the delight of Italian fans. Kimi Räikkönen and Sebastian Vettel are currently the drivers on record for Ferrari.

Adalgisa Bisbini Ferrari

While no written documentation exists regarding Ferrari's mother's demand that Piero Lardi be legitimized, my interviews with Mauro Forghieri, Romolo Tavoni, Franco Lini, Don Sergio Mantovani and others close to the family during the 1960s produced agreement on the subject. It is possible that Adalgisa Ferrari knew of Piero's existence well before Laura. Like most Italian sons, Ferrari was extremely close to his mother and deferential to her in most family matters. The exact genesis of the agreement is obviously lost to history unless Piero Lardi Ferrari chooses to discuss it at a later time, but it must be presumed that Enzo Ferrari complied with his mother's wishes when he legally acknowledged Piero as his legal son following Laura's death in 1978.

Ferrari and Agnelli

While Enzo Ferrari and Giovanni Agnelli knew each other on a business basis for thirty years, they were never what could be described as close friends. In the classic Italian manner, Agnelli was lavish in praise of Ferrari, especially following his death in 1988. But in reality the connection had been strained at best. In 1988, shortly after Ferrari's passing, Agnelli told Lee Iacocca in private conversation that his relationship and business dealings with Ferrari had been "extremely difficult." The source of this comment is a close friend of Iacocca's who requests anonymity. I consider his recollection to be absolutely accurate.

Porsche and Ferrari

While it is hardly the intent of this book to chronicle all of the racing wars in which Enzo Ferrari engaged, it should be noted that once Ford left the scene, Porsche replaced it as every bit as formidable a rival for the Scuderia. The Porsche 917 coupe, powered by a 4.5-liter flat-12 of the type that had been employed by Ferrari since the mid-1960s, was overwhelming in endurance racing. Ferrari countered with his 512 series, but they were simply unable to compete with the big machines from Stuttgart. Ferrari also developed the little 312PBs with 3-liter flat-12s in the early 1960s. But by then he was on the verge of leaving sports-car competition for good. Had he chosen to stay, he would have faced enormous challenges from both Porsche and the French aerospace firm of Matra. Again, this must serve as buttressing the theory that much of Ferrari's successes in endurance racing came at times when other major manufacturers were absent from the scene. When the likes of Ford or Porsche brought the full force of their technology to bear, it was extremely difficult for the Maranello operation to keep up. This only makes his decision to drop such competition in 1973 and concentrate on Formula One seem all the more prudent.

CHAPTER SEVENTEEN

During the four years that Niki Lauda drove for Ferrari he recorded fifteen wins, twelve second-place finishes and no fewer than twenty-three pole positions. He won the World Championship twice. While his tenure at Ferrari was difficult for all concerned, he claims that his affection for Italy was considerable and that he thoroughly enjoyed the people of Modena and Maranello. Like so many drivers, he found the political atmosphere at the factory to finally become unbearable and this alone led to his departure. His detailed version of the Ferrari years is recounted in two books, *For the Record: My Years with Ferrari* (Verlag Orac, 1977) and *Niki Lauda, Meine Story* (Verlag Orac, 1986). Lauda was to win yet another World Championship with McLaren in 1984 (after a two-year retirement in 1980–81). He founded and operated two successful charter airlines, Lauda Air and Niki, and continues to actively work in Formula One.

No one has had the time—much less the patience—to calculate the number of books, stories, pamphlets, technical treatises, etc., written about Enzo Ferrari and his automobiles. Suffice it to say that much of the material is repetitive and essentially derived from research done by such fine automotive journalists and historians as the Italians Luigi Orsini, Piero Casucci and Gianni Rogliati, Englishmen Doug Nye, Alan Henry, Denis Jenkinson and L. J. K. Setright, as well as the Americans Jonathan Thompson, Dean Batchelor, Stan Nowak, Karl Ludvigsen and Pete Lyons. However, much of their work involves classification and description of the various Ferrari automobiles and the logging of Ferrari's extensive racing record. Very little serious research has been done on Ferrari's personal life. Because many professional journalists were in a sense beholden to Ferrari for access to test cars, archive materials, factory tours, interviews, etc., few dared to be overly critical during Ferrari's lifetime. Ferrari was extremely protective of his public image and tolerated no deviation from his official biography. Therefore, almost nothing was written about the "private side" of the man, other than to occasionally note that he was willful, manipulative and fiercely—even ruthlessly—competitive. The Italian sporting press was often critical of his racing cars and his driver selections, but seldom if ever treated the man himself critically. While the existence of his mistress and Piero was widely known in the press from the mid-1960s onward, a decade would pass before any mention was made of Piero. He was not widely known until Niki Lauda mentioned him in his 1977 autobiography, *For the Record*.

Books dealing with the personal life of Enzo Ferrari are rare and include Gino Rancati's *Ferrari: A Memory* (1989), which is a polite and decidedly circumspect personal biography. Ferrari's early life and brief racing career are dealt with in detail in both Giulio Schmidt's *Roaring Races* (1988) and Valerio Moretti's *Enzo Ferrari, Pilota* (in Italian, 1987). Each writer has a slightly different version of Ferrari's racing efforts, but both are valuable sources. Before his death Gioachino Colombo, in *Origins of the Ferrari Legend* (1985), gave his version of how the early Ferrari automobiles were created. This is excellent, although Colombo downplays the acrimony between himself and Lampredi and his ultimate battle with Ferrari. Carlo Chiti's *Chiti Grand Prix* (1987), with Piero Casucci, deals with Ferrari in personal terms, although he too glosses over the reason for his participation in the great 1962 mutiny, surely for political reasons. The fine review of the Ferrari operation by *Automobile Quarterly*, titled *Ferrari: The Man, the Machines* (1975), edited by Stan Grayson, contains a number of excellent pieces, highlighted by Griffith Borgeson's two fine profiles of the man. Subsequent research has made some of the material obsolete, but it nonetheless remains valuable as an overview. Niki Lauda, who left Ferrari on unpleasant terms, produced some rather unflattering insights in his two autobiographies, *For the Record* (1977) and *Meine Story* (1986). Both books were among the first to reveal the cruder, more ruthless side of Enzo Ferrari. Derek Bell, who raced briefly for the factory team in the late 1960s and early 1970s, offers some anecdotal material in his autobiography (with Alan Henry), *My Racing Life* (1988). A number of personal interviews with some of Ferrari's early associates can be found in Angelo Tito Anselmi's excellent examination of the first production Ferraris, titled *Tipo 166: The Original Sports Ferrari* (1984). Of course, Ferrari's own works (see Notes for Chapter 2) are valuable, but they are useful as much for what is omitted as for what is included.

The Ferrari automotive history has been covered by a number of authors, but surely the seminal work is *Ferrari*, originally written by Hans Tanner (1959) and expertly updated by Doug Nye. For an overview of the operation, in terms of both passenger cars and racing machines, this is the primary source. However, Ferrari's Grand Prix racing history is best covered in *Ferrari: The Grand Prix Cars, 1948–89* (1989) by Alan Henry. The book is brightly written, contains a wealth of anecdotes and is a gem of automotive research. Several other Ferrari histories purport to cover the subject, but for the most part deal in polite generalities. However, they ought to be read by the serious student. They include: *Ferrari: Catalogue Raisonné* (1987) by Luigi Orsini and Augusto Constantino; *Ferrari: 40 Years* (1985) by Luigi Orsini; *Enzo Ferrari: 30 Years of Greatness* (1982) by Piero Casucci. An excellent introduction to the subject is Peter Lyons's *Ferrari: The Man and His Machines* (1989), while L. J. K. Setright's *Ferrari* (1975) offers some interesting insights.

Specific motor-racing history regarding Ferrari can be found in *La Ferrari alla Mille Miglia* by Giannino Marzotto (1987) and *Conte Maggi's Mille Miglia* (1988) by Peter Miller, as well as Dominic Pascal's *Ferraris at Le Mans* (1984). Surely the masterpiece in this group is Luigi Orsini's *The Scuderia Ferrari* (1979), which recounts the activities of the team from 1930 to 1938 and contains extraordinary photos from the Franco Zagari collection. Kevin Desmond's biography of Alberto Ascari, *The Man with Two Shadows* (1981), offers revealing peeks inside the Ferrari motorsports operation of the early 1950s.

Hundreds of books have been written about the Ferrari road cars and most of them are little more than collections of pretty pictures. Although they do not include recent models, Warren Fitzgerald and Richard F. Merritt's *Ferrari: The Sports and Gran Turismo Cars* (1979), Dean Batchelor's *Ferrari Gran Turismo and Competition Machines* (1977) and Gianni Rogliatti's *The Ferrari* (1973) are superb examinations of the period when Ferrari was building the most original and outrageous of his road machines. *Fantastic Ferraris* by Antoine Prunet and Peter Vann (1988) and Braden and Roush's *Ferrari 365GTB/4 Daytona* (1982) are but two of dozens of books that take interesting, if rather adoring looks at the passenger machines produced at Maranello.

Because Ferrari's fortunes in many cases overlapped those of other marques, a number of books regarding Porsche, Bugatti, Alfa Romeo, OSCA and Maserati are also of value. *The Alfa Romeo Catalogue Raisonné* is a mother lode. Certainly Luigi Orsini and Franco Zagari's excellent history of Maserati, *Maserati: A Complete History from 1926 to the Present* (1980), tops the list, although H. G. Conway's *Bugatti* (updated, 1989) and a history of Porsche by Richard Von Frankenberg and Karl Ludvigsen, *Porsche: Excellence Was Expected* are superb efforts. Orsini's history of OSCA and Griffith Borgeson's personal memoir regarding the Bugatti family offer insights into the mood and temperament of the people who created exotic cars in the 1930s, as does Mark Dees's superb history of Miller racing cars, *The Miller Dynasty* (1981).

Any student of Ferrari must seek background information in the plethora of books that deal with international motorsports. Any number of anthologies exist, including Mike Lawrence's pithy, broad-ranging *Directory of Grand Prix Cars, 1945–63* (1989); L. J. K. Setright's fine *The Pirelli History of Motorsport* (1981); *The Hamlin Encyclopedia of Grand Prix* by David Hodges (1988); *Formula One*, by Nigel Roebuck, with paintings by Michael Turner (1983); *The History of Motor Racing* by William Boddy and Brian Laban (1977); *Fifty Famous Motor Races* by Alan Henry (1988); *A Motor Racing Camera, 1894–1916* by G. N. Georgano (1979); and a personalized look at the sport, John Dugdale's *Great Motorsport of the '30's* (1977). Nigel Roebuck's *Grand Prix Greats* (1986) and his comparison piece, *Inside Formula One* (1988), offer excellent personalized views of the sport and its participants, while Joe Seward's *World Atlas of Motor Racing* (1989) provides excellent descriptions of the major racing circuits of the world. Gianni Lurani's *History of the Racing Car* (1972) is also recommended.

If any marque has been written about as much as or more than Ferrari, it may be Mercedes-Benz. Because of its major impact on racing (and on Ferrari) during the 1930s and again in the 1950s, a number of fine books offer important background. Among the best is Chris Nixon's *Racing the Silver Arrows* (1986), a flawlessly researched look at the Mercedes campaigns of 1934–39. Karl Ludvigsen's *The Mercedes-Benz Racing Cars* (1971) is a masterful technical treatise on the subject, while George Monkhouse's *Motor Racing with Mercedes-Benz* (1945), his *Motor Racing* (1947) and his later pictorial *Mercedes-Benz Grand Prix Racing, 1934–55* are monumental works of history and photography. In terms of photographers, Jesse Alexander was among the very best working in the 1950s and 1960s and his *Looking Back* (1986) offers a magnificent visual treat to the enthusiast.

For a look at the technical side of Grand Prix racing, a number of valuable treatises exist. The classic in the field is Lawrence Pomeroy's two-volume work, *The Grand Prix Car*, updated by L. J. K. Setright. *The Racing Car: Development and Design* by Cecil Clutton, Cyril Posthumus and Denis Jenkinson (1956) is a standard reference. *A History of the World's Racing Cars* by Michael Frostick and Richard Hough (1965) is also serviceable but not as technically detailed as the foregoing.

Driver biographies are useful, offering both background on the Ferrari operation and overall observations of the sport. *Grand Prix Driver* by Hermann Lang (1953) and *A Racing Car Driver's Life* by Rudolf Caracciola (1954) afford looks at the 1930s racing scene, as does the excellent *My Two Lives* by René Dreyfus and Beverly Rae Kimes (1983). Carroll Shelby's *The Cobra Story* (1965) deals in detail with his battles with Ferrari, as does Leo Levine's outstanding history of Ford's racing in the 1960s, *The Dust and the Glory* (1968). Michael L. Shoen's *The Cobra-Ferrari Wars, 1963–65* (1990) is a detailed examination of that rivalry. Gianni Lurani produced a useful biography of Tazio Nuvolari, *Nuvolari* (1959), but Cesare De Agostini's (in Italian) is more detailed. Gunther Molter's *Juan Manuel Fangio* (1969) is a fine work on the great Argentinian champion, as is Fangio's autobiography, *My Racing Life*. Chris Nixon's *Rosemeyer* (1989) is also recommended.

Solid biographical information on drivers can be found in *The Encyclopedia of Auto Racing Greats* (1973) by Bob Cutter and Bob Fendell and in *Winners* by Brian Laban (1981). *The Encyclopedia of Motor Cars, 1885 to the Present* (1982), edited by G. N. Georgano, is a standard in the field, as is the valuable *The Encyclopedia of the Motor Car* (1979), edited by Phil Drackett.

Doug Nye's *Racing Car Oddities* (1975) describes some of racing's more bizarre efforts, including the Ferrari-Alfa *bimotore*, while Chris Nixon's *Road Race* details some of the more famous international road races, including the Mille Miglia. Robert Daly's *Cars at Speed* (1962) is a lurid but essentially accurate reporter's look at the violent sport of motor racing in the late 1950s. A late but valuable entry for those interested in the Ferrari factory scene is Michael Dregni's *Inside Ferrari*.

Beyond this list of books, which is admittedly incomplete, Enzo Ferrari has been the subject of endless stories in such magazines as *Cavallino*, *Ferrari World* and the *Ferrari Owners Club Newsletter*, all of which are exclusively (and passionately) devoted to the subject, as well as *Car and Driver*, *Road & Track*, *Autoweek*, *Automobile Quarterly* and *Automobile Magazine*. Additionally, numerous books are available through Motorbooks International in Osceola, Wisconsin. Those periodicals, plus a number published in Great Britain, Germany and Italy, were all valuable adjuncts to the preparation of this book. The author is grateful to all the historians and journalists who have worked so diligently to record and classify this complex and often baffling subject.

BROCK YATES, a founding father of modern automotive journalism, was a prolific author of columns, books, and movies that span decades and embrace his love and respect for all things motorsports. Best known for creating the highly illegal Cannonball Run Sea-to Shining-Sea Memorial Trophy Dash, he was the longtime executive editor for *Car and Driver* and contributed to *The Wall Street Journal, Playboy, The American Spectator, Boating*, and *Vintage Motorsport*, along with countless other publications. He created the One Lap of America race and authored fifteen books and two screenplays, including *The Cannonball Run*. As a commentator for CBS Sports during the historic 1979 Daytona 500, Yates, along with Ken Squier, David Hobbs, and Ned Jarrett, ushered in a groundbreaking era in live coverage, airing the first ever "flag to flag" race, cementing NASCAR's place in live televised sports and changing the face of racing coverage forever. Nicknamed "the Assassin" by his peers, Yates was known for his incisive industry critiques and was the recipient of numerous awards, including the coveted Ace Award, the Golden Medallion Award from the International Automotive Media Competition (IAMC), and the Ken Purdy Award for excellence in automotive journalism, and was inducted into the Motorsports Hall of Fame in 2017.